A

HISTORY OF THE BRITISH ARMY

A History of
The British Army

BY

THE HON. J. W. FORTESCUE

FIRST PART—TO THE CLOSE OF THE SEVEN YEARS' WAR

VOL. I

The Naval & Military Press Ltd

Published by
The Naval & Military Press Ltd
Unit 10, Ridgewood Industrial Park,
Uckfield, East Sussex,
TN22 5QE England
Tel: +44 (0) 1825 749494
Fax: +44 (0) 1825 765701

www.naval-military-press.com

© The Naval & Military Press Ltd 2004

PREFACE

The civilian who attempts to write a military history is of necessity guilty of an act of presumption; and I am not blind to my own temerity in venturing to grapple with such a task as the History of the British Army. But England has waited long for a soldier to do the work; and so far no sign has been given of the willingness of any officer to undertake it beyond the publication, a few years since, of Colonel Walton's *History of the British Standing Army from* 1660 *to* 1700. Nor is this altogether surprising, for the leisure of officers is limited, the subject is a large one, and the number of those who have already toiled in the field and left the fruit of their labour to others is sadly small. A civilian may therefore, I hope, be pardoned for trying at any rate to make some beginning, however conscious of his own shortcomings and of the inevitable disadvantage from which he suffers through inexperience of military life in peace and, still more fatally, in war. His efforts may at least stimulate some one better qualified than himself to treat the subject in a manner better befitting its dignity and its worth.

My design is to write the history of the Army down to the year 1870, the two present volumes carrying the story down to the Peace of Paris in 1763, and two future volumes bringing it forward to the great reforms which virtually closed the life of our old Army and opened that of a new. It would have been easy to have filled a score of volumes with matters germane to the subject and of genuine interest to at least some groups of military students; nor would such treatment have been foreign to the methods of one school of British historians. There is indeed much to be said for it from the writer's standpoint, for it simplifies his task beyond belief. To me, however, rightly or wrongly, it seemed better to gather the story if possible into a smaller compass, even at the cost of omitting many instructive statistics and picturesque details. Accordingly I have compressed the six hundred years of our military history from Hastings to Naseby into one-third that number of pages, endeavouring only to set down such points and incidents as were essential to a coherent sketch of the growth of our military system. Even after Naseby and up to the reign of Queen Anne I have dealt with the history in a like arbitrary spirit, thus passing over, not I confess without regret, the Irish campaigns of Cromwell and King William, though entering with some detail into that of Schomberg. All could not be written down, as any one can bear me witness who has attempted to go below the surface of the Great Civil War alone. The reader must decide whether I have judged well or ill in that which I have left unwritten.

I must plead guilty also to deliberate omission of sundry small details which are rather of antiquarian than of true military interest, minute particulars of dress, armament and equipment and the like, the real place for which is rather in a military dictionary than in a military history. These I have sacrificed, not because I felt them to be trivial, but because I thought that the space which they demanded would be more profitably occupied by a sketch of the political relations between the Army and the country. I cannot, however, claim completeness for this sketch: and I am conscious that many questions of great constitutional importance are left unresolved, as I must frankly acknowledge, through my inability to cope with them. I have sought our acknowledged authorities on constitutional questions in vain; not one is of help. I confess that I have been amazed when reading our innumerable political histories to see how unconcernedly Army, Navy, and the whole question of National Defence are left out of account.

It is this, the political not less than the military aspect of the Army's history that I have endeavoured, however slightly and however unsuccessfully, to elucidate, at the sacrifice sometimes of purely military matters; and it is this which makes the subject so vast as to be almost unmanageable. The difficulties of tracing military operations are frequently trying enough, but they are insignificant compared to those presented by the civil administration of the Army, and by the intolerable complication of the finance. Here again the reader must judge whether or not I have chosen aright; and I would ask him only not to attribute to neglect omissions which have been made after mature deliberation.

My authorities from the reign of Queen Anne onward, and occasionally before, are quoted at the foot of the page; but in the earlier portion of the first volume I have been content to group them in a brief note at the close of each chapter or section;[1] and I have followed the same plan with some modification throughout. I must, however, mention that these notes rarely comprise the whole of the authorities that I have consulted, much less all that lie open to consultation. It would be a simple matter, for instance, to cover a page with works consulted on the subject of the Civil War alone; but while I have, as I trust, taken pains to make my work thorough, I have been content frequently to refer the reader to such authorities as will guide him to further sources of information, should he desire to pursue them. I have spared no pains to glean all that may be gleaned from the original papers preserved at the Record Office in reference to the military administration and to the various campaigns, and I have waded through many thousands of old newspapers, with and without profit. What unknown treasures I may have overlooked among the archives preserved by individual regiments, I know not, since with an army so widely dispersed as our own it seemed to me hopeless to attempt to search for them; but such regimental histories as exist in print I have been careful to study, sometimes

[1] I must mention here that where reference is made to Mr. Oman's *Art of War*, the volume alluded to is the short essay, published in 1885, not the larger and far more important work of the same author, which, to my great misfortune, appeared too late for me to avail myself of it.

with advantage but not always with profound respect for their accuracy.

Maps and plans have been a matter of extreme difficulty, owing to the inaccuracy of the old surveys and the disappearance of such fugitive features as marsh and forest. I have followed contemporary plans wherever I could in fixing the dispositions of troops, but in many cases I should have preferred to have presented the reader with a map of the ground only, and left him to fill in the troops for himself from the description in the text. Blocks of red and blue are pleasing indeed to the eye, but it is always a question whether their facility for misleading does not exceed their utility for guidance. Actual visits to many of the battlefields of the Low Countries, with the maps of so recent a writer as Coxe in my hand, did not encourage me in my belief in the system, although, in deference to the vast majority of my advisers I have pursued it.

It remains to say a few words on some minor matters, and first as to the question of choosing between Old Style and New Style in the matter of dates. Herein Lord Stanhope's rule seemed to be a good one, namely to use the Old Style in recording events that occurred in England, and the New for events abroad. But I have supplemented it by giving both styles in the margin against the dates of events abroad; lest the reader, with some other account in his mind, should (like the editor of Marlborough's Despatches) be bewildered by the arrival in England of news of an action some days before it appears to have been fought in the Low Countries. One difficulty I have found insuperable, which is to discover when the New Style was accepted in India; but finding that the dates given by French writers differ by eleven days from those of Orme I have been driven to the conclusion that the Old Style endured at any rate until 1753, and have written down the dates accordingly.

Another difficulty, more formidable than might be imagined, has been the choice of orthography for names of places abroad. Before the war of 1870 the French form might have been selected without hesitation; but with the rise of the German Empire, the decay of French influence in Europe and the ever increasing importance of German writings in every branch of literature, science and art, this rule no longer holds good. Finding consistency absolutely impossible, I have endeavoured to choose the form most familiar to English readers, and least likely to call down upon me the charge of pedantry. Even so, however, the choice has not been easy. Take for instance the three ecclesiastical electorates of the Empire. Shall they be Mainz, Köln and Trier, or Mayence, Cologne and Trèves? The form Cologne is decided for us by the influence of Jean Maria Farina; Trèves is, I think, for the present better known than Trier; but Mainz, a large station familiar to thousands of British travellers, seemed to me preferable to the French corruption Mayence, as reminding the reader of its situation on the Main. For German names of minor importance I have taken the German form, since, their French dress being equally unfamiliar to English readers, there seemed to be no reason why they should not be written down correctly; but the French form is adopted so exclusively in contemporary histories that possibly not a few instances of it may have escaped my vigilance. In Flanders again it is frequently necessary to choose between the French and the Flemish spelling of a name; and, where it has been possible without pedantry, I have preferred the Flemish as nearer akin to the English. Thus I have always written Overkirk rather than Auverquerque, Dunkirk rather than Dunquerque, Steenkirk rather than Estinquerque (the form preferred for some reason by Colonel Clifford Walton), since the French forms are obviously only corruptions of honest Flemish which is very nearly honest English. Actual English corruptions I have employed without scruple, though here again consistency is impossible. It is justifiable to write Leghorn for Livorno; but The Groyne, a familiar form at the beginning of this century, is no longer legitimate for Corunna, any more than The Buss for Bois-le-duc (Hertogenbosch) or Hollock for Hohenlohe. Then there is the eternal stumbling-block of spelling Indian names. Here I have not hesitated to follow the old orthography which is still preserved in the colours of our regiments. Ugly and base though the corruptions may be they are at any rate familiar, and that is sufficient; while they probably convey at least as good an idea of the actual pronunciation as the new forms introduced by Sir William Hunter. Here once more it would be confusing to write Ally for Ali or Caubool for Cabul, though possibly less so than to confront the reader with Machhlīpatan or Machlīpatan (two forms used indifferently by Colonel Malleson) for Masulipatam, and Maisur for Mysore. We are an arbitrary nation in such matters and very far from consistent. Even in such simple things as the names of West Indian Islands we have dropped the old form Martinico in favour of Martinique, though we still affect Dominica in lieu of Dominique. All that a writer can do is to study the prejudices of his readers without attempt either to justify or to offend them.

Lastly, I must give the reader warning that I have spoken of our regiments throughout by the old numbers instead of by their territorial titles. As I do not propose to carry the history beyond 1870 I may plead so much technically in justification; but apart from that I would advance with all humility that life is short, and that it is too much to ask a man to set down such a legend as "The First Battalion of the York and Lancaster Regiment" (in itself probably only an ephemeral title), when he can convey the same idea at least as intelligibly by writing the words Sixty-fifth. I have also called regiments by their modern appellations (so far as the numbers may be reckoned modern) throughout, ignoring the anachronism of denominating what were really regiments of Horse by the term Dragoon-Guards, for the sake of brevity and convenience. An Appendix gives the present designation of each regiment against its old number, so that the reader may find no difficulty in identifying it. I may add that I have written the numbers of regiments at full length in the text in all cases where such regiments have survived up to the present day, so that the reader need be in no doubt as to their identity; and I have carefully avoided the designation of disbanded regiments by the numbers which they once bore, in order to avoid confusion.

In conclusion, I have to express my deepest thanks to Mr. G. K. Fortescue at the British Museum and to Mr. Hubert Hall at the Record Office for their unwearied and inexhaustible courtesy in disinterring every book or document which could be of service to me.

J. W. F.

June, 1899.

CONTENTS

BOOK I

CHAPTER I

CHAPTER II

CHAPTER III

CHAPTER IV

CHAPTER V

BOOK II

CHAPTER I

CHAPTER II

CHAPTER III

CHAPTER IV

CHAPTER V

CHAPTER VI

CHAPTER VII

BOOK III

CHAPTER I

CHAPTER II

CHAPTER III

CHAPTER IV

CHAPTER V

BOOK IV

CHAPTER I

CHAPTER II

BOOK V

CHAPTER I

CHAPTER II

CHAPTER III

CHAPTER IV

BOOK VI

CHAPTER I

CHAPTER II

CHAPTER III

CHAPTER IV

CHAPTER V

CHAPTER VI

CHAPTER VII

CHAPTER VIII

CHAPTER IX

CHAPTER X

CHAPTER XI

MAPS AND PLANS

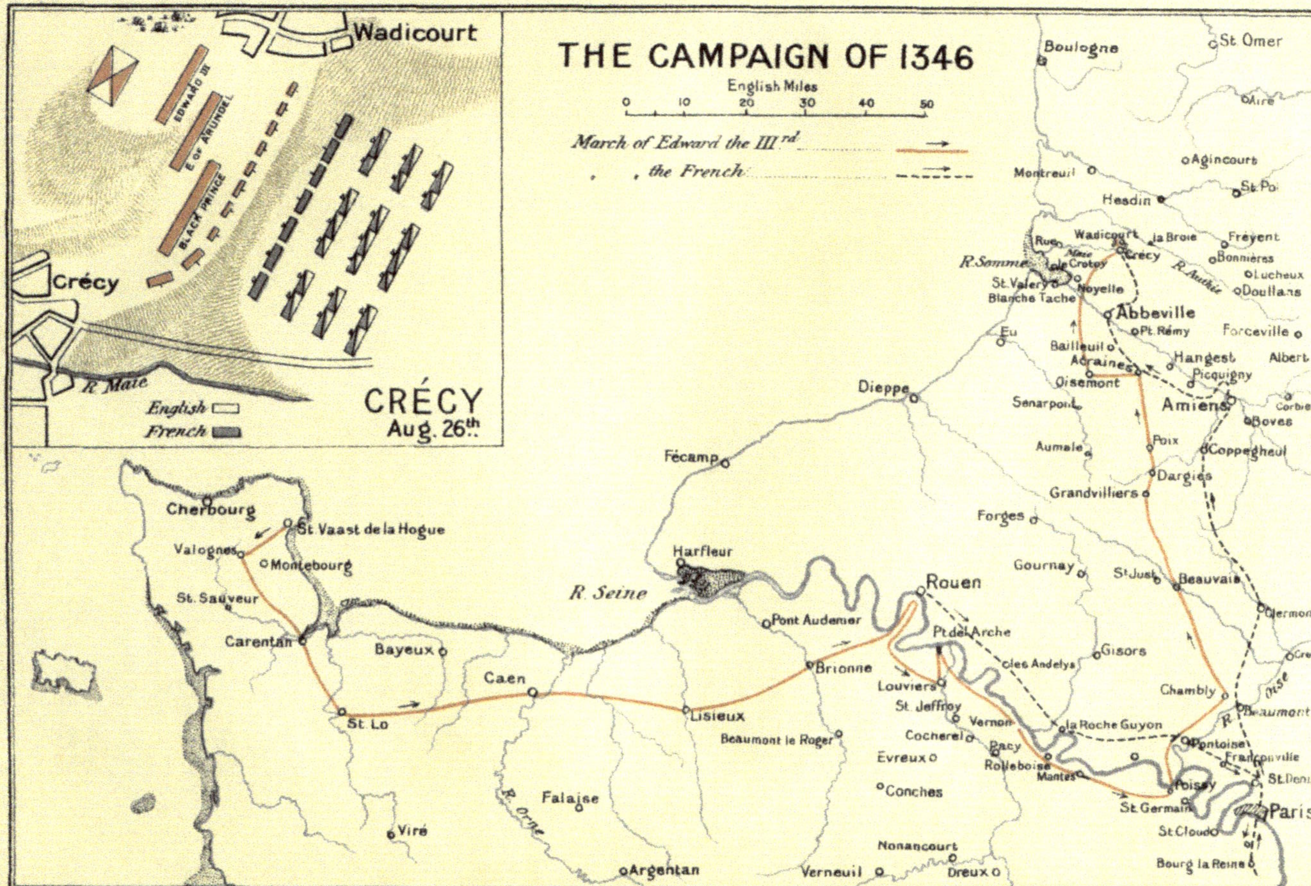

To face page 36.

THE CAMPAIGN OF 1356.

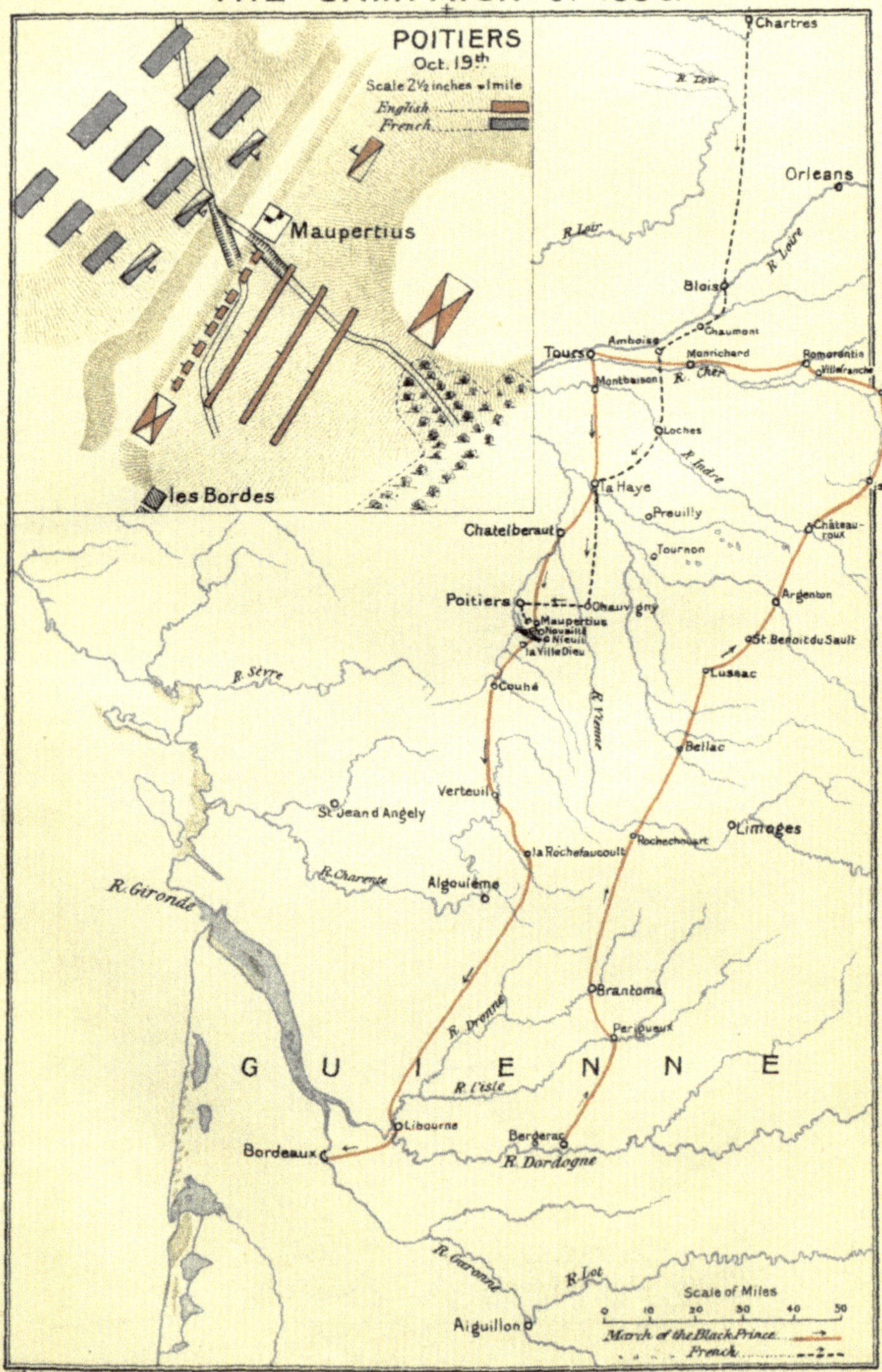

THE CAMPAIGN OF 1367.

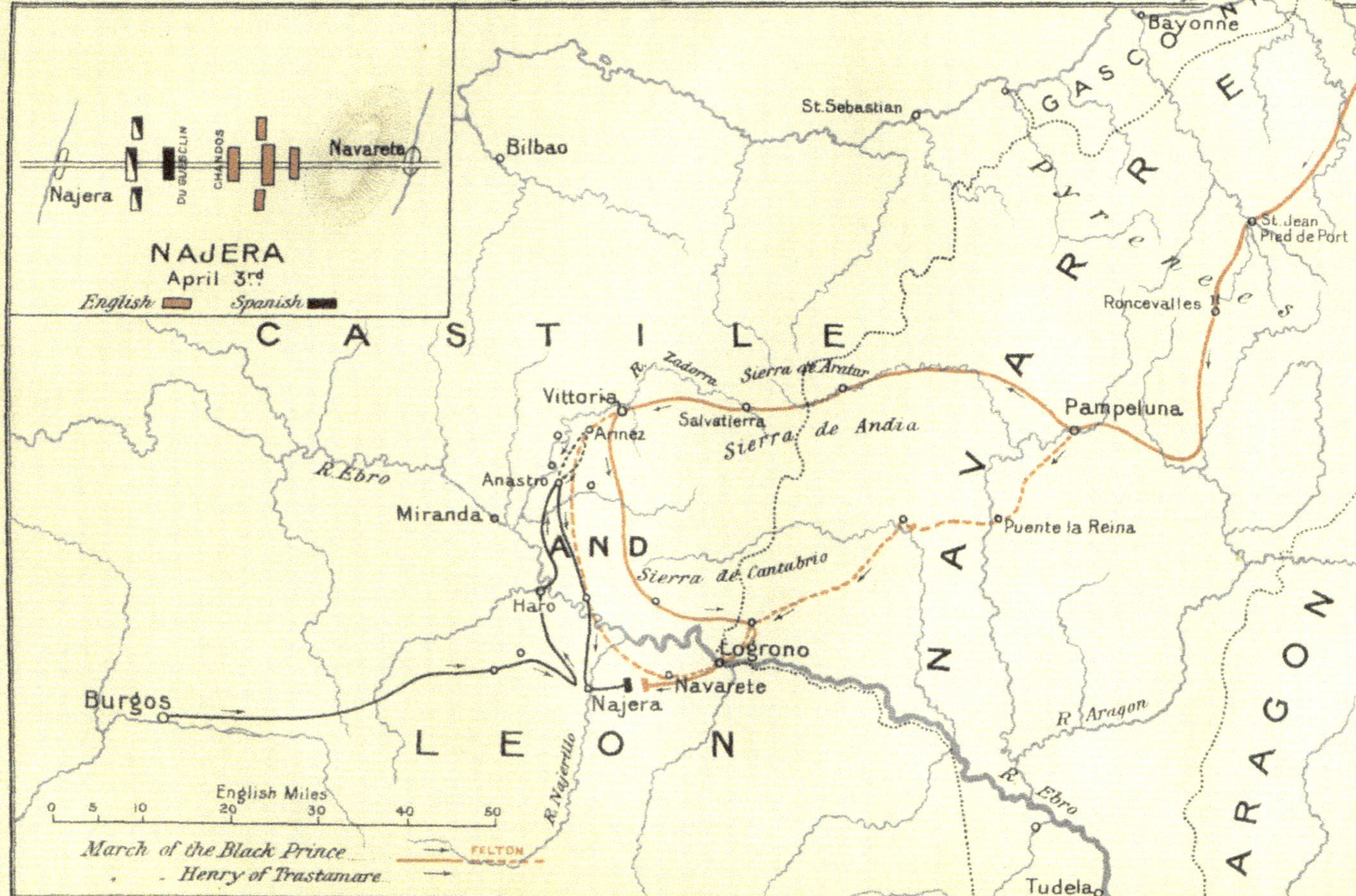

Walker & Boutall del.

To face page 46.

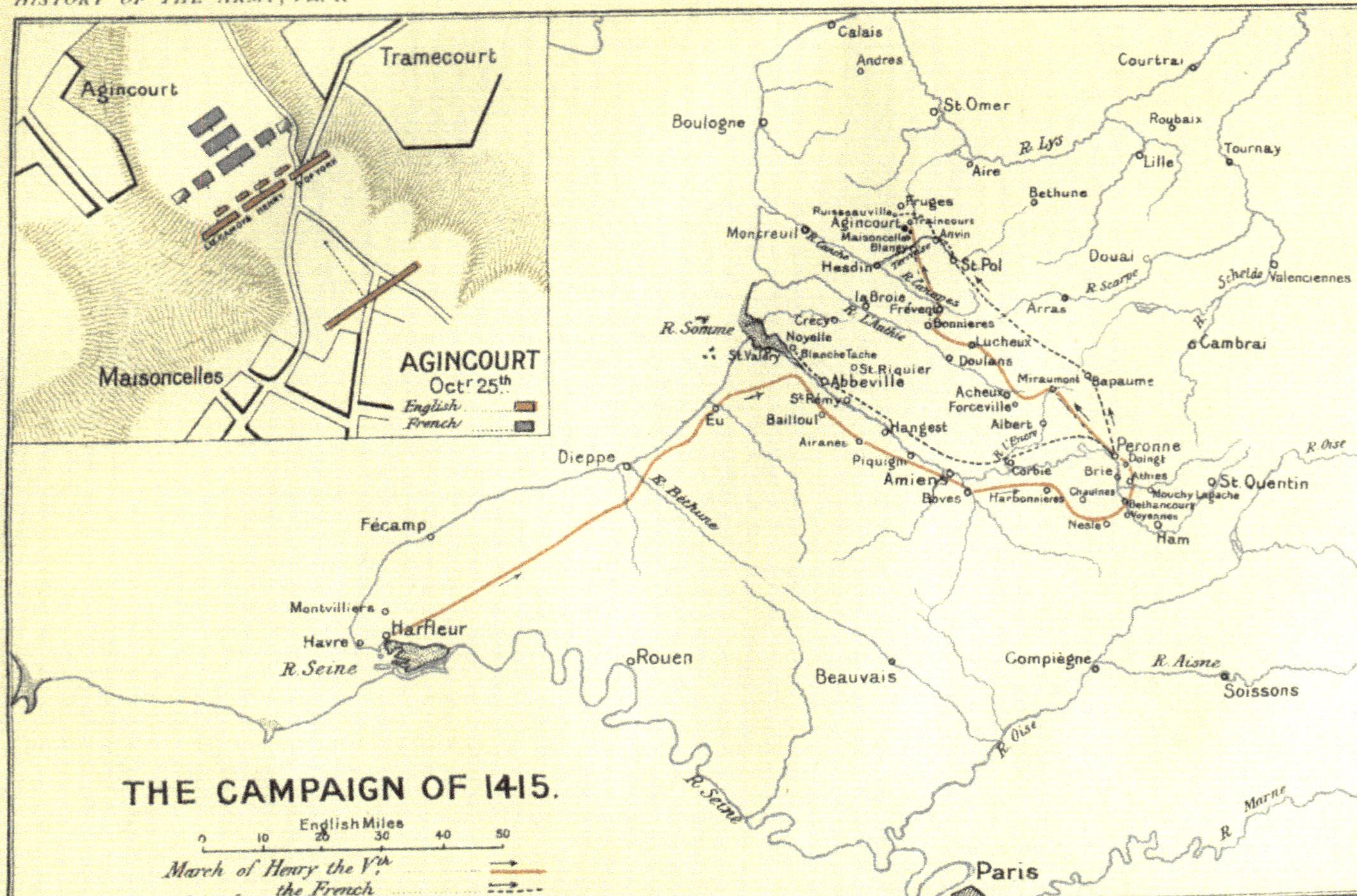

To face page 62.

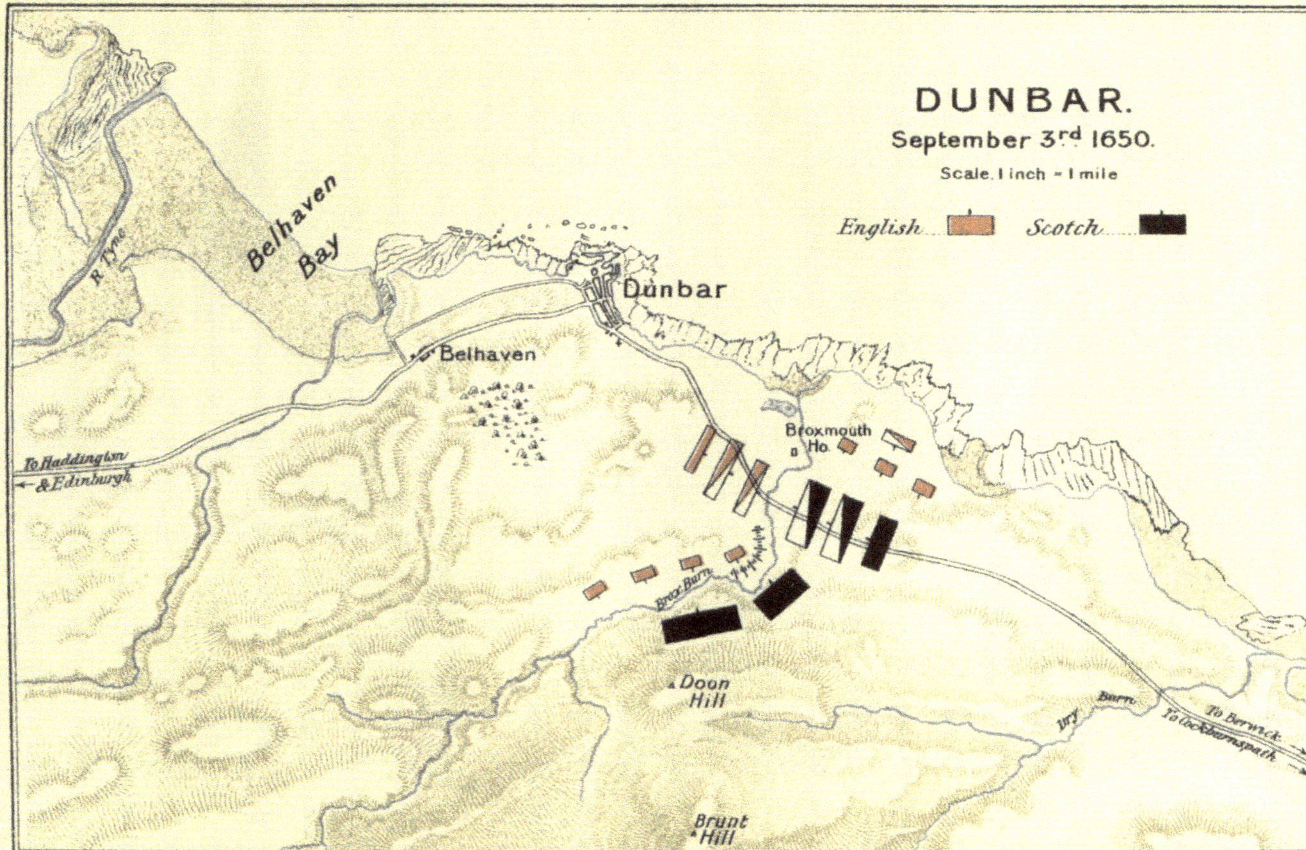

Walker & Boutall, del.

To face page 244.

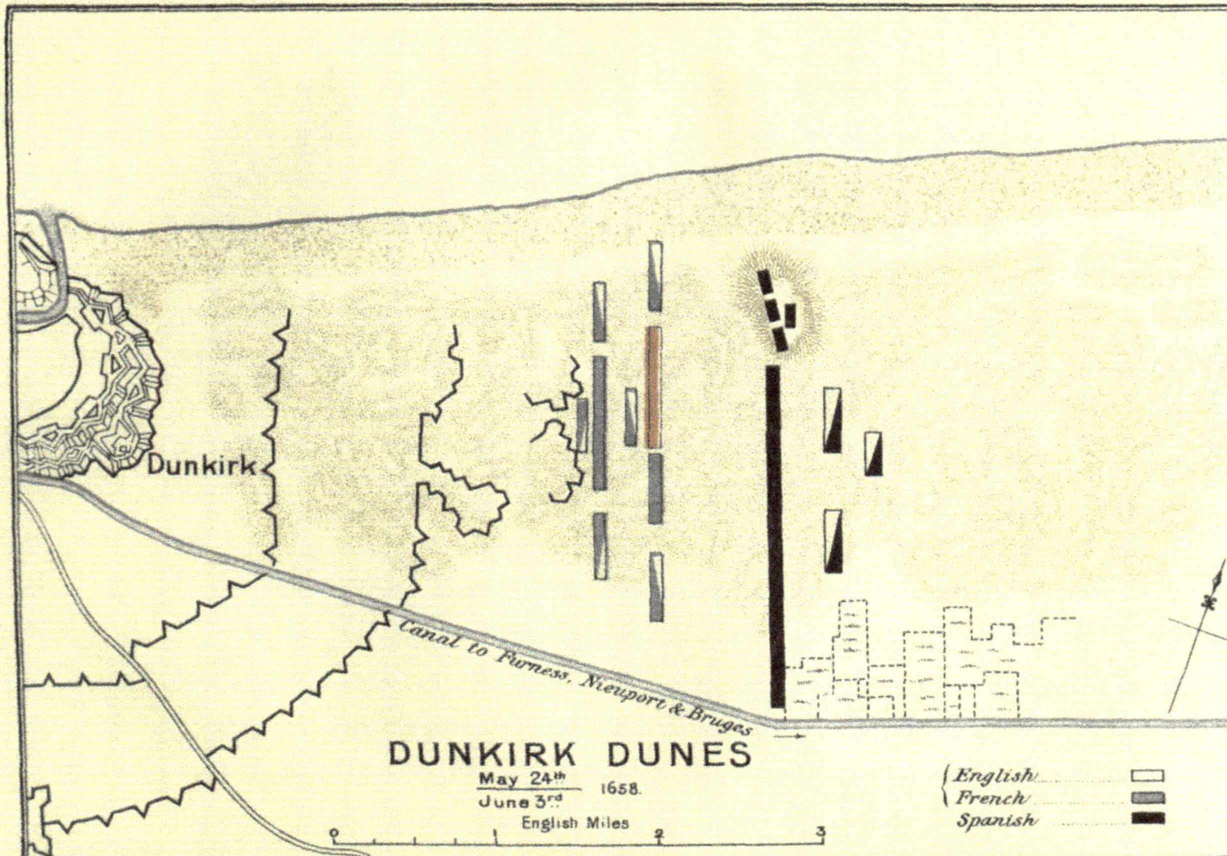

Walker & Boutall, del.

To face page 272.

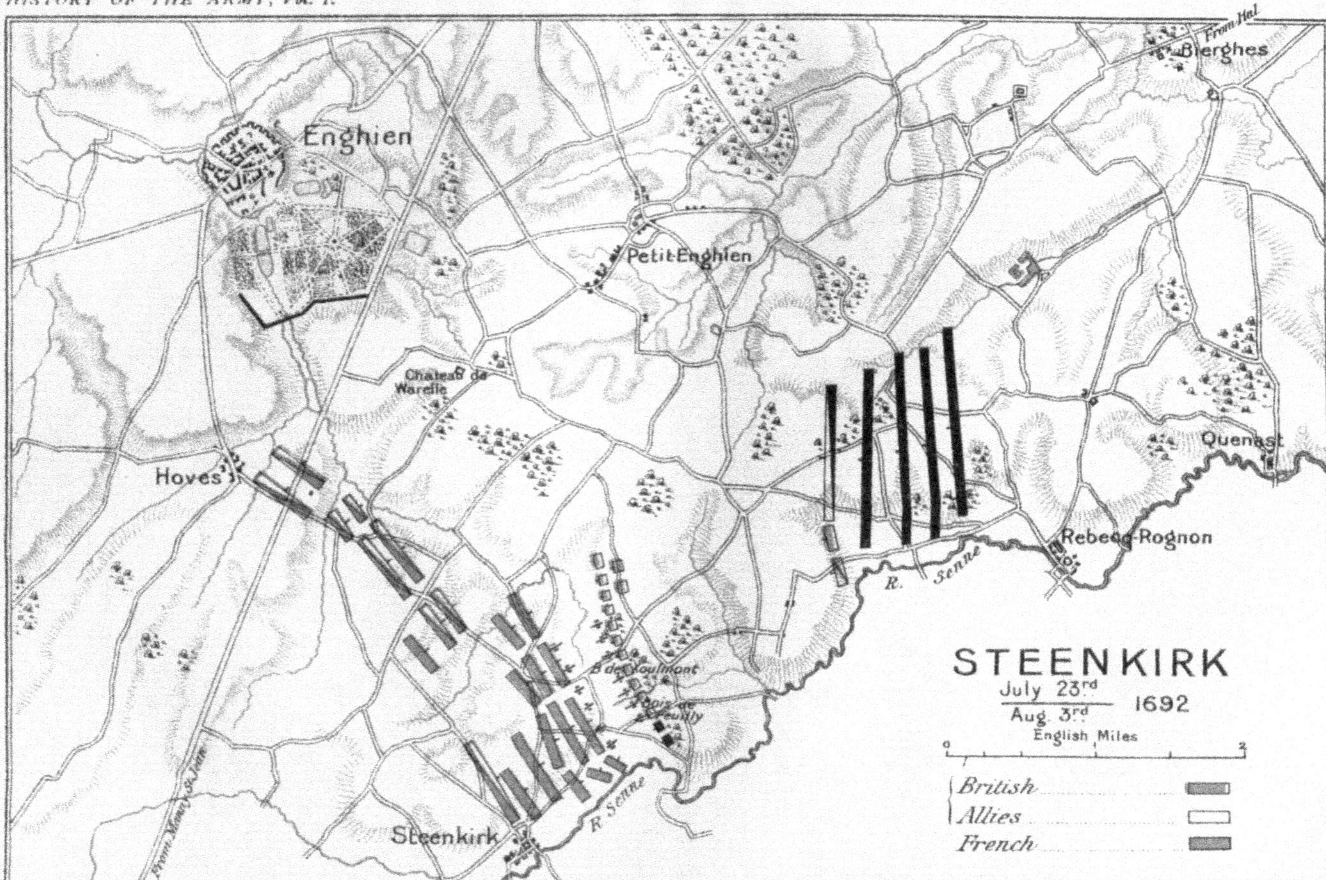

Walker & Boutall del.

To face page 366.

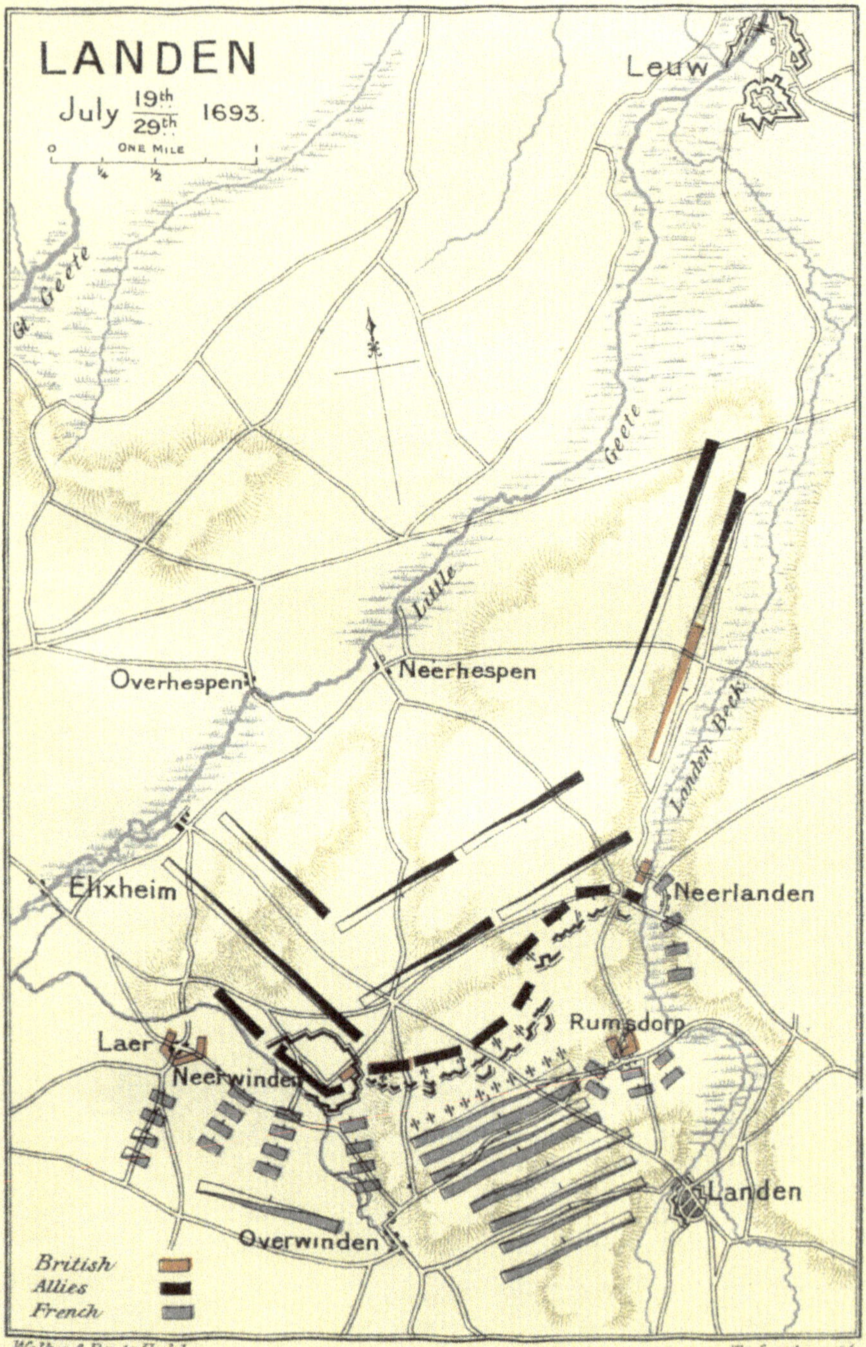
LANDEN
July 19th/29th 1693.
0 ONE MILE 1
¼ ½
Leuw
Gt. Geete
Geete
Little
Landen Beck
Overhespen
Neerhespen
Elixheim
Neerlanden
Rumsdorp
Laer
Neerwinden
Landen
Overwinden
British
Allies
French

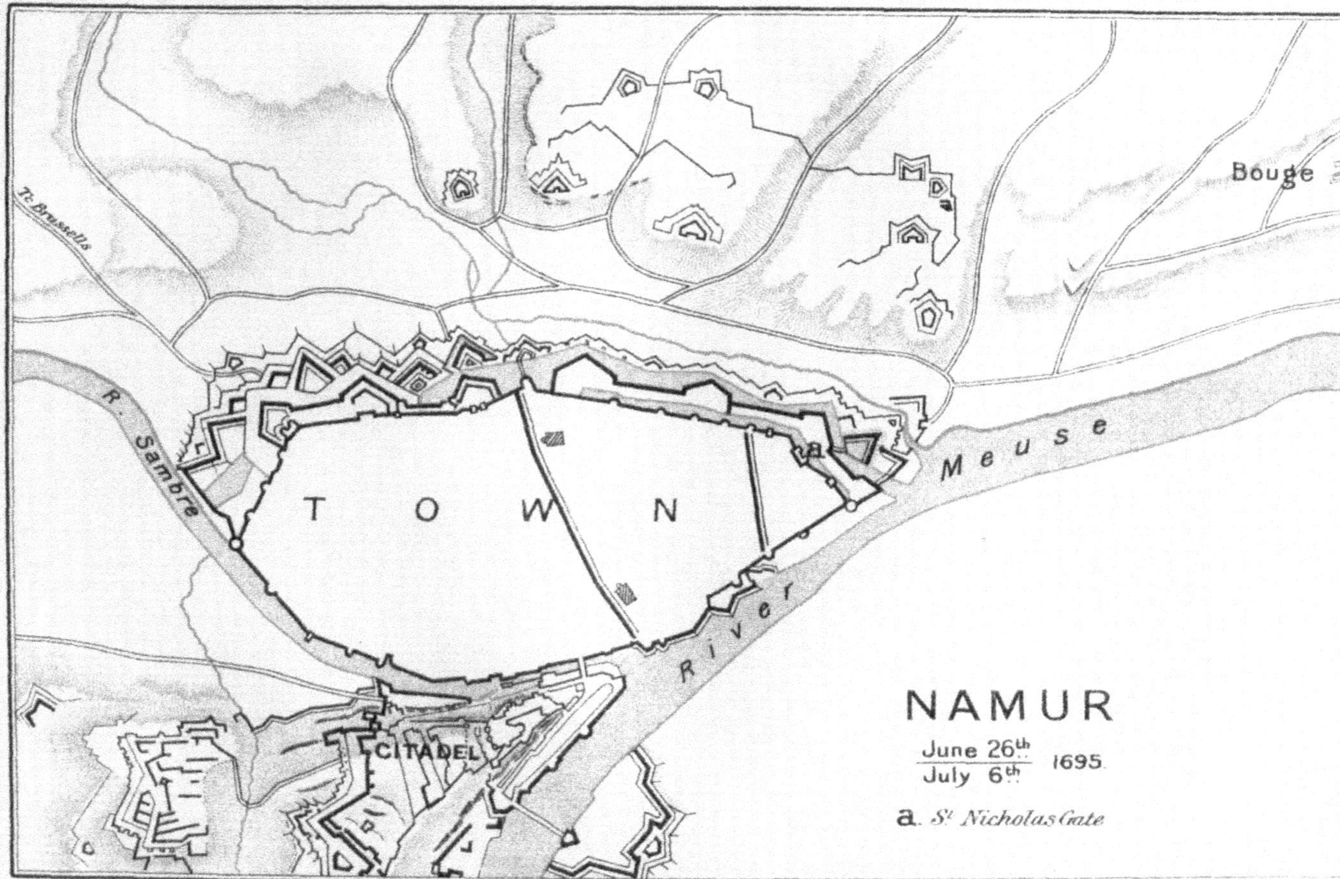

To face page 378.

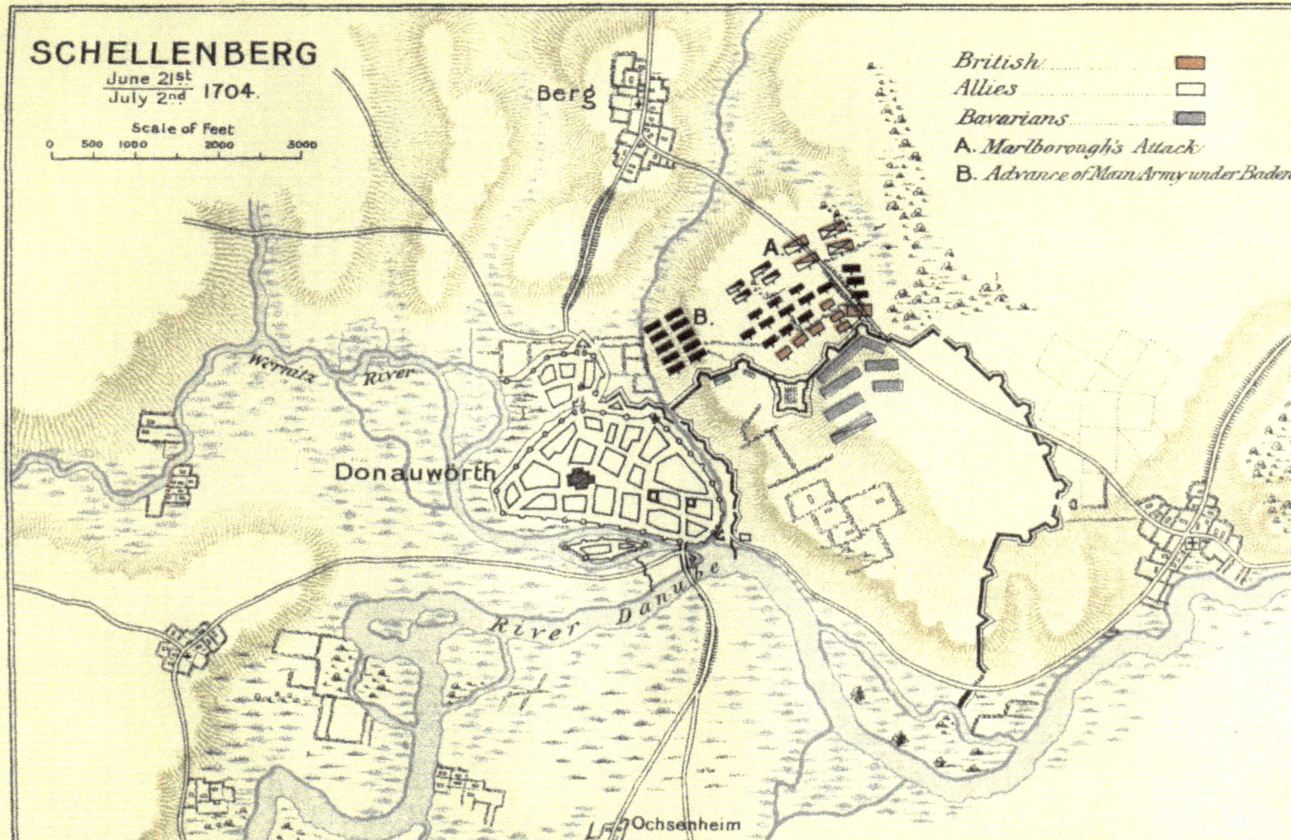

Walker & Boutall del.

To face page 426.

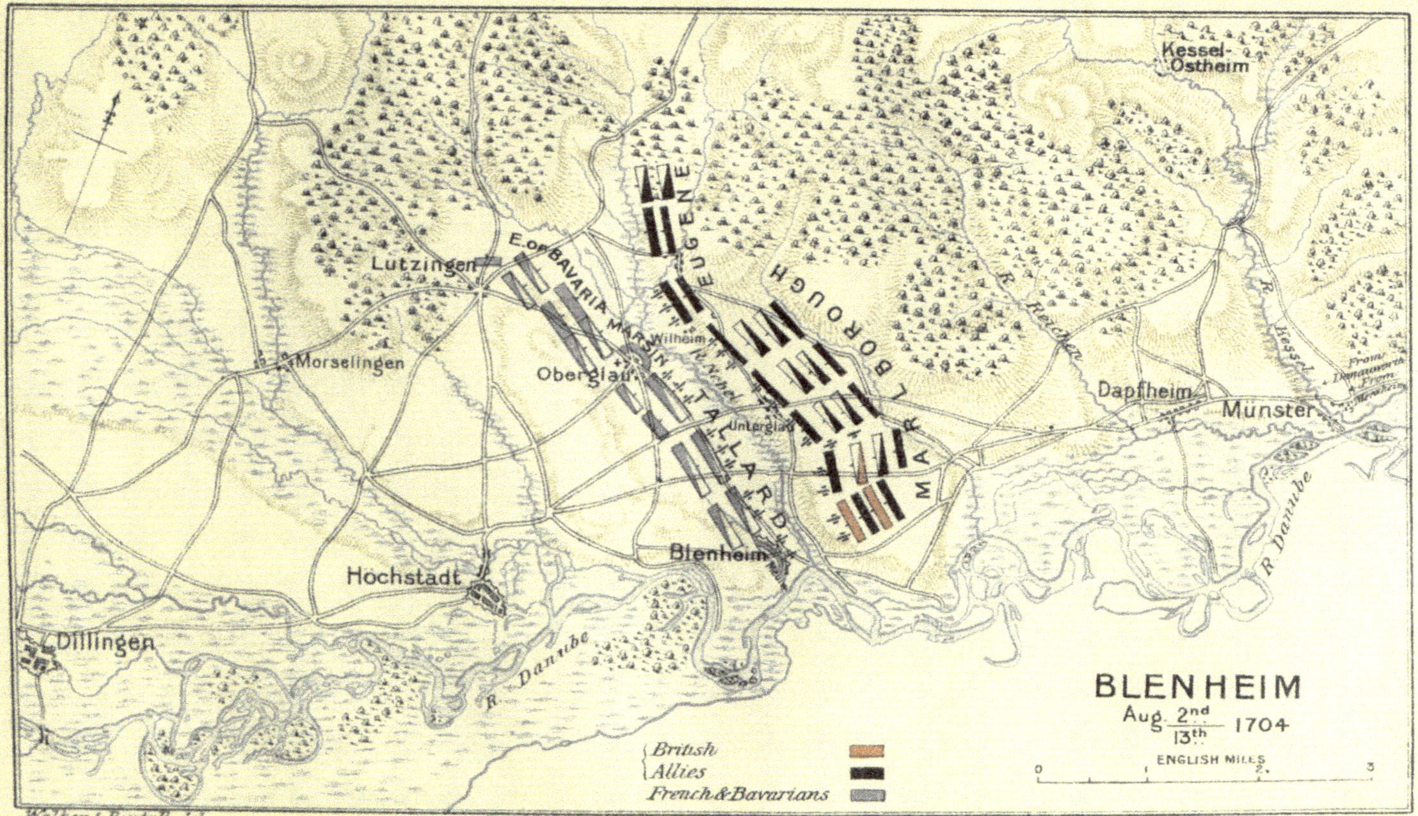

Walker & Boutall del.

To face page 442.

To face page 150.

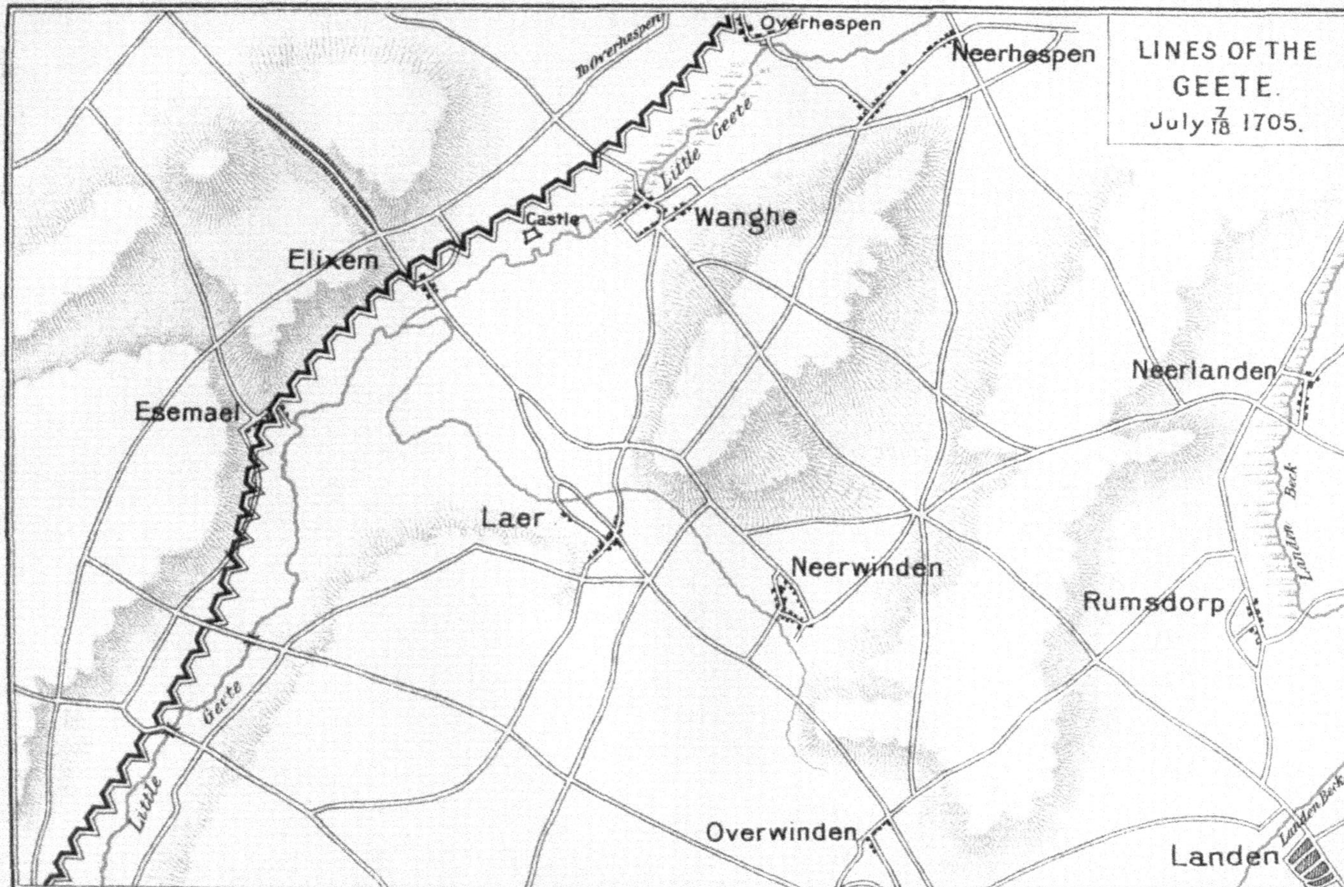

Walker & Boutall, del.

To face page 454.

Walker & Boutall. del.

To face page 402.

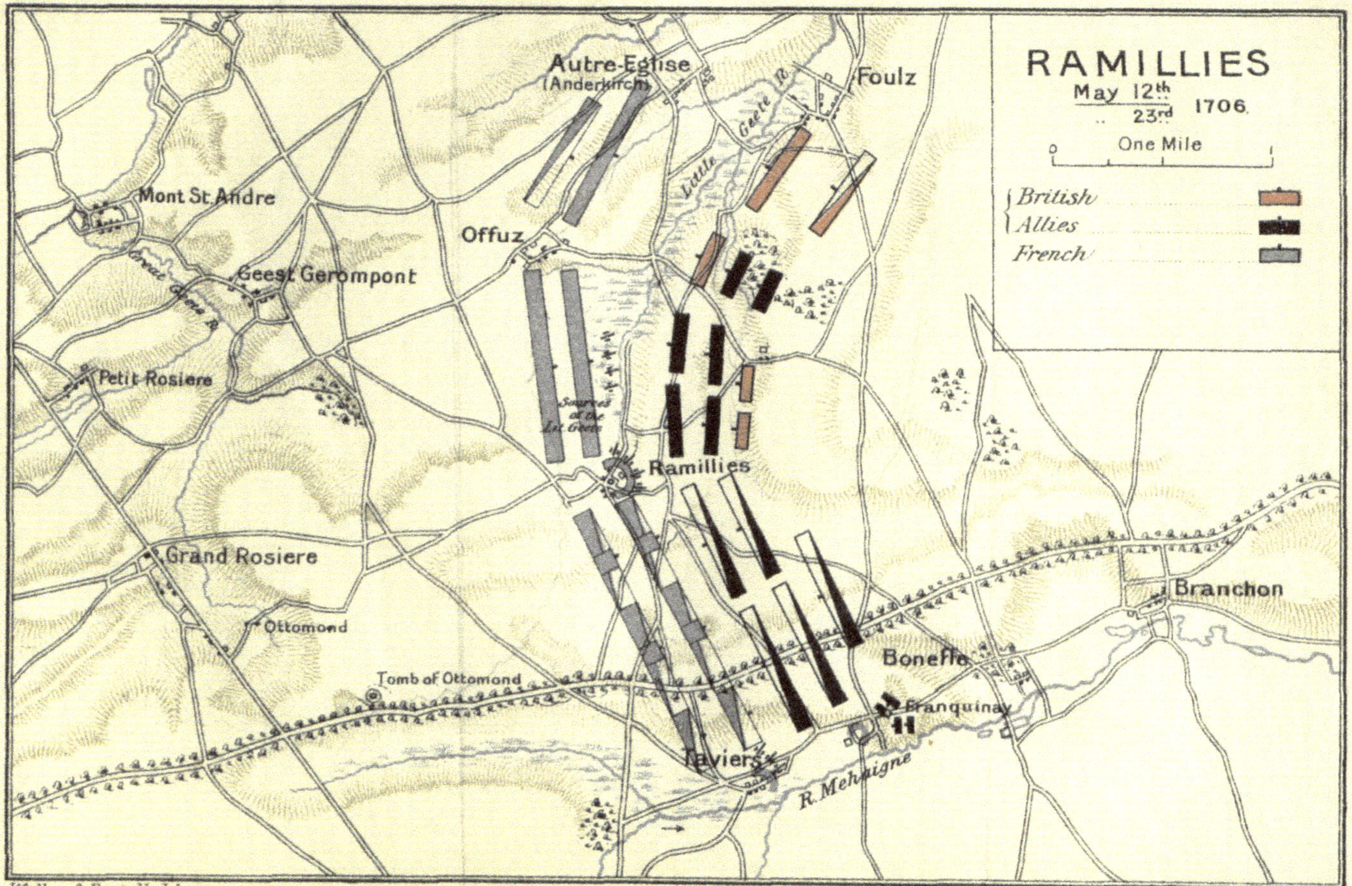

Walker & Boutall del.

To face page 472.

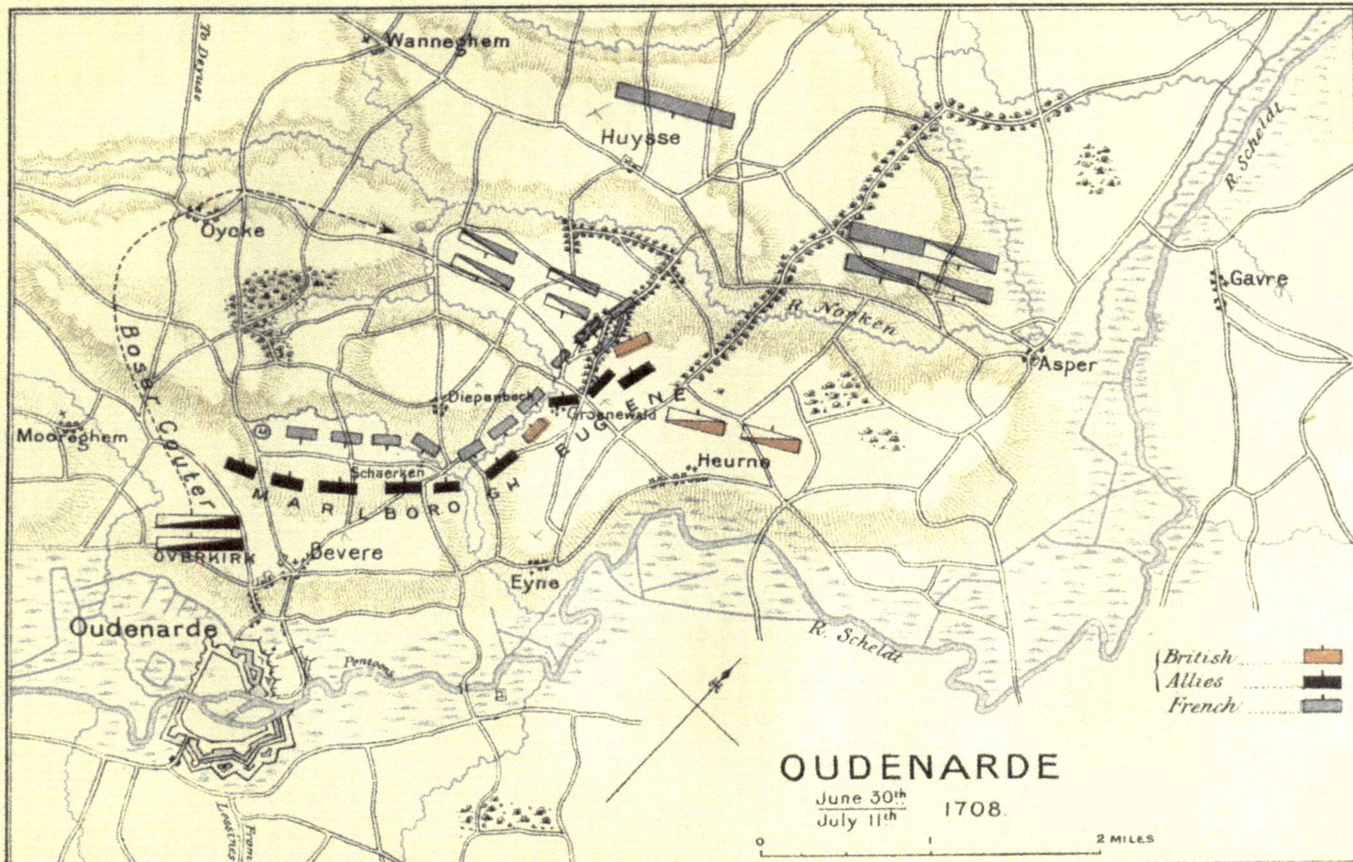

Walker & Boutall del.

To face page 500.

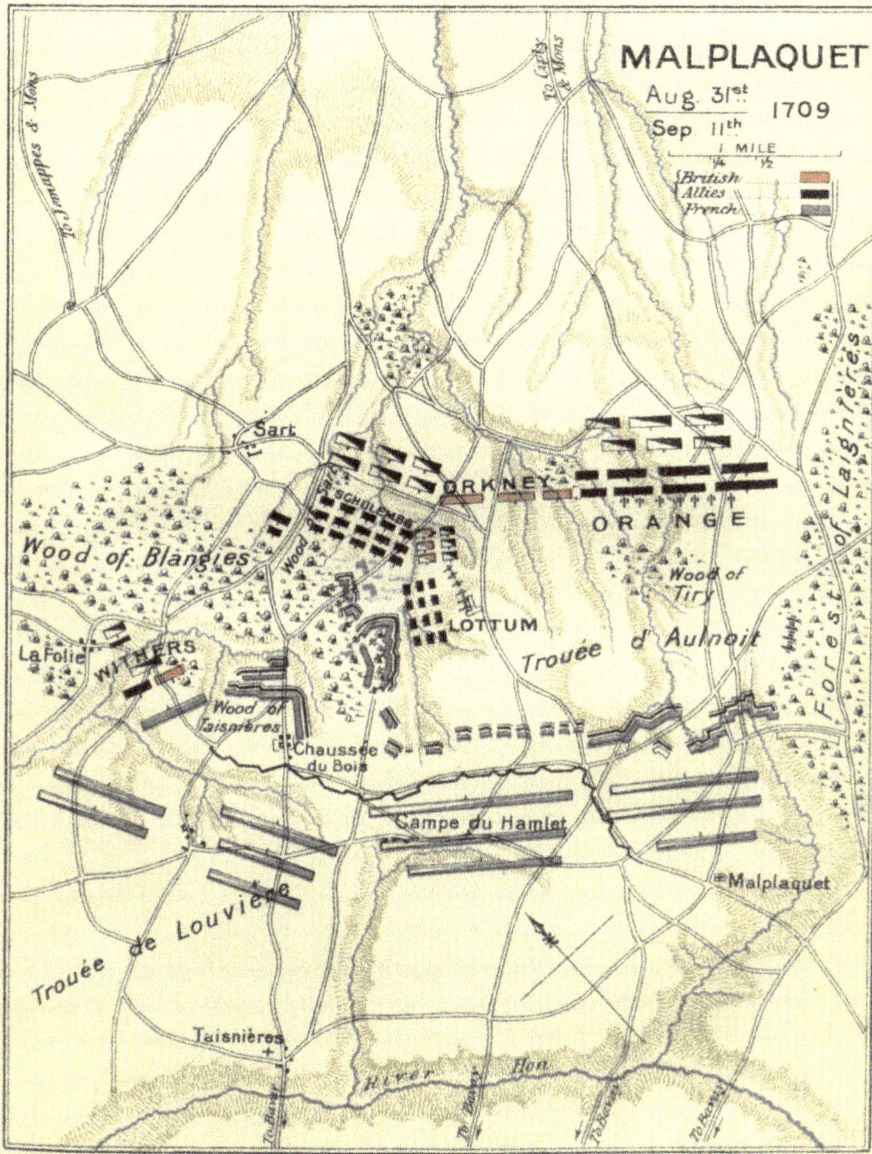

Walker & Boutall del.

To face page 524

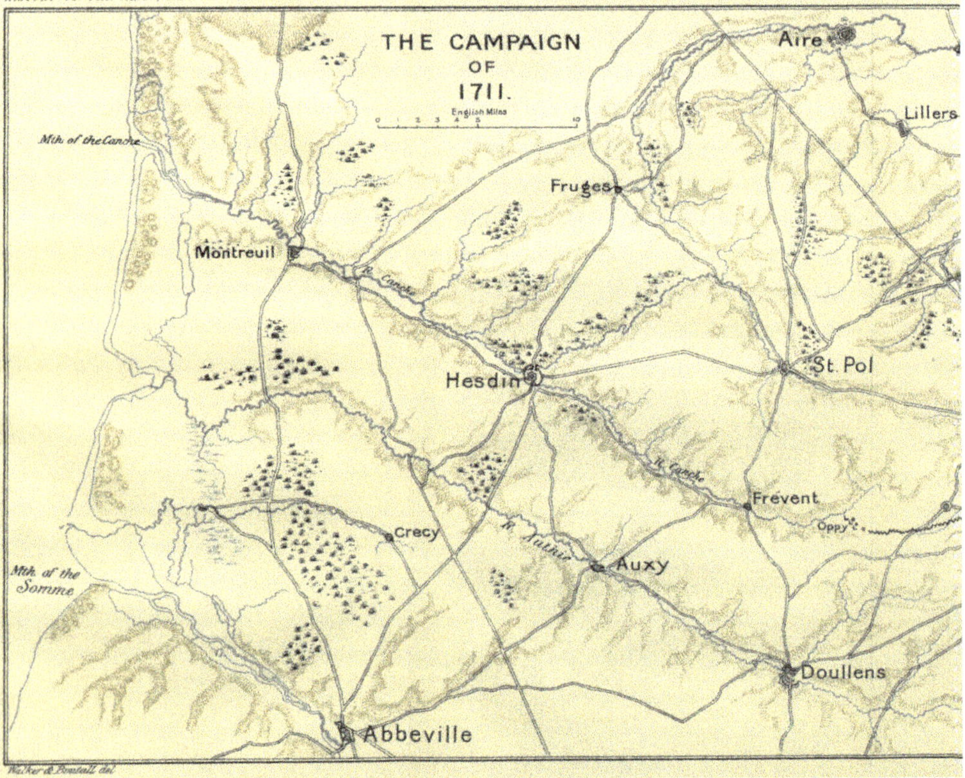

Walker & Boutall del.

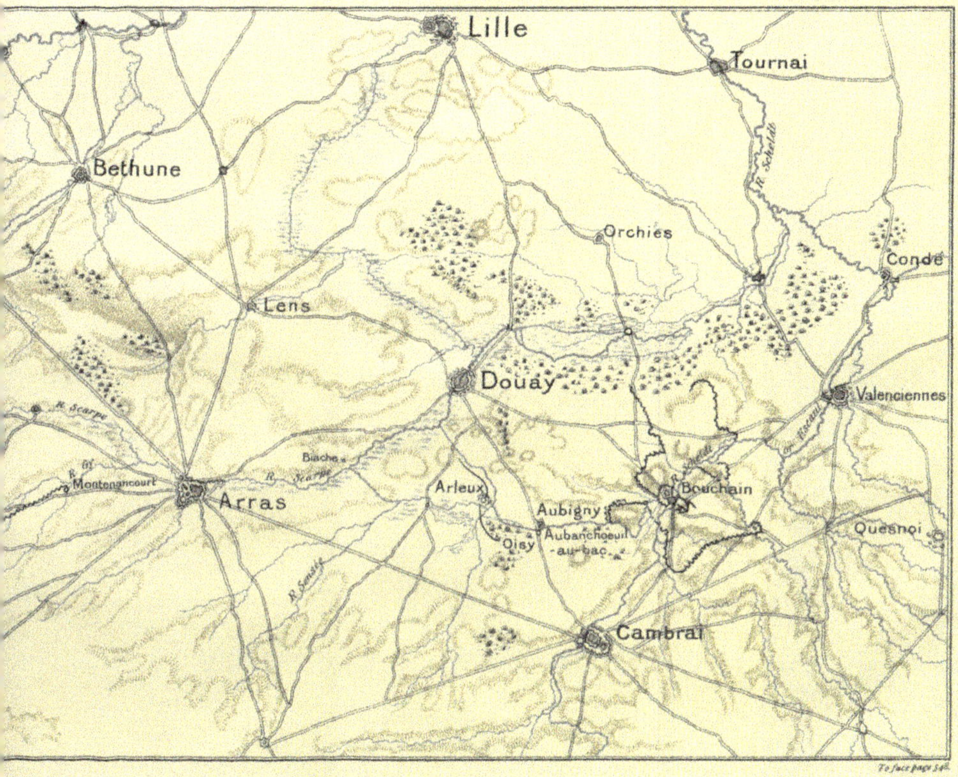

To face page 548.

THE BRITISH ISLES
AND
NORTHERN FRANCE.
English Miles
0 10 50 100
The Garrisons as they existed in 1760 are shown in red thus:- Governorships Lieutenant-Governorships
The Cinque Ports are underlined thus:- Dover
Moray Firth
Inverness
Aberdeen
Montrose
Firth of Tay
Perth
Firth of Forth
Stirling
Berwick
Ayr
Londonderry
Belfast
Enniskillen
Newry
Dundalk
Drogheda
Dublin
Galway
Athlone
Morpeth
Newcastle
Carlisle
Durham
Hull
York
Wigan
Liverpool
Warrington
Chester
Gainsborough
The Wash
R. Humber
R. Ribble

Walker & Boutall, del.

MAP I.

End of Vol. I.

THE NETHERLANDS
In the 18th Century
English Miles
0 10 20 30 40 50
The United Netherlands
Spanish (Austrian) Netherlands
French Netherlands
Names and Boundaries of the
Seven Provinces shown in red
Fortified Towns
3°
4°
5°
6°
7°
53°
52°
GRONINGEN
Groningen
Leeuwarden
FRIESLAND
DRENTHE
OVERYSSEL
Deventer
Zutphen
GELDERLAND
Doesburg
Arnheim
UTRECHT
Utrecht
Amsterdam
R. Rhine
The Hague
Ryswick
Rotterdam
Brill
R. Leck
R. Yssel
Fl. Schenk
Waal
Nimegen
Grave
Cleves
Gennep
Goch
Uden
Xanten
Wesel
R. Lippe
Rheinberg
Williamstadt
Geertruidenberg
Terheyden
Breda
Tilburg
Bergen op Zoom
DUTCH BRABANT
CLEVES
ZEALAND
N. Beveland
I. of Walcheren
Middleburg

FLANDERS
BRABANT
HAINAULT
NAMUR
BISHOPRIC OF LIEGE
LUXEMBURG
ARTOIS
FRANCE
Dunkirk
Ostend
Sluys
Damme
Bruges
Cadsande
Hulst
Stabroek
Antwerp
Ghent
Furnes
Dixmuyde
Bergues
Roulers
Ypres
Courtrai
Menin
Oudenarde
Dendermond
Mechlin
Aerschot
Diest
Alost
Brussels
Louvain
Grammont
Lessines
Ath
Tournay
Lille
Aire
St Venant
la Bassée
Bethune
Lens
Orchies
Douay
Arras
Bouchain
Cambray
Condé
Valenciennes
Mons
Maubeuge
Malplaquet
Charleroi
Nivelles
Genappe
Ramillies
Namur
Dinant
Philipville
Marienburg
Huy
Liege
Landen
Hasselt
Maestricht
Limburg
Aix la Chapelle
Düren
Cologne
Bonn
Venloo
Roermond
(Ruremonde)
Stevensweert
Maeseyk
Helchteren
Crefeld
Dusseldorf
Nuys
Luxemburg
Treves
Trarbach
To Coblenz
R. Lys
R. Scheldt
R. Scarpe
R. Sambre
R. Meuse
R. Roer
R. Rhine
R. Moselle
R. Saar
3°
4°
5°
6°
7°
50°
51°
Walker & Boutall del.
End of Vol. I.

MAP II.

Walker & Boutall, del.

MAP

SPAIN AND PORTUGAL
English Miles
0 100 200 300
Names and Boundaries of the Spanish Kingdoms in the time of the Black Prince are printed in red.
Roads
Fortified Towns
F R A N C E
Pyrenees Mountains
KINGDOM OF NAVARRE
KINGDOM OF ARAGON
CATALONIA
VALENCIA
MURCIA
Rousillon
Bayonne
Roncevalles
Pampeluna
Puente de la Reina
Navarete
Tudela
Saragossa
R. Ebro
Alfaraz
Almenara
Agramont
Lerida
Perpignan
Gerona
Barcelona
Tarragona
Tortosa
R. Ebro
Traguera
Vinaroz
San Mateo
Oropesa
Castellon de la Plana
Villa Real
Nules
Murviedro
Valencia
Cuenca
Requena
Almanza
Montalegre
Ontinyence
Villena
Denia
Alicante
Cartagena
Minorca
Mahon
Majorca
Iviça
Cifuentes
From Bordeaux
42
40
38
36

End of Vol. I.

Walker & Boutall, del.

MAP

12°
14°
16°
18°
20°
54°
52°
50°
48°
I. of Rügen
Rügenwalde
Danzig
Stralsund
Swedish Pomerania
P o m e r a n i a
ECKLENBURG
Strelitz
Stettin
Netze District
Thorn
Vistula
PRUSSIA
New Ruppin
BRANDENBURG
Werben
Oder
Zorndorf
Custrin
Berlin
Brandenburg
Potsdam
Spree
Frankfurt
Kunersdorf
Schwiebus
Kay
Warta
POLAND
Elbe
Wittenburg
Lubben
Cottbus
Glogau
Wohlau
Breslau
Warta
Torgau
Halle
SAXONY
Breitenfeld
Leipzig
Rossbach
Weimar
Jena
Hochkirch
Görlitz
Liegnitz
Leuthen
Dresden
Hohenfriedberg
Brieg
Mollwitz
Kesselsdorf
Pirna
Maxen
Königstein
Zittau
Chemnitz
Toplitz
Lobositz
Leitmeritz
Eger
Glatz
Kosel
Melnitz
Königgratz
Prague
Kolin
Chotusitz
Czaslau
GALICIA
Bohemia
Beraun
Moldau
Olmütz
AUSTRIA
HUNGARY
Passau
BAVARIA
Inn
Linz
Danube
Vienna
Munich
Salzburg

End of Vol. I.

BOOK I

CHAPTER I

THE history of the British Army is commonly supposed to begin with the year 1661, and from the day, the 14th of February, whereon King Charles the Second took over Monk's Regiment of Foot from the Commonwealth's service to his own, and named it the Coldstream Guards. The assumption is unfortunately more convenient than accurate. The British standing army dates not from 1661 but from 1645, not from Monk's regiment but from the famous New Model, which was established by Act of the Long Parliament and maintained, in substance, until the Restoration. The continuity of the Coldstream regiment's existence was practically unbroken by the ceremony of Saint Valentine's day, and this famous corps therefore forms the link that binds the New Model to the Army of Queen Victoria.

But we are not therefore justified in opening the history of the army with the birth of the New Model. The very name indicates the existence of an earlier model, and throws us back to the outbreak of the Civil War. There then confronts us the difficulty of conceiving how an organised body of trained fighting men could have been formed without the superintendence of experienced officers. We are forced to ask whence came those officers, and where did they learn their profession. The answer leads us to the Thirty Years' War and the long struggle for Dutch Independence, to the English and Scots, numbered by tens, nay, hundreds of thousands, who fought under Gustavus Adolphus and Maurice of Nassau. Two noble regiments still abide with us as representatives of these two schools, a standing record of our army's 'prentice years.

But though we go back two generations before the Civil War to find the foundation of the New Model Army, it is impossible to pause there. In the early years of Queen Elizabeth's reign we are brought face to face with an important period in our military history, with a break in old traditions, an unwilling conformity with foreign standards, in a word, with the renascence in England of the art of war. For there were memories to which the English clung with pathetic tenacity, not in Elizabeth's day only but even to the midst of the Civil War, the memories of King Harry the Fifth, of the Black Prince, of Edward the Third, and of the unconquerable infantry that had won the day at Agincourt, Poitiers, and Creçy. The passion of English sentiment over the change is mirrored to us for all time in the pages of Shakespeare; for no nation loves military reform so little as our own, and we shrink from the thought that if military glory is not to pass from a possession into a legend, it must be eternally renewed with strange weapons and by unfamiliar methods. This was the trouble which afflicted England under the Tudors, and she comforted herself with the immortal prejudice that is still her mainstay in all times of doubt,

> "I tell thee herald,
> I thought upon one pair of English legs
> Did march three Frenchmen."

The origin of the new departures in warfare must therefore be briefly traced through the Spaniards, the Landsknechts, and the Swiss, and the old English practice must be followed to its source. Creçy gives us no resting-place, for Edward the Third's also was a time of military reform; the next steps are to the Battle of Falkirk, the Statute of Winchester, and the Assize of Arms; and still the English traditions recede before us, till at last at the Conquest we can seize a great English principle which forced itself upon the conquering Normans, and ultimately upon all Europe.

This then is the task that is first attempted in this book: to follow, however briefly and imperfectly, the growth of the English as a military power to the time of its first manifestation at Creçy, and onward to the supreme day of Agincourt; then through the decay under the blight of the Wars of the Roses to the revival under the Tudors, and to the training in foreign schools which prepared the way for the New Model and the Standing Army. The period is long, and the conditions of warfare vary constantly from stage to stage, but we shall find the Englishman, through all the changes of the art of war unchangeable, a splendid fighting man.

The primitive national army of the English, as of other Teutonic nations, consisted of the mass of free landowners between the ages of sixteen and sixty; it was called in the Karolingian legislation by the still existing name of *landwehr*, and known in England as the fyrd. Its term of service was fixed by custom at two months in the year. The force was reorganised by King Alfred or by his son through the division of the country into military districts, every five hides of land being required to provide an armed man at the king's summons, and to provide him with victuals and with pay. Further, all owners of five hides of land and upwards were required to do thane's service, that is to say, to appear in the field as heavily-armed men at their own charge, and to serve for the entire campaign. The organisation of the thanes was by shires. With the conquest of England by Canute a new military element was introduced by the establishment of the royal body-guard, a picked force of from three to six thousand Danish troops, which were retained by him after the rest of the army had been sent back to Denmark, and were known as the house-carles.

It was with an army framed on this model—the

raw levies of the fyrd and the better trained men of the body-guard—that King Harold, flushed with the victory of Stamford Bridge, marched down to meet the invasion of William of Normandy. The heavily-armed troops wore a shirt of ringed or chain-mail, and a conical helmet with a bar protecting the nose; their legs were swathed in bandages not wholly unlike the "putties" of the present day, and their arms were left free to swing the Danish axe. They carried also a sword, five missile darts, and a shield, but the axe was the weapon that they loved, for the Teutonic races, unlike the Latin, have ever preferred to cut rather than to thrust. The light-armed men, who could not afford defensive armour, came into the field with spear and shield only. Yet the force was homogeneous in virtue of a single custom, wherein lies the secret of the rise of England's prowess as a military nation. Though the wealthy thanes might ride horses on the march, they dismounted one and all for action, and fought, even to the king himself, on their own feet.[1]

The force was divided into large bands or battalions, of which the normal formation for battle was a wedge broadening out from a front of two men to a base of uncertain number; the officers and the better armed men forming the point, backed by a dense column of inferior troops. It was with a single line of such wedges, apparently from five-and-twenty to thirty of them, that Harold took up his position to bar the advance of the Norman army. Having no cavalry, he had resolved to stand on the defensive, and had chosen his ground with no little skill. His line occupied the crest of a hill, his flanks were protected by ravines, and he had dug across the plain on his front a trench which was sufficient to check a rapid advance of cavalry. Moreover, he had caused each battalion to ring itself about

[1] An alien captain of the garrison of Hereford tried in 1055 to break through this custom. "Anglos *contra morem* in equis pugnare jussit" (see Hewitt, vol. i. p. 17).

with sharp stakes, planted into the ground at intervals with the points slanting outwards, as a further protection against the attack of horse.[1] The reader should take note of these stakes, for he will find them constantly reappearing up to the seventeenth century. There then the English waited in close compact masses, a wall of shields within a hedge of stakes, the men of nine-and-twenty shires under a victorious leader. There is no need to enter into details of the battle. The English, as has been well said,[2] were subjected to the same trial as the famous squares at Waterloo, alternate rain of missiles and charges of cavalry, and as yet they were unequal to it. Harold's orders had been that not a man should move, but when the Normans, after many fruitless attacks, at last under William's direction simulated flight, the order was forgotten and one wing broke its ranks in headlong pursuit of the fugitives. Possibly, if Harold had been equal to the occasion, a general advance might have saved the day, but he made no such effort, and he was in the presence of a man who overlooked no blunder. The pursuing wing was enveloped by the Normans and annihilated ; and then William turned the whole of his force against the fragment of the line that remained upon the hill. The English stood rooted to the ground enduring attack after attack, until at last, worn out with fatigue and choked with dead and wounded, they were broken and cut down, fighting desperately to the end. Indiscipline had brought ruin to the nation ; and England now passed, to her great good fortune, under the sway of a race that could teach her to obey.

But the English had still one more lesson to learn. Many of the nobles, chafing against the rule of a foreigner, forsook their country and, taking service with the Byzantine emperors, joined the famous Varangian

[1] This seems to be the simplest and likeliest solution of the problem of the palisade, which has provoked such acrimonious controversy (see Köhler, vol. i. p. 8).

[2] Oman.

Guard of the Emperor Alexius Comnenus. At Durazzo they for the second time met the Normans, under the command of Robert Guiscard. True to their custom, they dismounted and fought on foot, a magnificent corps, the choicest of the whole army. As at Hastings, the Normans attacked and were repulsed, and as at Hastings, the undisciplined English broke their ranks in pursuit. Robert Guiscard saw his opportunity, hurled his cavalry on to their flank, and then surrounding them on all sides cut them down, in spite of a furious resistance, to the very last man. So perished these untameable, unteachable spirits, the last of the unconquered English.

The Conquest was immediately followed by the institution of knight-service. But this system, as introduced into England, differed in many material respects from that which reigned on the continent of Europe. It was less distinctly military in character, and far less perfect as an organisation for national defence. The distribution of England into knight's fees, however clearly it might be mapped out on paper, was a work of time and not to be accomplished in a day. Moreover, there was disloyalty to be reckoned with; for the English were a stiff-necked people, and were not readily reconciled to the yoke of their new masters. We find, therefore, that in very early days the practice of accepting money in lieu of personal service crept in, and enabled the Norman kings to fight their battles with hired mercenaries. For this reason England has been called the cradle of the soldier; the soldier being the man who fights for pay, *solde*, *solidus*, or, as we may say by literal translation of the Latin, the man who fights for a shilling.

The sole military interest therefore of the reigns of the Norman kings is to follow the breakdown of the feudal system for military purposes, and the rapid reversion to the Saxon methods and organisation. William Rufus was the first to appeal to the English to arm in his cause, and he did so twice with success. But in the seventh year of his reign he played them a trick which lost him their confidence for ever. The fyrd had furnished twenty thousand men for service against the Norman rebels in France, and had provided every man, at the cost of his shire, with ten shillings for the expenses of his journey or, to use a later expression, for his conduct-money. William met them at the rendezvous, took their two hundred thousand shillings from them to hire mercenaries withal, and dismissed them to their homes. This Rufus has been selected by an historian of repute as the earliest example of an officer and a gentleman; he should also be remembered as the first officer who set the fashion, soon to become sadly prevalent, of misappropriating the pay of his men. The reader should note in passing this early instance of conduct-money, for we shall find in it the germ of the Queen's shilling.

The reign of Henry the First is interesting in that it shows us English knights serving in the field against
Robert of Normandy under the walls of Tenchbrai. 1106.
We find that the old order of battle, the single line of Hastings,[1] has disappeared and has given place to the three lines of the Byzantine school, but that, strange to say, the Saxons have forced their peculiar principle upon the Normans. Henry caused his English and Norman knights to dismount, formed them into a solid battalion and placed himself at their head, keeping but one small body still on their horses. The enemy's cavalry attacked Henry's mounted men and dispersed them; but the phalanx of the dismounted remained unbroken, pressed on against the rabble of hostile infantry, broke it down and almost annihilated it. The victory was hailed by the English as atonement for the defeat at Hastings, so bitter even then was the rivalry between ourselves and
our gallant neighbours across the channel. Ten years 1116.
later the English were again in France, fighting not only

[1] A single line of course must not be understood as a single rank. It was a line of wedges or, as we should now say, a line of columns.

against rebellious Norman barons but against their ally, the French King Louis the Sixth. A long and desultory
1125, war was closed by the action of Brenville. Again Henry
25th March. dismounted four hundred out of five hundred of his knights and following the tactics of Tenchbrai won, though not without hard fighting, a second victory. A third engagement, known as the battle of Beaumont, saw the old English practice repeated for the third time with signal success; but here must be noticed the entry of a new force, a company of archers, which contributed not a little to the fortunate issue of the day. For as the Norman cavalry came thundering down on the English battalion, the archers moved off to their left flank and poured in such a shower of arrows that the horsemen were utterly overthrown. These archers must not be confounded with the famous English bowmen of a later time, for most probably they were merely copied, like the order of battle, from the Byzantine model; but they taught the English the second of two useful lessons. Henry had already discovered that dismounted knights could hold their own against the impetuous cavalry of France; he now learned that the attack of horse could be weakened almost to annihilation by the volley of archers. This, at a time when cavalry held absolute supremacy in war, was a secret of vital importance, a secret indeed which laid the foundation of our military power. Henry, evidently alive to it, encouraged the practice of archery by ordaining that, if any man should by accident slay another at the butts, the misadventure should not be reckoned to him as a crime.

The miserable reign of Stephen, so unsatisfactory to the general historian, possesses through the continued development of English tactical methods a distinct military interest. The year 1138 is memorable for the Battle of the Standard, the first of many actions fought against the Scots, and typical of many a victory to come. The English knights as usual fought on foot, and aided by archers made havoc of the enemy. Here is already the germ of the later infantry; we shall find

lances and bows give way to pikes and muskets, but for five whole centuries we shall see the foot compounded of two elements, offensive and defensive, until the invention of the bayonet slowly welds them into one. At 1141.
the battle of Lincoln, on the other hand, we find the defensive element acting alone and suffering defeat, though not disgrace; for the dismounted knights who stood round Stephen fought with all the old obstinacy and yielded only to overwhelming numbers. Thus, though two generations had passed since the Conquest, the English methods of fighting were still in full vigour, and the future of English infantry bade fair to be assured.

Nor was the cavalry neglected; for amid all the earnest of this turbulent reign there was introduced the mimic warfare known as the tournament. This was an invention of the hot-blooded, combative French, and had been originally so close an imitation of genuine battle, that the Popes had intervened to prohibit the employment therein of any but blunt weapons. The tournament being not a duel of man against man, but a contest of troop against troop, was a training not only for individual gallantry, but for tactics, drill, discipline, and leadership; victory turning mainly on skilful handling and on the preservation of compact order. Thus by the blending of English foot and Norman horse was laid, earlier than in any other country of Europe, the foundation of an army wherein both branches took an equal share of work in the day of action.

The next in succession of our kings was a great soldier and a great administrator, yet the work that he did for the army was curiously mixed. Engaged as he was incessantly in war, he felt more than others the imperfection of the feudal as a military system. The number of knights that could be summoned to his standard was very small, and was diminished still further by constant evasion of obligations. He therefore regulated the commutation of personal military service for payment in money, and formed it, under the old name

of scutage, into a permanent institution. Advantage was generally taken of the system, and with the money thus obtained he took Brabançon mercenaries, the prototypes of the landsknechts of a later time, permanently into his pay. When he needed the feudal force to supplement these mercenaries, he fell back on the device of ordering every three knights to furnish and equip one of their number for service; and finally, driven to extremity, he re-established the old English fyrd as a
1181. National Militia by the Assize of Arms. This, the earliest of enactments for the organisation of our national forces, and the basis of all that followed down to the reign of Philip and Mary, contained the following provisions:—

Every holder of one knight's fee shall have a coat of mail,[1] a helmet, a shield, and a lance; and every knight as many coats of mail, helmets, shields, and lances as there are fees in his domain.

Every free layman having in chattels or rent to the value of sixteen marks shall keep the same equipment.

Every free layman having in chattels or rent ten marks, shall keep an habergeon,[2] a chaplet[3] of iron, and a lance.

All burgesses and the whole community of freemen shall have a wambais,[4] a chaplet of iron, and a lance.

It is noteworthy that neither the bow nor the axe appear in this list of the national weapons, an omission for which it is difficult to account, since the bow was evidently in full use at the time. Possibly the temptation to employ it for purposes of poaching may have been so strong as to make the authorities hesitate to enjoin the keeping of a bow in every poor freeman's house. The influence of the poacher will be found

[1] The coat of mail was made of rings or scales of iron sewn on to leather.

[2] The habergeon was a similar but smaller coat without sleeves.

[3] The chaplet was an iron scull-cap without vizor.

[4] The wambais was a doublet padded with cotton, wool or hair, and generally covered with leather.

equally potent when the time comes for the introduction of firearms.

Richard the Lion-Heart, like his predecessors, preferred to employ mercenaries for his wars, while even the knights who accompanied him to the Crusade were in receipt of pay. Were it not that his achievements in the Holy Land had left little mark on English military history they would be well worthy of a detailed narrative, for Richard was beyond dispute a really great soldier, a good engineer, and a remarkably able commander. The story of his march from Joppa to Jerusalem and of his victory at Arsouf is known to few, but it remains to all time an example of consummate military skill. A mixed force compounded of many nations is never very easy to control, and it was doubly difficult when the best of it was composed of knights who hated the very name of subordination. Yet it was with such material, joined to a huge body of half-disciplined infantry, that Richard executed a flank march in the presence of the most formidable of living generals, and repulsed him brilliantly when he ventured, at an extremely trying moment, to attack. The plan of the campaign, the arrangements and orders for the march, the drill and discipline imposed on the knights, and the handling of the troops in the action are all alike admirable. Yet, as has been already stated, the lessons of the Crusades wrought little influence in England, mainly because she had already learned from her own experience the value of a heavily armed infantry, and of the tactical combination of missile and striking weapons. In the rest of Europe they were for a time remembered but very soon forgotten ;[1] and England was then once more left alone with her secret.

Two small relics of the Crusades must however find mention in this place. The first is the employment of the cross as a mark for distinguishing the warriors of different nations, which became in due time the recog-

[1] The mortality among horses and the difficulty of obtaining remounts frequently forced the crusading knights to fight afoot.

nised substitute for uniform among European soldiers. Each nation took a different colour for its cross, that of the English being at first white, which, curiously enough, is now the regular facing for English regiments of infantry. The second relic is the military band which, there seems to be little doubt, was copied from the Saracens. In their armies trumpets and drums, the latter decidedly an Oriental instrument, were used to indicate a rallying-point ; for though at ordinary times the standards sufficed to show men the places of their leaders, yet in the dust of battle these were often hidden from sight ; and it was therefore the rule to gather the minstrels (such was the English term) around the standards, and bid them blow and beat strenuously and unceasingly during the action. The silence of the band was taken as a proof that a battalion had been broken and that the colours were in danger ; and the fashion lasted so long that even in the seventeenth century the bandsmen in all pictures of battles are depicted, drawn up at a safe distance and energetically playing.

The reign of King John accentuated still further the weak points of the English feudal system as a military organisation. The principle introduced by the Conqueror had been to claim for the sovereign direct feudal authority over every landholder in the country, suffering no intermediate class of virtually independent vassals, such as existed in France, to intercept the service of those who owed duty to him. Of the advantages of this innovation mention shall presently be made elsewhere, but at this point it is necessary to dwell only on its military defects. The whole efficiency of the feudal system turned on the creation of a caste of warriors ; and such a caste can obviously be built up only by the grant of certain exclusive privileges. The English knights possessed no such privileges. There were no special advantages bound up with the tenure of a fief. Far from enjoying immunity from taxation, as in France and Germany, the knights were obliged to pay not only the imposts

required of all classes, but scutage into the bargain. Again the winning of a knight's fee lay open to all ranks of freemen, so that it could not be regarded as the hereditary possession of a proud nobility. Yet again, the grant of the honour of knighthood was the exclusive right of the sovereign, who converted it simply into an instrument of extortion. Briefly, there was no inducement to English knights faithfully to perform their service; the sovereign took everything and gave nothing; and at last they would endure such oppression no longer. When John required a feudal force, in the year 1205, he was obliged to arrange that every ten knights should equip one of their number for service. Moreover, the knights who did serve him showed no merit; the English contingent at Bouvines having covered itself with anything but 1214.
glory. Finally, came mutiny and rebellion and the Great Charter, wherein the express stipulation that fiefs should be both alienable and divisible crushed all hopes of an hereditary caste of warriors for ever.

After the Charter the national force was composed nominally of three elements, the tenants in chief with their armed vassals, the minor tenants in chief, and the freemen subject to the Assize of Arms, the last two being both under the orders of the sheriffs. It made an imposing show on paper, but was difficult to bring efficient into the field. No man was more shameless than Henry the Third in forcing knighthood, for the sake of the fees, upon all free landholders whom he thought rich enough to support the dignity; yet, when the question became one not of money but of armed men, he was forced to fall back on the same resource as his greater namesake. He simply issued a writ for the enforcement of the Assize of Arms, and 1252.
ordered the sheriffs to furnish a fixed contingent of men-at-arms, to be provided by the men of the county who were subject thereto.

The defects of feudal influence in military matters were now so manifest, that Edward the First tried

hard to do away with them altogether. Strictly speaking the feudal force was summoned by a special writ addressed to the barons, ordering them to appear with their due proportion of men and horses, and by similar directions to the sheriffs to warn the tenants in chief within their bailiwicks. The system was however, so cumbrous and ineffective that Edward
1282. superseded it by issuing commissions to one or two leading men of the county to muster and array the military forces. These Commissions of Array, as they were called, will come before us again so late as in the reign of Charles the First.

But, like all his predecessors, Edward was careful to cherish the national militia which had grown out
1285. of the fyrd. The Statute of Winchester re-enacted the Assize of Arms and redistributed the force into new divisions armed with new weapons. The wealthiest class of freemen was now required to keep a hauberk [1] of iron, a sword and a knife, and a horse. The two lower classes were now subdivided into four, whereof the first was to keep the same arms as the wealthiest, the horse excepted; the second a sword, bow and arrows, and a knife; the third battle-axes, knives, and "other less weapons," in which last are included bills; [2] and the rest bows and arrows, or if they lived in the forest, bows and bolts, the latter being probably less deadly to the king's deer than arrows. Here then was the axe of Harold's day revived, and the archers established by statute. It is evident, from the fact that they wore no defensive armour, that the archers were designed to be light infantry, swift and mobile in their limbs, skilful and deadly with their weapons. The name of Edward the First must be ever memorable in our history for the encouragement that he gave to

[1] The hauberk was a complete suit of mail, a hood joined to a jacket with sleeves, breeches, stockings, shoes and gauntlets of double chain-mail.

[2] A bill was a broad curved blade mounted at the end of a seven-foot shaft, sometimes with a point and a hook added.

the long-bow; but we seek in vain for the man, if such there was, who founded the tradition, still happily strong among us, that the English whatever their missile weapon shall always be good shots. Even at the siege of Messina by Richard the First the archers drove the Sicilians from the walls; "for no man could look out of doors but he would have an arrow in his eye before he could shut it."

The bowmen had not long been a statutory force before they were called upon for active service. The defeat of the English by William Wallace at Cambus- 1297.
kenneth had summoned Edward from France to take the field in person against the Scots; and he met them on the field of Falkirk. The Scottish army consisted 1298.
for the most part of infantry armed with pikes, not yet the long pikes of eighteen feet which they were to wield so gallantly under Gustavus Adolphus, but still a good and formidable weapon. Wallace drew them up behind a marsh in four circular battalions ringed in with stakes, posting his light troops, which were armed principally with the short-bow, in the intervals between them, and his one weak body of horse in rear. The English knights were formed as usual in column of three divisions, vanguard, battle and rear-guard, and with them was a strong force of archers. Untrue to its old traditions, the English cavalry did not dismount, but galloped straight to the attack. The first division plunged headlong into the swamp (for the mediæval knight, in spite of a hundred warnings, rarely took the trouble to examine the ground before him), did no execution, and suffered heavy loss. The second division, under the Bishop of Durham, then skirted the swamp and came in sight of the Scottish horse. The Bishop hesitated and called a halt. "Back to your mass, Bishop," answered one contemptuous knight. His comrades charged, dispersed the Scottish cavalry, and drove away the archers between the pikemen; but the four battalions stood firm and unbroken, and the knights surged round them in vain.

Then the king brought up the archers and the third division of horse. Pushing the archers forward, he held the cavalry back in support until an incessant rain of arrows had riddled the Scottish battalions through and through, and then hurling the knights forward into the broken ranks, he fairly swept them from the field. It was the old story, heavy fire of artillery followed by charges of cavalry, the training of the Scots as Hastings had been of the English, for the trial of Waterloo.

It is interesting to note that Edward made an effort even then for the constitutional union of the two countries which had so honourably lost and won the day at Falkirk, but he was four centuries before his time. The war continued with varying fortune during the ensuing years. The maker of the English archers died, and under his feeble son the English
1314. army learned at Bannockburn an ignominious lesson in tactics. The Scotch army, forty thousand strong, was composed principally of pikemen, who were drawn up, as at Falkirk, in four battalions, with the burn in their front and broken ground on either flank. Their cavalry, numbering a thousand, a mere handful compared to the host of the English men-at-arms, was kept carefully in hand. Edward opened the action by advancing his archers to play on the Scottish infantry, but omitted to support them; and Bruce, seeing his opportunity, let loose his thousand horse on their flank and rolled them up in confusion. The English cavalry then dashed in disorder against the serried pikes, failed, partly from want of space and partly from bad management, to make the slightest impression on them, and were driven off in shameful and humiliating defeat. So the English learned that their famous archers could not hold their own against cavalry without support,[1] and they took the lesson to

[1] Mr. Oman (*Art of War in the Middle Ages*, p. 104) holds the opinion that to force a line of long-bowmen by a mere front attack was a task almost as hopeless for cavalry as the breaking of

heart. The old system of dismounting the men-at-arms had been for the moment abandoned with disastrous results; the man who was to revive it had been born at Windsor Castle just two years before the fight.

Thirteen years later this boy ascended the throne 1327.
of England as King Edward the Third, and almost immediately marched with a great host against the Scots. The campaign came to an end without any decisive engagement, but on the one occasion when an action seemed imminent, the English men-at-arms dismounted and put off their spurs after the old English fashion. Peace was made, but only to be
broken by the Scots, and then Edward took his revenge 1333.
for Bannockburn at Halidon Hill. The English men-at-arms alighted from their horses, and were formed into four battalions, each of them flanked by wings of archers, the identical formation adopted two centuries later for the pikemen and musketeers. The Scots, whose numbers were far superior, were also formed on foot in four battalions, but without the strength of archers. "And then," says the old historian,[1] "the English minstrels blew aloud their trumpets and sounded their pipes and other instruments of martial music, and marched furiously to meet the Scots." The archers shot so thick and fast that the enemy, unable to endure it, broke their ranks, and then the English men-at-arms leaped on to their horses for the pursuit. The Scotch strove gallantly to rally in small bodies, but they were borne down or swept away; they are said to have lost ten thousand slain out of sixty thousand that entered the battle.

The mounting of the men-at-arms for the pursuit gave the finishing touch to the English tactical methods, and the nation was now ready for war on a grander

a modern square, and would have it that archers needed support on their flanks only. With all respect I must reject this view, as opposed alike to history and common sense.

[1] Barnes.

scale. Moreover, there was playing round the knees of good Queen Philippa a little boy of three years old who was destined to be the victor of Poitiers. It is therefore time, while the quarrel which led to the Hundred Years' War is maturing, to observe the point to which two centuries and a half of progress had brought English military organisation.

AUTHORITIES.—By far the best, so far as I know the only, account of the rise of English tactics and of English military power is to be found in *Die Entwickelung des Kriegswesens in der Ritterzeit*, by Major-General Köhler, vol. ii. pp. 356 sq., and vol. v. pp. 97 sq., a work to which my obligations must be most gratefully acknowledged. The authorities are faithfully and abundantly quoted. Freeman's *Norman Conquest*, Mr. J. H. Round's *Feudal England*, Hewitt's *Ancient Armour*, Oman's *Art of War in the Middle Ages*, Grose's *Military Antiquities*, and Rymer's *Fædera* are authorities which will occur to every one, as also the Constitutional Histories of Hallam, Stubbs, and Gneist.

CHAPTER II

ATTENTION has already been called to the defects of the feudal system for military purposes, and to the shifts whereby successive sovereigns sought to make them good. With Edward the Second resort was made to a new device. Contracts, or as they were called indents, were concluded by the King with men of position, whereby the latter, as though they had been apprentices to a trade, bound themselves to serve him with a force of fixed strength during a fixed term at a fixed rate of wages. In some respects this was simply a reversion to the old practice of hiring mercenaries; but as Edward the Third placed his contracts for the most part within his kingdom, the force assumed a national character. The current ideas of organisation were still so imperfect that the contractors generally engaged themselves to provide a mixed force of all arms; but as they naturally raised men where they could most easily get hold of them, that is to say in their own neighbourhoods, there was almost certainly some local or personal feeling to help to keep them together. For the rest the contractor of course made his own arrangements for the interior economy of his own particular troops, and enjoyed in consequence considerable powers, which descended to the colonels of a later day and have only been stripped from them within the last two generations. It is not difficult to imagine that men thus enlisted should presently, when released from national employment, have sold their services to the highest bidder and become, as they presently did become, *condottièri*. It is characteristic of the commercial genius of our race that England should be the cradle not only of the soldier but of the *condottière*;[1] in other words, that she should have set the example in making warfare first a question of wages, and next a question of profit. But her work did not end here; for these reforms created the race of professional soldiers and through them the renascence of the Art of War. In short, with the opening of the Hundred Years' War the British army quickens in the womb of time, and the feudal force sinks into ever swifter decay.

But there is another side to this picture of feudal inefficiency. Moral not less than physical force is a mighty factor in war; and it was precisely the military defects of the English feudal system that first made her a military power. Though the growth of a caste of warriors was checked, it was to make room for that which was worthy to overshadow it, a fighting nation. For in England there was not, as in other countries, any denial of civil rights to the commons of the realm. Below the ranks of the peerage all freemen enjoyed equality before the law; nay, the peerage itself conferred no privilege except on those who actually possessed it, the sons of peers being commoners, not as elsewhere noble through the mere fact of their birth. In England there were and are nobility and gentry: in other countries nobility and gentry were merged in a single haughty exclusive caste, and between them and other freemen was fixed a great and impassable gulf. Thus the highest and the lowest of the freemen were in touch with each other in England as nowhere else in Europe. More than two centuries later than Creçy, so great and gallant a gentleman as Bayard could refuse with disdain to fight by the side of infantry. In England, whatever the pride of race, the son of the noblest peer in the land stood shoulder to shoulder with his equal when the archer fell in by his side, and where the son stood the

[1] William of Ypres, who came to England in the pay of Stephen in 1138, is reckoned the first of the *condottièri*.

father could feel it no shame to stand. No other nation as yet could imitate this; no other could recall a Hastings where all classes had stood afoot in one battalion. Other nations could indeed, when taught by experience, dismount their knights and align cross-bowmen with them, just as at this day they can erect an upper and lower chamber and speak of a constitution on the English model; but then as now it was the form only, not the substance, that was English.

So far for the commercial and political influence that helped to mould our military system; there remains yet another great moral force to be reckoned with. Chivalry, which had been growing slowly in England since the Third Crusade, burst in the fourteenth century into late but magnificent blossom. The nation woke to the beauty of a service which gave dignity to man's fighting instincts, which taught that it was not enough for him to be without fear if he were not also without reproach, and that though the government of the world must always rest upon force, yet mercy and justice may go hand in hand with it. The girding on of the sword was no longer a social but a religious act; it marked not merely the young man's entrance into public life, but his ordination to a great and noble function. Concurrently there had arisen a sense of the charm of glory and adventure. Hitherto the English knights had gained no repute in Europe. Hatred and jealousy had held the Saxon aloof from his Norman master; now there was no more Saxon and Norman, but the English, united and strong, a fighting people that thirsted for military fame.

Let us now briefly consider the composition and organisation of the armies that were to work such havoc in France. The cavalry was drawn for the most part from the wealthier classes, though, as has been seen, there was one division of the freemen under the statute of Winchester which was called upon to do mounted service. The more important branch, the men-at-arms, was composed of two elements, knights and squires.

From the first institution of the feudal system, the number of men required from the greater vassals had forced them to equip their sons and serving-men, who after many changes were finally in the thirteenth century merged together under the generic name of *servientes*, a term which was soon corrupted into its present form of sergeants. In the year 1294 these *servientes* were dignified by the higher title of *servientes equites*, mounted sergeants, which was six years later abandoned for the familiar name of squires. These squires must not, however, be confounded with a different class of the same appellation, namely, the apprentices who were the personal attendants of the knights. The squire of which I now speak was rather a knight of inferior order corresponding to the *bachelier* (*bas chevalier*) of France. The word knight itself gives us a hint of this inferiority, being the same as the German *knecht*, whereas *ritter* is the German term that expresses what is generally understood as a knight in English. The inner history of chivalry is the story of the struggle of the sergeants to rise to an equality with the knights of the first order, and in the fourteenth century they were not far from their goal. Even now they were considered the backbone of the English army, and were equipped in all points like the class above them.

Men-at-arms, an expression derived from the French, were so called because they were covered with defensive armour from top to toe; but as the middle of the fourteenth century is a period of transition in the development of armour, it is difficult to describe their equipment with any certainty. Their offensive arms were the lance, sword, dagger, and shield. Trained from very early youth in the handling of weapons they were doubtless proficient enough with them; but they do not seem to have been great horsemen, and indeed it is recorded that they were sometimes tied to the saddle. Monstrelet, writing in the year 1416, tells us of the astonishment which certain Italians created among the French because they could actually turn their horses at the gallop. It is probable that the bits employed were too weak, and that the cumbrousness of the saddle and the weight carried by each man were sad obstacles to good horsemanship; but it is worth remembering in any case that, as this passage plainly shows, men-at-arms in the saddle were reduced to one of two alternatives, to move slowly and retain control of their horses, or to gallop for an indefinite period wherever the animals might choose to carry them.

The favourite horses, alike for speed, endurance, and courage, were the Spanish, which, as they could only reach England by the journey overland through France, were not always very easily obtained. Philip the Bold in 1282 refused to allow one batch of eighty such horses to be transhipped to England; but from a contract still extant, of the year 1333, it appears that Edward the Third still counted on Spain to provide him with remounts. These horses, however, were only bestridden for action, being committed on the march to the care of the shield-bearers or squires, who led them, as was natural, on their right-hand side, and thus procured for them the curious name of *dextrarii*.[1] The usual allowance of horses for a knight was three, besides a packhorse for his baggage, the smallest of which, named the palfrey, was that which he rode on ordinary occasions; in fact, to put the matter into modern language, a knight started on a campaign with a first charger, a second charger, and a pony. The first charger was always a stallion; the rest might be geldings or mares. From the year 1298 the practice of covering horses with defensive armour was introduced into England, an equipment which soon came to be regarded as so essential that one branch of the cavalry, and that the most important, was reckoned by the number of barded horses.

The personal retinue of the knights was made up of apprentices or aspirants to the rank which they held. The squire or shield-bearer took charge of the knight's armour on the march, and was responsible for maintaining

[1] Whence the French word *destrier*.

it in proper order; and it is worth remarking that the English squire took a pride in burnishing the metal to the highest pitch of brilliancy, thus early establishing those traditions of smartness which are still so strong in our cavalry. It was also the squire's duty, among many others, to help his master to don his harness when the time for action came, beginning with his iron shoes or sollerets, and working upwards till the fabric was crowned by the iron headpiece, and the finishing touch added by the assumption of the shield. The reader will readily understand that a really efficient squire must have been invaluable, for if an engagement came in any way as a surprise there was an immediate rush for the baggage, and a scene of confusion that must have beggared description. Fortunately, the fact that both sides were generally alike unready, and the punctiliousness of chivalric courtesy, permitted as a rule ample time not only for the equipment of all ranks, but for the marshalling of the host.

In the matter of administrative organisation the men-at-arms were distributed into constabularies, being commanded by officers called constables. The strength of a constabulary seems to have varied from five-and-twenty to eighty; and this variety, together with the absence of any tactical unit of fixed strength, makes it impossible to state how many constabularies were included in the next tactical division. This was called the banner, and was commanded by a banneret, a rank originally conferred only upon such as could bring a certain number of followers into the field. Promotion to the degree of banneret was marked by cutting off the forked tail of the pennon which was carried by the ordinary knight, and leaving the remnant square. So at the present day, the pennons of lances are forked, the square being reserved for the standards of squadrons and regiments.

The independent employment of small bodies in action was almost unknown, the rule being to pack an indefinite number of men-at-arms, hundreds or even thousands, into a close and solid mass, its depth almost if not quite as great as its frontage. The *haye*, or thin line, is of much later date. Ordinarily some modification of the wedge was the formation preferred; that is to say, that the frontage of the front rank was somewhat less than that of the rear; the mass of that particular shape being judged to be less liable to disorder and better adapted for breaking into a hostile phalanx. The relative strength of the front and rear ranks depended entirely on the numbers that were packed in between them, and it may readily be supposed that the evolutions which so unwieldly a body could execute were very few. Probably, until the moment of action came, sufficient space was maintained to permit every horse to turn on his own ground, after the Roman fashion, to right, left, or about; but for the attack ranks and files were closed up as tightly as possible, and all other considerations were sacrificed to the maintenance of a compact array. It was said of the French knights who marched with Richard the Lion-heart that an apple thrown into the midst of them would not have fallen to the ground. We must therefore rid ourselves of the popular notion of the knight as a headlong galloping cavalier. The attack of men-at-arms could not be very rapid unless it were made in disorder; and though it comes strictly under the head of shock-action, the shock was rather that of a ponderous column moving at a moderate pace than of a light line charging at high speed. By bearing these facts in mind it will be easier to understand the failure of mounted men-at-arms to break a passive square of infantry.

Next after the men-at-arms came a species of cavalry called by the name of pauncenars,[1] which was less fully equipped with defensive armour, but wore the habergeon [2] and was armed with the lance.

Lastly came the light cavalry of the fyrd, originally established to patrol the English coast. These were called hobelars, from the hobbies or ponies which they

[1] From the German *panzer*, a coat of mail.
[2] A sleeveless coat of chain-mail.

rode, and were equipped with an iron helmet, a heavily padded doublet (*aketon*), iron gloves, and a sword.

Turning next to the infantry, there were Welsh spearmen, carrying the weapon which gave them their name, but without defensive armour. Indeed it should seem that they were not overburdened with clothes of any kind, for they were every one provided at the King's expense with a tunic and a mantle, which were by express direction made of the same material and colour for all. These Welsh spearmen therefore were the first troops in the English service who were dressed in uniform, and they received it first in the year 1337.[1] The colour of their clothing unfortunately remains unknown to us.

Next we come to the peculiar strength of England, the archers. Though a certain number of them seem generally to have been mounted, yet, like the dragoons of a later day, these rode for the sake of swifter mobility only, and may rightly be reckoned as infantry. As has been already stated, the archers wore no defensive armour except an iron cap, relying on their bows alone. These bows were six feet four inches long; the arrows, of varying length but generally described as cloth-yard shafts, were fitted with barb and point of iron and fledged with the feathers of goose or peacock. But the weapon itself would have gone for little without the special training in its use wherein the English excelled. "My father," says Bishop Latimer (and we may reasonably assume that in such matters there had been little change in a hundred and fifty years), "My father was diligent in teaching me to shoot with the bow; he taught me to draw, to lay my body to the bow, not to draw with strength of arm as other nations do, but with the strength of the body. I had my bows bought[2]

[1] The earliest instance of uniform in modern Europe is found in the militia of the Flemish towns at the battle of Courtrai, 1302 (Köhler).

[2] The contract price of a bow in 1341 was, unpainted 1s., painted 1s. 6d.; of a sheaf of twenty-four arrows 1s. 2d. An archer's pay was 3d. a day.

me according to my age and strength; as I increased in these my bows were made bigger and bigger." The principle was in fact analogous to that which is taught to young oarsmen at the present day. The results of this training were astonishing. The range of the long-bow in the hands of the old archers is said to have been fully two hundred and forty yards, and the force of the arrow to have been such as to pierce at a fair distance an inch of stout timber. Moreover, the shooting was both rapid and accurate. Indeed the long-bow was in the fourteenth century a more formidable weapon than the cross-bow, which had been condemned by Pope Innocent the Second as too deadly for Christian warfare so far back as 1139. It was at no disadvantage in the matter of range, while it could be discharged far more quickly; and further, since it was held not horizontally but perpendicularly to the ground, the archers could stand closer together, and their volleys could be better concentrated. Thus the long-bow, though the cross-bow was not unknown to the English, was not only the national but the better weapon. In action the archers were ranked as deep as was consistent with the delivery of effective volleys, the rear ranks being able to do good execution by aiming over the heads of the men before them. It may be imagined from the muscular training undergone by the archers that they were physically a magnificent body of men.

Strictly speaking the archers were the artillery of the army, according to the terminology of the time,[1] the word *artillator* being used in the time of Edward the Second to signify the officer in charge of what we now call the ordnance-stores. But to avoid confusion we must use the word in its modern sense, the more so since we find among the stores of the custodian[2] of the King's artillery in 1344 the items of saltpetre and

[1] See 1 Samuel xx. 40.

[2] As the historian of the Royal Artillery has ignored this gentleman we may give his name, Thomas de Roldeston (see Hewitt, vol. ii. p. 289).

sulphur for the manufacture of powder, and among his men six "gonners." Gun, it should be added, was the English, cannon the French name for these weapons from the beginning. It will presently be necessary to notice their first appearance in the field.

As to the general organisation of the army, the whole was divided into thousands under an officer called a millenar, subdivided into hundreds, each under a centenar, and further subdivided into twenties, each under a vintenar. The commander-in-chief was usually the King in person, aided by two principal officers, the High Constable and the Marshal, whose duties were, roughly speaking, those of Adjutant and Quartermaster-General. For tactical purposes the army was distributed into three divisions, called the vanguard, battle and rearguard, which kept those names whatever their position in the field or on the march, whether the host was drawn up, as most commonly, in three lines, or in one. Trumpets were used for purposes of signalling, though so far as can be gathered they sounded no distinct calls, and were dependent for their significance on orders previously issued. The failing in this respect is the more remarkable, inasmuch as the signals of the chase with the horn were already very numerous and very clearly and accurately defined.

The pay of all ranks can fortunately be supplied from the muster-roll of Calais in 1346, and although I shall not again encumber these pages with a pay-list I shall for once print it entire :

The Prince of Wales	20s.	a day.
The Bishop of Durham	6s. 8d.	" "
Earls	6s. 8d.	" "
Barons and Bannerets	4s.	" "
Knights	2s.	" "
Esquires, Constables, Captains, and Leaders	1s.	" "
Vintenars	6d.	" "
Mounted Archers	6d.	" "
Pauncenars	6d.	" "

Hobelars	6d.	a day.
Foot-Archers	3d.	,, ,,
Welsh Spearmen	2d.	,, ,,
,, Vintenars	4d.	,, ,,

Masons, Carpenters, Smiths, Engineers, Miners, Gunners, 10d., 6d., and 3d.

It is melancholy to have to record that even so early as in 1342 corruption and fraudulent dealing had begun in the army. The marshals were ordered to muster the men-at-arms once a month, and to refuse pay for men who were absent or inadequately armed or indifferently mounted. We shall see the practice of drawing pay for imaginary men and the tricks played on muster-masters increase and multiply, till they demand a special vocabulary and a certain measure of official recognition. A favourite abuse among men-at-arms was the claim of extortionate compensation for horses lost on active service, leading to an order in this same year that all horses should be valued on admission to the corps, and marked to prevent deception. Thus early was the road opened that leads to the broad arrow. The taint of corruption, indeed, clings strongly to every army, with the possible exception of the Prussian, in Europe. War is a time of urgency and stress, which does not admit of strict audits or careful inspections, and poor human nature is too weak not to turn such an opportunity to its profit. It is an unpleasant thought that dishonesty and peculation should be inseparably associated with so much that is noble and heroic in human history, but the fact is indisputable, and must not be lightly passed over. Moreover the days when English cavalry shall go to war on their own horses may not yet be numbered; and it may be useful to remember that the mediæval man-at-arms would mount himself on his worst animal in order to break him down the quicker, and claim for him the price of his best. It is only by constant wariness against such evils that there can be built up a sound system of military administration.

Authorities.—As for previous chapter.

CHAPTER III

HAVING now sketched the composition of the English forces, let us move forthwith to the scene of action.
1339. We must omit the early incidents of the war, and the assumption by Edward of the famous motto wherein he consecrated his claim to the crown of France, *Dieu et mon droit*. We must pass by the famous naval action
1340, June 24. of Sluys, where the English commanders in their zeal to follow the precepts of Vegetius, thought it more important to have the sun in the enemy's eyes than the wind in their own favour, and where the archers, acting as marine sharp-shooters, were the true authors of the English victory. We must overlook likewise the innumerable sieges, even that of Quesnoy, where the English first came under the fire of cannon, merely remarking that owing to their ignorance of that particular branch of warfare, the English were uniformly unsuccessful; and we must come straight to the year 1345, when Henry of Lancaster, Earl of Derby, landed at Bayonne with a force of three thousand men for a campaign in Gascony and Guienne. The name of our first artillery-officer has been given; attention must now be called to our first engineer, this same Earl of Derby, who had lately been recalled from service with the Spaniards against the Moors at the siege of Algesiras, and was the first man who taught the English how to take a fortified town.

Derby then with his little army harried Gascony and Guienne for a time, until the arrival of a superior French force compelled him to retire and gave him

much ado to defend himself. Accordingly, in June 1346, June.
1346 Edward the Third impressed a fleet of innumerable small vessels, none of them exceeding sixty tons burden, embarked thereon four thousand men-at-arms, ten thousand archers and five or six thousand Welsh spearmen, and sailed for the coast of France. On the 12th of July he put into St. Vaast de la Hogue, a little to the east of Cherbourg, dispersed a French force that was stationed to oppose him, and successfully effected his landing. Six days were allowed to recruit men and horses after the voyage, and the army then moved eastward to the Seine, leaving a broad line of ruin and desolation in its wake, and advanced up the left bank of the river. King Philip of France had meanwhile collected an army at Rouen, whence he marched parallel to the English along the right bank of the Seine, crossed it at Paris, and stood ready to fall upon Edward if he should strike southward to Guienne. But Edward's plans were of the vaguest; his diversion had already relieved Derby, and he now crossed the Seine at Poissy and struck northward as if for Flanders. Philip no sooner divined his purpose than he too hastened northward, outmarched the English, crossed the Somme at Amiens, gave orders for the occupation of every bridge and ford by which the English could pass the river, and then recrossing marched straight upon Edward's right flank.

The position of the English was now most critical, for they could not cross the Somme and were fairly hemmed in between the river and the sea. At his wits' end Edward examined his prisoners, and from them learned of the ford of Blanche Tache in the tidal water about eight miles below Abbeville. Thither accordingly he marched, and after waiting part of a night for the ebb-tide, forced the passage in the teeth of a French detachment that had been stationed to guard it, and sending six officers to select for him a suitable position pursued his way northward through the forest of Creçy. On the morning of the 26th of August he crossed the

1346, August 26. river Maie, and there swinging his front round from north to south-east he turned and stood at bay.

The position was well chosen. The army occupied a low line of heights lying between the villages of Creçy and Wadicourt, the left flank resting on a forest, the right on the river Maie. Edward ordered every man to dismount, and parked the horses and baggage waggons in an entrenched leaguer[1] in rear. The army was too weak to cover the whole line of the position, so the archers were pushed forward and extended in a multitude of battalions along the front, and backed with Welsh spearmen. Echeloned in rear of them stood the three main divisions of the army; foremost and to the right the vanguard of twelve hundred men-at-arms under the Black Prince, next to it the battle of as many more under the Earl of Arundel, and behind it, covering the extreme left, the rearguard, consisting of fifteen hundred men-at-arms and six thousand mixed archers and infantry under the King. The country being rich in provisions Edward ordered every man to eat a hearty meal before falling into his place, for he knew that the Englishman fights best when he is full. When the host was arrayed in order he rode round the whole army to cheer it; and then the men lay down, the archers with their helmets and bows on the ground before them, and waited till the French should come.

Philip meanwhile had crossed the Somme at Abbeville on the morning of the 26th, and turned eastward in the hope of cutting off the English. Finding that he was too late, he countermarched and turned north, at the same time sending forward officers to reconnoitre. The afternoon was far advanced, and the French were wearied with a long, disorderly march when these officers returned with intelligence of the English. Philip ordered a halt, but the indiscipline and confusion were such that the order could not be obeyed. The noblest blood in France was riding on in all its pride to make

[1] What since the Zulu war we have called a *laager*, forgetting the English word that lay ready to our hand.

an end of the despised English, and a mass of rude 1346,
infantry was waiting to share the slaughter and the spoil. August 26.
So they blundered on till they caught sight of the English lying quietly down in order of battle; and therewith all good resolutions vanished and Philip gave the order to attack.

It was now nearly five o'clock, and the heaven was black with clouds, which presently burst in a terrific thunderstorm. The English archers slipped off their bowstrings to keep them dry, and waited; while six thousand Genoese cross-bowmen, jaded by the long march, drenched and draggled with the rain that beat into their faces, conscious that they were almost disarmed by the wetness of their bowstrings, shuffled wearily into their stations along the French front. Their leaders complained that they were unfairly treated. "Who cares for your rabble?" answered the Count of Alençon. "They are nothing but useless mouths, more trouble than help." So the cross-bowmen sulkily took their position, and the rest of the French army, from twelve to twenty thousand men-at-arms and some fifteen thousand infantry, ranged themselves in three massive lines behind them. A vast flight of ravens flew over the opposing arrays, croaking loudly over the promised feast of dead men.

Then the storm passed away inland into France, and the sun low down in the west flashed out in all his glory full in the faces of the French. The Genoese advanced and raised a loud cry, thrice repeated, to strike terror into the English: the archers over against them stood massive and silent. The loud report of two or three cannon, little more harmful than the shouts of the Genoese, was the only answer; and then the archers stepped forward and drew bow. In vain the Genoese attempted to reply; they were overwhelmed by the torrent of shafts; they shrank back, cut their bowstrings and would have fled, but for a line of French mounted men-at-arms which was drawn up in their rear to check them. The proud chivalry of France was chafing im-

46, August 26.

patiently behind them, and Philip would wait no longer. "Slay me these rascals," he said brutally; and the first line of men-at-arms thundered forward, trod the hapless Genoese under foot, and pressed on within range of the arrows. And then ensued a terrible scene. The great stallions, maddened by the pain of the keen barbed shafts, broke from all control. They jibbed, they reared, they swerved, they plunged, striking and lashing out hideously, while the rear of the dense column, carried forward by its own momentum, surged on to the top of the foremost and wedged the whole into a helpless choking mass. And still the shower of pitiless arrows fell swift as snow upon the thickest of the press; and the whole of the French fighting line became a confused welter of struggling animals, maimed cross-bowmen, and fallen cavaliers, crippled by the weight of their armour, an easy prey to the long, keen knives of the Welsh.

Nevertheless some few of the French men-at-arms had managed to pierce through the archers. The blind king of Bohemia had been guided by two faithful knights through the centre, Alençon had skirted them on one flank, the Count of Flanders on the other, and all had fallen upon the Black Prince's battalion. The danger was greatest on the left flank; but the Earl of Arundel moved up the second line of the echelon to his support, and the English held their own. Then the second line of the French advanced, broke through the archers, not without heavy loss, and fell likewise upon the English men-at-arms. The Prince of Wales was overthrown, and was only saved by the devotion of his standard-bearer, but the battalion fought on. It was probably at this time that Arundel sent a messenger to the King for reinforcements. "Is my son dead or hurt?" he asked. "No, sire, but he is hard beset." "Then return to those who sent you and bid them send me no more such messages while my son is alive; tell them to let the boy win his spurs." The message was carried back to the battalion, and the men-at-arms fought on stoutly as ever. The archers seem also to have rallied and closed on the flank and rear of the attacking French. Alençon's banner could still be seen swaying behind a hedge of archers, and Philip, anxious to pour his third and last line into the fight, had actually advanced within range of the arrows. But the power of the bowmen was still unweakened, the ground was choked with dead men and horses, and the light was failing fast. He yielded to the entreaties of his followers and rode from the field; and the first great battle of the English was won. 1346, August 26.

When morning dawned the country was full of straggling Frenchmen, who from the sudden change in the direction of the advance had lost all knowledge of their line of retreat; the few that retained some semblance of organised bodies were attacked and broken up. Never was victory more complete. The French left eleven great lords, eighty-three bannerets, over twelve hundred knights and some thousands of common soldiers dead on the field. It was a fortunate issue to a reckless and ill-planned campaign. It is customary to give all credit for the victory to the archers, but this is unjust. Superbly as they fought they would have been broken without the men-at-arms, even as the men-at-arms would have been overwhelmed without the archers. Both did their duty without envy or jealousy, and therein lay the secret of their success.

The siege and capture of Calais followed, and then by the mediation of the Pope peace was made, and for a time preserved. Petty hostilities however never ceased in Brittany, and finally in 1355 the war broke 1355.
out anew. Three armies were fitted out,—one of a thousand men-at-arms under the Black Prince for operations in Guienne, a second under the Earl of Derby for Brittany, and a third under the personal command of the King. Little, however, was effected in the campaign of 1355. The King was recalled to England by an invasion of the Scots, and the operations of 1356 in Brittany were checked by the appearance of 1356.
the French King in superior force. But at the close

1356, July. of July the Black Prince suddenly started on a wild raid from the Dordogne in the south to the Loire. His object seems to have been to effect a junction with Derby's forces at Orleans; but it is difficult to see how August 28. he could have hoped for success. He had reached Vierzon on the Cher when he heard that the King of France was on his way to meet him in overwhelming strength. Unable to retreat through the country which he had laid waste on his advance, he turned sharp to the west down the Cher and struck the Loire at Tours. There for four days he halted, for what reason it is difficult to explain, since the delay enabled the French to cross the Loire and seriously to threaten his retreat.

There was now nothing for the Prince but to retire southward with all haste. The French were hard on his track, and followed him so closely that he was much straitened by want of supplies. On the 14th of September the English were at Chatelheraut and the French at La Haye, little more than ten miles apart, and on the 15th the French made a forced march which brought them fairly to southward of the Prince, and between him and his base at Bordeaux. All contact however had been lost; and the French King, making sure that the Prince had designs on Poitiers, swung round to the westward and moved straight upon the town. On the 17th, while in full march, his rearguard was suddenly surprised by the advanced parties of the Prince. As in the movements after the Alma, each army was executing a flank march, quite unconsciously, in the presence of the other. The French rearguard pursued the reconnoitring party to the main body of the English, and after a sharp engagement was repulsed with heavy loss. The French army had actually marched across the line of the Black Prince's retreat, and left it open to him once more.

Edward lost no time in looking for a suitable position, and presently found it at Maupertuis some fifteen miles south-west of Poitiers. There to the

north of the river Miosson is a plain seamed with deep 1356,
ravines running down to that stream ; and behind one Sept. 18.
of these he took his stand, facing north-east. The sides of the ravine were planted with vineyards and blocked by thick hedges, so that it was impossible for cavalry to cross it except by a track which was broad enough for but four horsemen abreast ; and these natural advantages the Prince improved by repairing all weak places in the fences and by digging entrenchments. One exposed spot on his left flank he strengthened by a leaguer of waggons as well as with the spade. He then told off his archers to line the hedges which commanded the passage across the ravine, and drew up his men-at-arms, all of them dismounted, in three lines behind it. The first line he committed to the Earls of Warwick and Suffolk, the rearmost to the Earl of Salisbury, and the centre he reserved for himself. His whole force, augmented as it was by a contingent of Gascons, did not exceed six or seven thousand men, half of whom were archers.

So passed the day of the 18th of September on the English side. The French on their part, instead of blocking up their retreat to the south and reducing them by starvation, simply moved down from Poitiers to within a league of the English position and halted for the night. Their force amounted to sixty thousand men, and they might well feel confident as to the issue of an action. Indeed, when the Black Prince, fully alive to the desperate peril of his situation, negotiated for an evacuation of the country, they imposed such terms that he could not in honour accept them. They therefore reconnoitred the English position, and laid their plans for the morrow. Three hundred chosen men-at-arms, backed by a column of German, Italian, and Spanish knights, were to charge down the ravine upon the archers, disperse them, and attack the English men-at-arms on the other side. Three lines, each of three massive battalions containing from three to four thousand men-at-arms, with lances shortened to a

1356, length of five feet, were to follow them afoot, and the English were to be crushed by their own tactics.

It is hardly surprising that in the night the Black Prince's heart failed him. He resolved while he could to place the Miosson between him and the French, and
Sept. 19. at dawn began his retreat, leaving the rearguard, however, still in the position at Maupertuis in case withdrawal should be impossible.[1] He also sent two knights to watch the French army, who however approached too closely to it and were captured. His first line had already crossed the Miosson when intelligence reached him that the French had advanced, and that the rearguard was engaged. He at once ordered the vanguard to return, and himself hastening back with his own division, despatched three hundred mounted men-at-arms and as many mounted archers without delay to strengthen his right wing. The French meanwhile had moved forward, gaily singing the song of Roland, to find the way blocked by the hedges and vineyards of the ravine. Undismayed they plunged down into the narrow track; and then the English archers behind the hedges opened at close range a succession of frightfully destructive volleys. The foremost of the horsemen fell headlong down, the rear plunged confusedly on the top of them, and the pass was blocked with a heaving, helpless crowd, on which the arrows hissed down in an eternal merciless shower. The supporting column of foreign cavalry was unable to act in the confusion; it was already under the fire of the archers, and before it could move the English mounted men on the right wing came down full upon its left flank, and killed or captured every man.

And now the wounded French horses, mad with pain and terror, many of them riderless and all beyond control, dashed back on to the first line of the dis-

[1] The only authority for this is the rhymed chronicle of the Chandos herald, but, as Köhler observes, the proceeding was so natural, and, I may add, the invention of such a story so improbable, that it is difficult not to accept it.

mounted French men-at-arms. It was a charge of mad 1356.
animals, the most terrible of all charges, and the huge Sept. 19.
battalion fell into confusion before it. Edward was watching the battle keenly from his position; he had already ordered his men-at-arms to mount, and now Sir John Chandos, whose name must always be linked to Edward's as that of Collingwood to Nelson, broke out aloud with, "Forward, sire, forward, and the day is yours!" "Aye, John," answered the Prince, with a thought perhaps of the morning's retreat, "No going backward to day. Forward banner, in the name of God and St. George!" The preliminary attack of the mounted men on the right had already cleared the way for them. The English cavalry scrambled in haste down into the ravine on the right, and fell upon the French men-at-arms. The front and centre divisions, already much shaken, were easily broken and dispersed; the third and strongest still remained, and against this, which resisted desperately, the whole force of the English was turned. The lesson of Falkirk was remembered. The mounted archers made the gaps and the men-at-arms rode into them. The division was broken, the King was captured, and the mass of the fugitives making for Poitiers found the gates closed against them and were cut down by hundreds. The action began at six in the morning, and lasted till late into the afternoon. The French losses were enormous. Over and above the King and many great lords two thousand men-at-arms were captured, and two thousand five hundred more were left dead on the field; the number of the unhappy foot-men that were slain it is impossible to state. The English loss is variously set down, the reports ranging from half the force to sixty-four men. The battle, from the disparity between the strength of the two sides, must remain ever memorable in the annals of war. To the English, who had but lately risen above the horizon as a military power, it gave a prestige that has never been lost.

The peace of Brétigny closed the war, and the 1360.

1360. English army was disbanded. But the soldiers, like the ten thousand Greeks who returned from Cunaxa, were too deeply bitten with their profession to abandon it for the tedium of peace. They therefore formed themselves into independent bodies, or Free Companies, and for years were the scourge of France, their chamber as they called it, which they plundered and ravaged at their pleasure. The greatest of their leaders was John Hawkwood, of whom something more must presently be said, but these bands, in less or greater numbers, were constantly to be found fighting for hire against the French. Thus three hundred of them fought for the King of Navarre against the King of
1364, May 16. France at Cocherel. The numbers engaged were little more than fifteen hundred on each side, but the action is interesting as showing the efforts of the French to meet the peculiar tactics of the English. In order to have no more trouble with unruly horses the French men-at-arms dismounted and fought on foot, and now for the first time the archers found themselves outdone. The armour of the French was so good that it turned the cloth-yard shafts; and being slightly superior in numbers the French men-at-arms forced their enemy off the field. It was but a slight success, but a defeat even of a small body of English was such a rarity in those days that it gave the French great hopes for the future, hopes which were soon to be dashed to the ground.

1365. In the following year a quarrel as to the succession to the Duchy of Brittany between Charles of Blois and John of Montfort brought the English again into the field. The French King Charles the Fifth sent assistance to support the former, whereupon John of Montfort at once appealed to the English. John Chandos and several more of the garrison in France, eager for fresh battle against their old enemies, asked permission to join Montfort as volunteers. "You may go full well," answered the Black Prince. "Since the French are going for Charles of Blois, I give you good

leave." The English, both volunteers and mercenaries, 1365,
accordingly hurried to the scene of war; and at Auray Sept. 29.
they fought the action which decided the campaign. The numbers engaged did not exceed four thousand in either army. Both sides dismounted, and the French men-at-arms discarding the lance as unfit for fighting afoot equipped themselves with battle-axes, so that there promised to be a stubborn fight. The English archers as usual opened the engagement, but as at Cocherel their shafts could not penetrate the armour of the French; whereupon with great deliberation they threw down their bows, and boldly advancing to the French men-at-arms plucked their axes from their hands and plied the weapons against their astonished owners with terrible effect. The whole proceeding furnishes so good an example of the thoughtless, thick-headed gallantry of the English soldier, that one can only marvel that the battle of Auray should be practically unknown to Englishmen. The intensely ludicrous picture that can be conjured up of a series of detached struggles between the brawny active Englishmen in their doublets and hose, and the unhappy Frenchmen cased stiffly in their mail, the panting, the staggering, and the rattling, the agonised curses from behind the vizor, and the great broad laugh on the honest English face—this alone should have saved it from oblivion. The English men-at-arms came quickly to the support of the bowmen, and after a long and desperate engagement, for the noble and gallant Bertrand du Guesclin was in command of the French, the English drove their enemy from the field and as usual finished the pursuit on horseback. There was no question in the action of superior archery or advantage of position, though Chandos indeed handled his reserve in a masterly fashion, but it was simply a matter of what the Duke of Wellington called bludgeon-work; and at this too the English proved themselves the better men.

By this time the oppression of the Free Companies had become so insufferable that, in order to rid the

1366. country of them, Charles the Fifth ordered Bertrand du Guesclin to take a certain number of them into service and march with them to fight for the bastard Henry of Trastamare against Pedro the Cruel of Castile. It would be a mistake, we must note in passing, to look upon these companies as composed simply of low ruffians; they seem on the contrary to have been made up largely of the class of esquires, while there were poor noblemen serving even among the archers. On entering Spain they took to themselves a white cross, the old English colour of the Crusades, as their distinctive mark, and were apparently the first English troops that introduced this substitute for uniform. Further, they called themselves the White Company, and were in this respect the forerunners of the Buffs and Blues. They did little profitable work under du Guesclin, and were presently dismissed, just in time to be re-enlisted to the number of twelve thousand by the Black Prince, who, dreading an alliance of France with Spain, was preparing an expedition for the rescue of Peter the Cruel. The vassals of Aquitaine and Gascony were also summoned to the Prince's standard, a reinforcement under the Duke of Lancaster was sent from England to Brittany, whence it marched overland to the south, and by December 1366 thirty thousand mounted troops were concentrated on the frontier of Navarre. It was by general consent admitted to be the finest army that had ever been seen in Europe; so rapid had been the growth of military efficiency in England under the two great Edwards. It was organised in the usual three divisions, the vanguard being under command of the Duke of Lancaster, with Sir John Chandos at his side. The battle was under the command of the Prince himself, and the rearguard under a Gascon noble and famous soldier, the Captal de Buch. Every man wore the red cross of St. George on a white surcoat and on his shield, a badge which henceforth became distinctive of the English soldier for two centuries. The Spaniards, it is worth noting, wore a scarf, a fashion which, already two generations old, was destined to last through our great Civil War, and to survive, in the form of a sash, to the present day.

On Monday the 22nd of February 1367 the first 1367. division crossed the Pyrenees by the Pass of Roncesvalles. The next two followed it on the two succeeding days, and the whole force was reunited at Pampeluna. The Prince had now two lines of operations open to him, both leading to his objective, Burgos; the one by Vittoria and Miranda on the Ebro, the other by Puente la Reyna and Logrono. He chose the former, the identical line followed in the contrary direction by Wellington in chase of the beaten French, and sent only a small detachment of volunteers under Sir Thomas Felton along the latter route. This party of Felton's deserves mention as the first body of English irregular cavalry under a reckless and daring officer. No exploit was too hare-brained for them and they did excellent service, for they were the first to find contact with the Spanish army, at Navarete, and having obtained it they preserved it, keeping the Prince admirably informed of the enemy's movements. Henry of Trastamare, on learning the advance of the English, crossed the Ebro and marched on Vittoria, but finding that the Black Prince had been beforehand with him fell back on Miranda. Felton's volunteers stuck to him so persistently and impudently during this retreat that the Spaniards at last lost patience and attacked them in overwhelming force. The English, a mere hundred men, were too proud to retire but stood firm on the hill of Arinez, the very spot where Picton broke the French centre in the battle of the 21st of June 1813, and were killed to a man. Henry then recrossed the Ebro to his first position at Navarete; the Black Prince crossed the same river at Logrono and on the 3rd of April the two hosts stood face to face on the plain between Navarete and Najera.

It is not easy to ascertain the force engaged on each side, but it is certain that the Black Prince, with about

1367, April 3. ten thousand men-at-arms and as many archers, was superior in numbers and very decidedly superior in the quality of his troops. Nevertheless the force had suffered much hardship, and the men were individually enfeebled by want of food. The Spanish army was distributed into four divisions. The first of these, consisting of dismounted knights, was placed under the command of Bertrand du Guesclin and formed the first line. The remaining three formed the second line; the largest of them, composed of mounted men-at-arms and a rabble of rude infantry, being drawn up in rear of the vanguard, while the other two, made up chiefly of light cavalry copied from the Moorish model, were drawn up on either flank slightly in advance of the second and in rear of the first line. The arrangement of the Black Prince's army was similar but more massive; first came the vanguard under John Chandos, then a second line with two flanking divisions pushed slightly forward, as in the Spanish army, and lastly the third line in reserve. Every man in the English host was dismounted. The battlefield was a level plain; and the sight of the two armies advancing against each other, armour and pennons glancing under the morning sun was, in Froissart's words, great beauty to behold.

The English archers as usual opened the engagement, and then the divisions of Chandos and du Guesclin, the two most gallant and chivalrous soldiers of their day, met in full shock. In spite of a furious resistance the English, weakened by privation, were for a moment borne back. Chandos was overthrown and went near to lose his life. But meanwhile the English archers in the flanking divisions had driven off the light horse that stood before them, and now wheeling inward enveloped du Guesclin's devoted band on both flanks. The bastard Henry strove gallantly to save the day with the second line, but the Black Prince brought up not only a second line but a third, and the battle was soon over. Then the English men-at-arms flew, as at Poitiers, to their horses, and the defeat was turned into a rout. A rapid torrent, spanned by but a single bridge, barred the retreat of the fugitives; the narrow passage was choked by the press of the flying, and thousands were taken or slain. 1367.

This battle marks the zenith of early English military power. But the campaign was after all a failure. The ill faith of Pedro the Cruel forced the Black Prince to tax Gascony heavily for the expenses of the war; the province appealed to the King of France, and the Prince was summoned to be judged before his peers at Paris as a rebellious vassal. He shook his head ominously when he received the message. "We will go," he said, "but with helmet on head and sixty thousand men at our back." The war with France broke out anew, and petty operations were soon afoot all over the country; but now noble after noble in Aquitaine and Gascony forsook his allegiance and revolted to the French. Disaster came thick upon disaster. The Earl of Pembroke, a new commander, disdaining the help of the veteran Chandos, was defeated, and Chandos himself, while advancing to his relief, was slain in a skirmish, to the grief alike of friend and of foe. The Prince, already sickening of a mortal disease, turned in fury upon the insurgent town of Limoges, besieged it, took it, and ordered every soul in it to be put to the sword. Three thousand men, women, and children were cut down, crying "Mercy, mercy!" but the stern man, too ill to ride, looked on unmoved from his litter, till at the sight of three French knights fighting gallantly against overwhelming odds his heart softened, and he gave the word for the slaughter to cease.

A few weeks later his little son, but six years old, the boy upon whom the great soldier had lavished all that was tender in his nature, died suddenly at Bordeaux. The blow aggravated the Prince's sickness, and the physicians ordered him to England, in the faint hope that he might get better at home. He returned, hid himself in strict seclusion in his house at Berkhampstead, and waited for the end. Meanwhile things in France went from bad to worse. A great naval defeat before Rochelle cost England the command of the sea, and with the loss of the sea Guienne and Gascony were lost likewise. An expedition under John of Gaunt landed at Calais and marched indeed to Bordeaux, but lost four-fifths of its numbers through sickness on the way. By 1374 the English possessions in France were reduced to Calais, Bordeaux, and Bayonne; so swiftly had victory passed away with the withdrawal of the master's hand.

1376. At length, in 1376, the Prince came up to Westminster to attend, even in his sick-bed, the deliberations of Parliament. This was his last effort. Two months later, on the 8th of June, he summoned his faithful comrades to his chamber to bid them farewell, and as they filed past he thanked them for their good service and asked their pardon for that he could not reward them as he wished. Then he entreated them to be faithful to his son as they had been to himself: and they swore it, weeping like women, with all their hearts. The end came with a flash of the imperious soldier's spirit. Observing that a knight who had offended him had come in with the rest, the Prince instantly bade him begone and see his face no more; and then the noble heart cracked, and with a last ejaculation that he forgave all men as he hoped to be forgiven, the Black Prince, the hope and pride and treasure of England, sank back and died. Two months later he was buried with military pomp in the cathedral at Canterbury; and over his tomb were hung, and still hang, his helmet, his surcoat, his gauntlets, his crest, his shield, and his sword,[1] the veritable arms worn by the first great English soldier.[2] For a great soldier he was and a great commander. He could be stern and he could be merciless, but those were stern and merciless times, and the man whose last thoughts were for his comrades-in-arms was a chief who could hold men to him and a leader whom they would

[1] The sword is gone, but the scabbard remains.

[2] See for the whole scene Dean Stanley's *Memorials of Canterbury*.

follow to the death. Men no longer pray for his soul in the chapel which he founded in the crypt of the cathedral; but morning and evening the voice of the trumpet, calling English soldiers to their work and dismissing them to their rest, peals forth from the barracks without and pierces faintly into the silence of the sanctuary, no unfitting requiem for the great warrior who, waiting for the sound of a louder trumpet, sleeps peacefully beneath the shadow of his shield.

AUTHORITIES.—The principal authority for the period is of course Froissart, whose narrative has been elucidated, by the help of minor authorities, by Köhler with his usual care and pains. See his vol ii. pp. 385-523, and in particular the list of authorities on pp. 385 and 417.

CHAPTER IV

THE works of the Black Prince lived after him. Not that we must look for them immediately in England, where we now enter on forty years of intestine division and civil strife. We do indeed find that Richard the Second, on his invasion of Scotland in 1385, adopted for his army the organisation that had been taught by his father at Navarete ; but we discover no trace of military progress. Far more instructive is it to look to the continent of Europe and watch the spread of English military ideas there. It has already been seen that the French, not daring to meet the English archers on horseback, adopted the English system of dismounting for action ; and it is interesting to note that the same fashion spread to Germany and Italy, steadily tending to overthrow the supremacy of cavalry wrought by the feudal system, and to make a revolution in the art of war. Not one of the nations, however, seems to have grasped the pith of the English tactics, the combination of the offensive and defensive elements in the infantry. The French indeed, under King Charles the Sixth, strove to raise up archers, and with all too good success, for they became so efficient that they were esteemed a menace to the nobility, and were soon effectively discouraged out of existence. Perhaps the most striking example of the misapplication of the English system is
1382. the conduct of the Austrian commander at Sempach, who by dismounting his knights deliberately gave away every advantage to the Swiss, and thus helped forward that nation on the way to make its infantry the model

of Europe; a very significant matter in the history of the art of war.

But the truest disciples of the Black Prince were the English Free Companies, from whom there descended to England, and indeed to Europe, a legacy of a remarkable kind. These companies were military societies framed very much on the model of the ancient trade-guilds, and had as good a right to the name as they. A certain number of adventurers invested so much money in the creation of a trained body of fighting men, and took a higher or lower station of command therein, together with a larger or smaller share of the profits, according to the proportion of their venture. If any man wished to realise his capital he could sell out, provided that he could find a buyer; if any one partner seemed to the rest to be undesirable they would buy him out and take in another. Thus grew up what was known as the purchase-system. The abuse of their monopoly by these companies drove the sovereigns of Europe after a time to issue commissions to their subjects to raise companies for their own service only; but even so the commercial basis of the company remained unchanged, being only widened when the time came for the amalgamation of companies into regiments. These military adventurers taught the nations the new art of war, and the nations could not but follow their model.

The greatest leader of free companies was an Englishman, a pupil of the Black Prince but greater even than his master, John Hawkwood. It is true that he did his work for foreign nations and in a foreign land, but even so his name must not be omitted from a history of the British Army. The company which he commanded, English almost to a man, was the terror of Italy, and not only the most formidable in the field but the smartest to the eye, for its arms were burnished till they shone like silver. Hawkwood, though a mercenary, was celebrated as the only one who never broke faith, and as a general his reputation was

1387. European. The action which he fought at Castagnaro, when, in spite of great inferiority in numbers, he deliberately laid his plans for a sudden counterstroke, after the manner of Poitiers, extorts the admiration even of modern generals. Still more remarkable is his once famous retreat in the face of an overwhelming
1391. force from the Adda to the Adige, and perhaps greatest of all was the closing scene of that retreat. For, as he lay encamped in the plains by the Adige, the enemy broke the dykes of the river and turned the whole flood of its waters upon his army. It was night, and the men were encamping, weary after a hard day's march, when the deluge came upon them. Everything conspired to create a panic, but Hawkwood's coolness and confidence were equal to the danger. He bade every horseman take up one of the foot-men behind his saddle, and then placing himself at their head he led them through ten miles of the trackless waste of water, never less than girth-deep, and brought them out by sheer sagacity, not indeed without loss but without heavy loss, to the dry bed of the river. This was in his last campaign, when he was past seventy years of age; and Florence, the state which he had long faithfully served, voted him a pension for life and a monument even during his lifetime. He was making arrangements to return to England when he died; and King Richard the Second begged the city of Florence that the bones of so famous a warrior might be returned to his native land. The request was gracefully granted by the citizens, but the last resting-place of Hawkwood is now unknown. His monument in the Cathedral at Florence records that he was the most skilful general of his age, a height of military fame that has been reached by one other Englishman only, John, Duke of Marlborough.

Yet another action must be briefly noticed to show the value set on English military skill. During the invasion of Portugal by the King of Castile, in 1385, the Portuguese were joined by a party of about five hundred English adventurers, whose leaders appear to have directed most of the operations. It was under their guidance that the decisive battle of Aljubarotta, 1385, August 14. of which the Portuguese are still proud, was finally fought; and it is worthy of remark that, finding no advantageous position to hand, they deliberately constructed by means of abattis an imitation of the position of Poitiers, making it unassailable from the front except through a narrow strait, which was purposely left open and lined with archers. Marvellous to relate, the Spaniards and the French, who were fighting with them, rushed straight into the trap, and were of course utterly overthrown; whereupon, in due accordance with precedent, the Portuguese made their counter-attack and won a complete victory.[1] All this was due, as Froissart says, to the counsel of the English; and indeed, little though we may be conscious of it, it is doubtful whether even after Waterloo the prestige of English soldiers was greater than at the end of the fourteenth century.

But while the English military doctrines were thus spreading themselves over Europe, fresh innovations, which were destined to render them obsolete, were already making rapid progress. Artillery in the hands of the Germans was tending more and more to lose its cumbrous character and to take new form in mobile and practicable weapons. The heavy bombards, which could be neither elevated nor traversed, had before the close of the fourteenth century given place to lighter guns of smaller bore fixed on to the end of a shaft of wood and supported on a fork or hook, whence they derived their name of *Hakenbüchse*, a word soon corrupted by the English into hackbut, hagbush, and finally harquebus. A later improvement had fitted guns with a stock like that of the cross-bow, which could be brought up to the shoulder, thus more readily aligning the barrel to the eye. The step from this to the hand-gun, which could be served out as the individual weapon of a single man,

[1] Sir Arthur Wellesley occupied the Spanish position on his march to Roliça (*Conversations of the Duke of Wellington*, p. 3).

was but a short one and was soon to be taken. But as the traditions of Wellington and the Peninsula were to be tried once more at Alma and Inkerman before they finally perished, so the system of the two great Edwards was to be revived forty years after Navarrete at Agincourt.

1415. It is unnecessary to dwell on the pretensions which were put forward to excuse the wanton aggression of Henry the Fifth against France. Ambitious, like Frederick the Great, of military glory he made his will the true ground for his action, counting on the spirit of a people that was never strongly averse from a French war. The military devices introduced by the Edwards, the commissions of array,[1] and the system of indentures, were still in good working order, while the discipline of the Black Prince, like his order of battle, was stereotyped in a written code of Ordinances of War. All the old machinery was therefore to hand; and perhaps the most noteworthy change that had come over the English military world was the doubling of the archers' wages from threepence to sixpence a day. Parliament voted the King a large sum of money, which however proved to be insufficient, for, significantly enough, not a contractor would furnish his contingent of men without security for the repayment of his expenses. The crown jewels were pledged in all directions, ships were hired in Holland and in England, seamen were impressed, artisans of every trade, from the miner to the farrier, were engaged, and on the 7th of August 1415 the army embarked at Southampton and the adjacent ports, and sailed for the Seine. The whole fleet numbered some fourteen hundred vessels, and the army is reckoned at thirty thousand men, men-at-arms with their attendants, and archers both mounted and afoot, all distinguished by the red cross of St. George. Further, there was a great train of the newest and best artillery, great guns called by pet names such as the

[1] These had been recognised by a statute of 5 Henry IV., the enactment relied on later by Charles I.

London and the King's Daughter, the whole under the charge of four German gunmasters. 1415.

On the second day out the fleet anchored before Harfleur. A day was taken up by the disembarkation, which was unhindered by the French; and by the 19th of August the town was fully invested. Then came a month of siege, wherein the art that was dying blended strangely with that which was just coming to birth; wooden towers and quaint engines that might have been employed by the Romans plying side by side with sap and mine and countermine and the latest patterns of German artillery. The French made a most gallant defence, and dysentery breaking out in the English camp swept off thousands of the besiegers; but at length the heavy guns prevailed. The garrison begged for terms, praying that the King would make his gunners to cease, "for the fire was to them intolerable." On the 22nd of September the capitulation was agreed on, and Harfleur received an English garrison. It was the first town that the English had reduced by the fire of cannon.

But Henry was not yet satisfied. Two-thirds of his force had melted away, dead or invalided, but he had no intention of re-embarking at Harfleur. He devoted a fortnight to the repair of the defences of the captured town, and then collecting provisions for eight days he marched northward for Calais with an army, or, as we should now call it, a flying column, of nine thousand men.

Meanwhile the French, disorganised though they were by the insanity of their king, Charles the Sixth, began to bestir themselves, and collecting an army of sixty thousand men, fourteen thousand of them men-at-arms and several thousand archers and cross-bowmen, determined to hold the line of the Somme and bar Henry's passage of the river. Henry's idea, dictated like the whole of his campaign by the precedent of Edward the Third, had been to cross the Somme by the ford of Blanche Tache. He now learned that the

1415. passage was defended by the French in force. He wheeled at once to the right, and following the left bank of the river upward, tried in vain to find a crossing-place. Every bridge was broken down and every ford beset. It was plain that he was more effectually entrapped even than his predecessor Edward.

October. The eight days' supply of provisions was now consumed, and the position of the English became most critical. Retreat Henry would not, force the passage of the Somme he could not. He decided to follow the river upward to its head-waters, and on reaching Nesle learned from a countryman of a ford, the access to which lay across a morass. Two causeways that provided a footing over it had been broken down by the French, but these were quickly repaired with wood and faggots and straw till they were broad enough to admit three horsemen abreast. Henry himself was indefatigable in the work. He took personal charge of one end of the passages, and appointed special officers to attend to the other. The baggage was carried over along one causeway, and the men by the second. Thus the passage both of morass and river was accomplished between eight in the morning and an hour before dusk of an October day. The French, who were lying in force at Peronne, now for some unexplained reason retreated towards the north-west, but sent, according to custom, a challenge to Henry to fix time and place for battle. "I am marching straight to Calais through open country," he replied. "You will have no difficulty in finding me." And he continued his advance.

At Peronne the English struck the line of the French march and looked for an immediate engagement. The force moved in order of battle, every man armed and ready for action, while the archers by Henry's order carried a stake, eleven feet long and pointed at both ends, to make them defence against cavalry. To their surprise no enemy appeared; and Henry was presently able to disperse his force along a wider front, with the advantage alike of obtaining easier supply of victuals and surer information of the enemy. The English were much distressed by want of bread: other provisions were abundant, but grain was absolutely undiscoverable. Nevertheless discipline was most strictly enforced, and the order of the columns, as the speed of the march can avouch, was quite admirable. Robbery of churches or peasants, the slightest irregularity on the march or in the camp, the presence of women in the camp, all offences alike were visited with the severest punishment. One man, whom Shakespeare has immortalised as Bardolph, was detected in the theft of a pyx: he was paraded through the army as a criminal and hanged. Even French writers admit that the English dealt more mercifully with them than their own countrymen. The King himself avoided anything that might seem to indicate the slightest discouragement. One night he missed the camping-ground assigned to his division and took up that of the vanguard. "God forbid that in full armour I should turn back," he said; and pushing the vanguard further forward, he halted for the night where he stood. 1415, October.

On the 24th of October, Henry, who was lying at Frevent on the river Canopes, was informed by his scouts that the French were moving forward from St. Pol and must inevitably get ahead of him. He pushed on to Blangy, crossed the river Ternoise there, and advancing to Maisoncelle drew up his army in battle order before it. The whole French army was before him at Ruisseauville, but as dusk fell without an attack he withdrew for the night to Maisoncelle, and conscious of his desperate situation opened negotiations with the French, offering to restore Harfleur and make good all injuries if he might be permitted to evacuate France in peace. His overtures were rejected and he was warned to fight on the morrow. On the same evening the French moved down to a narrow plateau between the villages of Tramecourt and Agincourt, and there, cramped into a space far too narrow for sixty thousand

1415. men, they halted till the morrow within less than a mile of the English position.

The night was spent in very different fashion in the two camps. The French, doubtless much inconvenienced by the straitness of their quarters, were shouting everywhere for comrades and servants as noisily as a mob of sheep; while some, forgetting the lesson of Poitiers, gambled for the ransom of the prisoners that they were to take in the morrow's battle. Huge fires were kept burning round their banners, for the rain was incessant, and the English could see everything that passed among them. They too began shouting like the French till sternly checked by the King; and then the English camp fell silent, and the men, forbidden to forget their situation in the din of their own voices, sat down to face it in all its stern reality. They could be excused if they felt some misgiving. They had covered over three hundred miles in a continuous march of seventeen days, often in hourly expectation of a fight; for four days they had not tasted bread; and now, after a few short hours more of waiting in the ceaseless pattering rain, they were to meet a host outnumbering them by five to one. Arms and bowstrings were overhauled and repaired; and the priests had little rest from the numbers that came to them for shrift. But in the discipline of that silence lay the promise of success.

October 25. At dawn of the next morning Henry was astir, fully armed but bare-headed, riding a gray pony. Presently he led the army out of Maisoncelle to a newly-sown field, which was the position of his choice, and drew it up for battle. Every man was dismounted, and horses and baggage were parked in the rear under the protection of a small guard. But the numbers of his army were so weak that the favourite formation of the Black Prince could not be followed. The vanguard under the Duke of York became the right, the battle under the King the centre, and the rearguard under Lord Camoys the left of a single line, which even then was ranked but four men deep. It was a first example of English line against French column. Henry made the men a short speech, recalling to them the deeds of their fathers, and then the whole host kneeled down, thrice kissed the ground, and rose upright again into its ranks. 1415, October 25.

Meanwhile not a sign of attack came from the French. Their order of battle had been determined many days before, but it was ill adapted to so narrow a position. It was evident that only the vanguard could possibly come into action, and such was the indiscipline that every man of rank wished to command it. Finally the whole of the magnates were placed in the vanguard, and its strength was made up to about seven thousand men-at-arms, every one of them dismounted. On each flank was a wing of twelve hundred more dismounted men, and on their flanks again two small bodies of cavalry, three hundred on the right, and eight hundred on the left, which were designed to gallop down upon the archers. This was the first French line. The second was also made up of about eight thousand dismounted men-at-arms; while the remainder, who were ordered to dismount but would not, composed the third line. The whole stood on ploughed ground, soaked by the rain of the previous night and poached deep by the trampling of innumerable feet.

The French took advantage of the delay to give their men breakfast, an example which Henry immediately followed. Then seeing that the enemy remained motionless he prepared to attack. A gray old warrior, Sir Walter Erpingham, galloped forward with two aides-de-camp to make the necessary changes of formation. The archers were deployed in front and flanks, and when all was ready old Sir Walter tossed his baton into the air and sang out "Now strike." Then galloping back to the King's battalion he dismounted and took his place in the ranks. The King, already dismounted, gave the word "Forward banner," and the English answered with a mighty cry, the forerunner of that "stern and appalling shout" which four centuries later was to strike hesitation into so fine a soldier even as Soult. Then

15, tober 25.

the whole line advanced in close array, with frequent halts, for the ground was deep, and the archers in their leathern jackets and hose, ragged, hatless, and shoeless after two months of hard work, could easily wear down the men-at-arms in their heavy mail. Artillery in such a sea of mud could not be brought into position on either side, and the German gunners took no part in the fight. The French on their side stood firm and closed up their ranks. They were so heavily weighted with their armour, always heavier than that of the English, that they could hardly move, and their front was so much crowded that they could not use their archers; so they broke off their lances as at Poitiers to the length of five feet, and stood in dense array, thirty-one ranks against the English four.

Arrived within range the archers struck their stakes slantwise into the ground, and drew bow. The French vanguard then shook itself up and advanced slowly, while the cavalry on their flanks moved forward against the archers. The division of three hundred lances on the right made but a poor attack; little more than half of them really came on, and even these their horses, maddened as at Creçy by the pain of the arrows, soon carried in headlong confusion to the rear. The stronger division on the left charged home, and the leader and one or two others actually reached the line of stakes; but the stakes had no firm hold in the mud; the horses tripped over them and fell, and not one rider ever rose again. The remainder had as usual been carried back by their wounded horses upon their comrades in rear, and thence with them upon the wings of dismounted men-at-arms in which they tore terrible gaps. The centre of the French vanguard fared little better. Dazzled by the eastern sun that shone full in their eyes, and bending their heads before the sleet of arrows, they lost all idea of their direction, and became so clubbed together that they could not use their weapons. By sheer weight they forced back the English men-at-arms a lance's length, and for a time they fought hard.

King Henry was twice struck heavily on the helmet, one blow lopping a branch from the crown that encircled it. But meanwhile the archers had noted the gaps torn by the horses in the wings of the French fighting line. They dropped their bows, and with whatever weapon —axe, hammer, or sword—that hung at their girdle, they fell, light and active, upon the helpless, hampered men-at-arms and made fearful havoc of them. The French centre, exposed by the defeat of the wings to attack on both flanks, gave way before the King's battalion, and their first line was utterly defeated. There was no question of flight among the French men-at-arms, for the unhappy men could not move. The English simply took off the helmets of their prisoners, and, leaving them thus exposed, pressed on against the second line. This, however, was already shaken by the defeat of the vanguard; and though one leader who had arrived late in the field, the Duke of Brabant, set a gallant example, he was quickly cut down, and the defeat of the second line followed quickly on his fall. The third line still remained, but being mounted, contrary to orders, had no mind to stay and fight, but turned and fled, leaving some few of their leaders alone to redeem French honour by a hopeless struggle and a noble death.

1415, October 25.

This battle was hardly won when word was brought to Henry that his baggage, with all his treasure as well as all the horses, was in the hands of plunderers. The guard in fact had been unable to resist the temptation to join in the fight, and had left the baggage to take care of itself. The momentary confusion hereby caused gave some of the French time to rally, and Henry, not knowing how great the danger might be, ordered every man to kill his prisoners. The English hesitated, less possibly from humanity than from reluctance to lose good ransom, whereupon Henry told off two hundred archers for the duty, which was promptly carried out. He can hardly be blamed, for the fight had been won less by the slaughter than by the capture of the men-at-

415, October 25. arms; and the risk of undertaking a new attack in front with some thousands of unwounded prisoners in rear, was serious. Be that as it may, the deed was done. Henry then advanced against the rallied French and quickly broke them up; and at four o'clock, the victory being at last complete, he left the field. The French loss in nobles alone numbered from five to eight thousand men killed, exclusive of common men. A thousand prisoners and a hundred and twenty banners were taken. The losses of the English are uncertain, but probably did not exceed a few hundreds, the most distinguished of the fallen being the Duke of York.

So ended the great fight which King Harry himself decreed to be called by the name of Agincourt.[1] It sums up in itself the leading features of Creçy, Poitiers, and Cocherel, in a word of all the finest actions of the Edwards. But it was, as fate ordained, but the afterglow of the glory of the Plantagenets, not the light of a sun new risen like a giant to run his course.

To attempt to follow the later campaigns of Henry the Fifth in France would be alike tedious and unprofitable. To the last he stuck to the principles of the Black Prince, but his military talents ripened year after year, and while he lived France trembled under his
1420. sword. Finally, torn to pieces by the strife of Burgundian and Armagnac, France by the Treaty of Troyes surrendered her kingship into his hand. The contempt of the English for their enemy was such that the men once assaulted and captured a town without orders. But in the very next year came a reverse that boded ominously for the future. The Duke of Clarence was defeated at Beaugé, less by the French than by a body of Scottish auxiliaries, who had been sent to their assistance under the Earl of Buchan. Henry had hoped that the Scots would not fight against him, and ordered them henceforth to be treated as rebels, but it was to no purpose. The reader should take note of this fateful

[1] More correctly Azincourt.

year 1421, for it marks the permanent entrance of the Scots into the service of France, a fact full of import for both countries. Moreover, he will in due time see a regiment, still called the Royal Scots, withdrawn from the French army to become the first of the English Line.

The return of King Henry to France after Beaugé soon re-established the ascendency of the English arms; and then, while still in the prime of life, he sickened even in the midst of his operations and died. He was 1422. but thirty-four years of age, a great administrator, a great captain, and above all a grand disciplinarian. Yet he was no brutal martinet; nay, when once he had cast his wild days behind him he never even swore. "Impossible," or "It must be done," was the most that he said. But "he was so feared by his princes and captains that none dared to disobey his orders, however nearly related to him, and the principal cause was that if any one transgressed his orders he punished him at once without favour or mercy."[1] He and the army that fought with him at Agincourt are the true precursors of Craufurd and the Light Division. His body, borne with mournful pomp from the castle of Vincennes, still rests among us in Westminster Abbey, and above it still hang his saddle, his shield blazoned with the lilies of France, and the helmet, deeply dinted by two sword-cuts, which he wore at Agincourt. Not for three centuries was another soldier to rise up in England of equal fame with the Black Prince, John Hawkwood, and King Harry the Fifth.

AUTHORITIES.—For the life of Hawkwood see Temple Leader's *Sir John Hawkwood*. For the campaign of Agincourt, *Gesta Henrici Quinti* and Monstrelet's Chronicles are the chief authorities, while Sir Harris Nicholas's *Agincourt* furnishes a quantity of supplementary information. Other authorities will be found enumerated in Köhler, who is always the best guide in respect of military operations.

---

[1] Monstrelet.

CHAPTER V

It is now our sad duty to watch the military glory of the Plantagenets wane fainter and fainter, until it disappears, to be followed by a period of darkness until the light is slowly rekindled at the flame of foreign fires. The decline of our supremacy in arms was not at first rapid. John, Duke of Bedford, possessed a combination of military and administrative talent little less remarkable than that of his brother the late King, and as Regent of France he took up the reins of government and command with no unskilful hand. Everything turned upon the maintenance of existing factions in France. England working with Burgundy, the red cross of St. Andrew with the red cross of St. George, could preserve the English dominion; otherwise that dominion must inevitably fall. The French, after the lull created by Henry's death, gathered an army together of which the kernel was three thousand Scots, and marched into Burgundy to besiege Crevant. A body of four thousand picked English and Burgundians at once hastened after them, and although outnumbered, and compelled, by the advance of a second French army in their rear, to fight their battle and win it at whatever cost, they defeated the enemy completely and cut the Scots to pieces almost to a man. All was still done as King Harry had done it. English tactics were forced, on pain of death, upon English and Burgundians[1] alike, and discipline was most strictly preserved. It was not

[1] See Philippe de Commines, bk. i. chap. iii. "[At the battle of Montlhéry, 1464] the most honourable persons fought on foot among the archers . . . which order they learned of the English, who are the best shot in the world."

a promising beginning for the French, but Scotland was ready to furnish more men, and France not less ready to receive them ; and so the extraordinary struggle of French against French, and English against Scots was renewed once more.

Early in 1424 ten thousand Scottish men-at-arms, under Archibald, Earl of Douglas, arrived at Rochelle, and were welcomed with eagerness by the French. Douglas was created Duke of Touraine, and all went merrily until on the 17th of August French and English, with their allies, met under the walls of Verneuil. The French and Scots numbered close on twenty thousand men, the English twelve thousand, of whom eight thousand were archers. Contrary to the hitherto accepted practice, the French formed their army into a single huge central battalion of dismounted men, with cavalry on each wing, the mounted men being designed to fall upon the English flanks and rear. Bedford, who commanded the English, imitated the enemy in forming only a single battalion, but dismounted the whole of his force, covering his front and flanks with archers, who as at Agincourt carried stakes as a defence against the attack of horse. His baggage he parked in rear, the horses being tied collar to tail that they might be the less easily driven off; and he appointed as baggage-guard no fewer than ten thousand archers. 1424.

For the whole morning the two armies stood opposite to each other in order of battle, each waiting for the other to attack ; but at last, at three in the afternoon, the French advanced and were received by the English with a mighty shout. The French cavalry on the wings charged, broke through the archers, and sweeping round the English rear fell upon the baggage. They were greeted by the guard with a shower of arrows, but contrived none the less to carry off some quantity of spoil, with which they galloped away, feeling sure that the day was won.[1] But meanwhile the

[1] The reader will observe how early cavalry fell into the fault which caused the loss of Naseby.

two battalions of dismounted men-at-arms, those on the French side being exclusively Scots, had closed and were fighting desperately. For a moment the English were beaten back by superior numbers; but Salisbury, John Talbot, and other tried leaders were with them, and they soon recovered themselves. The archers on the wings rallied to their aid, while those of the baggage-guard, freed from all further alarm of cavalry, hurried up with loud shouts in support. The Scots wavered, and the English pressing forward with one supreme effort broke through their ranks, split up the battalion, and threw the whole into helpless confusion. And then began a terrible carnage, for the Scots had told Bedford that they would neither give nor receive quarter, and they certainly received none. Five thousand men, mostly Scots, were killed on the French side, John Stewart, Earl of Buchan, the Earl of Douglas and James his son being among the slain, and two hundred more were taken prisoners. Of the English some sixteen hundred only went down.

To France Verneuil was a disaster little less crushing than Agincourt, and indeed it seemed as though she had passed irrevocably under English dominion. All was however spoiled by Bedford's brother Humphrey, Duke of Gloucester, who, having made a match with a rich heiress, Jacqueline of Holland, carried away English troops to take possession of her dower-lands, and, worst of all, gave the deepest offence to Burgundy. At home Humphrey was equally troublesome, so much so that in 1425 Bedford was compelled to return to
1428. England to set matters right. It was not until three years later that he took the field again, well reinforced with men and with a powerful train of artillery. So far we have rarely found artillery employed except for sieges, but henceforth we see gunners regularly employed at the high wage of a man-at-arms, one shilling a day, and "hand-cannons" and "little cannons with stone shot of two pounds weight," playing ever a more prominent part in the field.

Against his better judgment Bedford now resolved to carry the war across the Loire, and detached the Earl of Salisbury with ten thousand men to the siege of Orleans. The operations opened unfortunately with the death of Salisbury, who was mortally wounded by a cannon-shot while examining the enemy's works; but the investment was carried on with spirit by the Earl of Suffolk, and a little action at the opening of 1429 1429.
showed that the English superiority still held good. This, the battle of Roveray, better known as the action of the Herrings, has a peculiar interest, though the occasion was simple enough. Lent was approaching; and as, among the many complications of mediæval warfare, the observance of the fast was by no means forgiven to fighting men,[1] it was necessary to send provisions of "Lenten stuff," principally herrings, to the besieging force round Orleans. The convoy being large was provided with an escort of sixteen hundred men under command of Sir John Falstolfe. The French and Scots decided to attack it on the march, but unfortunately could not agree as to their plan; the Scots insisting that it was best to dismount, the French preferring to remain in the saddle. Meanwhile Falstolfe with great dexterity drew his waggons into a leaguer, leaving but two narrow entrances defended by archers. It was the trap of Poitiers once more. The French and Scots after long discussion agreed to differ, and attacked each in their own fashion. The English archers shot with admirable precision; the Scots lost very heavily, the French after a short experience of the arrows rode out of range, and Falstolfe led his herrings triumphantly into Orleans, having killed close on six hundred of the enemy with trifling loss to himself. This was the last signal employment of the tactics of

[1] "The same difficulty of a Lenten campaign cropped up at the siege of Orleans a century later. It was surmounted by the general's insisting that the papal legate, who was in the camp, should grant a dispensation, which he very unwillingly did; whereupon every man in the army 'pria Dieu fort pour M. le legat'" (Brantôme, ed. Elzev. vol. i. p. 225).

Poitiers, the last brilliant success of the English in the Hundred Years' War, the first glimpse of a lesson learnt by England from the military genius of a foreign power. For the tactics of the waggon were those of John Zizka, the greatest soldier of Europe in the fifteenth century.

From this point the story is one of almost unbroken failure for the English in France. They were now about to pass through the experience which later befell the Spaniards in the Low Countries, and the French themselves in the Peninsula. The turning-point is of course the appearance in the field of Joan of Arc, a phenomenon so extraordinary that it has become the exclusive property of the votaries of poetry and sentiment, and is, perhaps rightly, not to be rescued from their hands. It is certain that her military talents were of the slightest; but, on the other hand, she possessed the magic of leadership and the amazing power of restoring the moral strength of her countrymen, which had been impaired as never before by an endless succession of defeats. The English not unnaturally attributed this power to witchcraft: for by what other agency could a peasant girl have checked the ever-victorious army? and the punishment of witchcraft being the fire they burnt her to death. Any other nation would have done the same in their place then, and there are still a few folks both in France and the United Kingdom who would do so now. But the fire in the market-place of Rouen availed the English little. "The French," as Monstrelet says, believed that "God was against the English"; and the English began to believe it themselves.

For the woman's quick instinct and the pure insight of a saintly soul had guided the maid aright. The moral quality of the English force was corrupted, and needed only to meet some loftier spirit to fall into decay. The chivalrous character of the war was gone. Hostile commanders no longer laid each other friendly wagers on the success of their next operations. The army too was ceasing to be national; the English element was growing smaller and smaller in number, and fast sinking to the level of the lawless adventurers who furnished the majority in the ranks. Long contempt of the enemy had bred insolence and carelessness, and the old discipline was almost gone. The sight of a deer or a hare sufficed to set a whole division hallooing, sometimes, as at Patay, with disastrous results. On 1430. that day the French scouts, who were feeling for the enemy, roused a stag, which ran towards the English array, and was greeted with such a storm of yells as told the French all that they wanted to know. The English force blundered on, without advanced parties of any kind, till it suddenly found itself on the verge of an engagement. Then the leaders wrangled as to the question of fighting in enclosed or open country, and, having finally in overweening confidence selected the open, were surprised and routed before the archers could plant their stakes in the ground. Worst of all, an officer in high command, Sir John Falstolfe, seeing that defeat was certain, disobeyed the order to dismount and galloped away. He was disgraced by Bedford, but was afterwards for some reason reinstated, though had Harry been king he would assuredly have lost his head.[1]

Among the French the revival of the military spirit soon showed itself in a remarkable development of new ideas. They had long copied, though with a bad grace, the English practice of dismounting men-at-arms and furnishing archers with a palisade of stakes, but in 1434 at Gerberoy they used the three arms, cavalry, infantry, and artillery, in combination, with signal success. Artillery was still so far a novelty in the field that only three years before a whole army collected by the Duke of Bar Sandacourt, had flung itself howling to the ground at the first dis- 1431. charge; but the English archers, though they knew better than to behave thus, were sadly dismayed when the round stone shot came bounding within their trusted

[1] He remains gibbeted, however, in the pages of Shakespeare, which is perhaps the worst fate that could have befallen him.

palisade. It was just after this, too, that two fatal blows were struck at the English by the shifting of Burgundy to the French side, and by the death of their ablest leader, John, Duke of Bedford.

Still the war, wantonly and foolishly continued by an inefficient Government, dragged on and on, and, though not unbroken by occasional brilliant exploits, turned steadily against the English. The behaviour of the soldiers was sullied more and more by shameful barbarity; and gradually but surely their hold on Normandy and Guienne slipped from them. Truce was made at last in 1444, and Charles the Seventh seized the opportunity to execute a series of long-meditated reforms in the French army. He established a national militia of fifteen companies of men-at-arms and archers, each six hundred strong, organised garrisons of trained men for the towns, took the greatest pains for the equipment, discipline, and regular payment of the troops, and formed the finest park of artillery thitherto seen. In a word, he laid the foundation of the French standing army, with the Scottish archers and Scottish men-at-arms at its head, two famous corps that remained in their old place on the army-list until the French Revolution. Thus French military organisation, spurred by a century of misfortune, made one gigantic bound ahead of English, and may be said to have kept the lead ever since.

In England there had been no such improvement. A feeble effort had been made to check by statute fraudu-
1440. lent enlistment and the still graver abuse of embezzlement of the soldiers' pay by the captains, but this was of little help when the enforcement of the Act[1] was entrusted to so corrupt and avaricious a commander as the Duke of Somerset. Throughout the truce the soldiers on the English side behaved abominably; but, since they were robbed of their wages by their officers, it is hardly surprising that they should have repaid
1449. themselves by the plunder of the country. When finally the truce was broken, and the French invaded

[1] 18 Henry VI. cap. 18.

Normandy, the English dominion fell before them like a house of cards. Town after town, their garrisons depleted to fill Somerset's pocket, surrendered to superior force, and the English as they marched forth had the mortification to see the Normans gleefully doff the red cross of St. George for the white cross of France. An attempt to save the province was foiled by the rout of the English reinforcements at Fourmigny, and Normandy 1450, April 18.
was lost. Anjou and Maine had been already made over to the father of Henry the Sixth's Queen, and Guienne and Gascony, which had been English since the reign of Henry the Second, alone remained. Next year they too went the way of Normandy and were lost.

Gascony, however, notwithstanding her hot southern blood, was in no such anxiety as Normandy to be quit of the English, and sent messages to England that, if an army were sent to help her, she would revolt against the French to rejoin her old mistress. England lent a willing ear, and John Talbot, the veteran Earl of Shrewsbury, was sent out to this, his last campaign. The decisive battle was fought under the walls of 1453, July 20.
Chatillon. The French were strongly entrenched, with three hundred pieces of artillery in position, a striking testimony to their military progress. The English fought with the weapon which for a century had won them their victories, and for the last as for the first battle of the Hundred Years' War, every man alighted from his horse. John Talbot alone, in virtue of his fourscore years, remained mounted on his hackney; and with the indomitable old man at their head the English hurled themselves upon the entrenchment. It was a mad, desperate, hopeless venture, but they stormed forward with such impetuosity that they went near to carry the position. For a full hour they persisted, until at last, riddled through and through by the fire of the artillery, they fell back. Then the French sallied forth and turned the defeat into a rout. Old John Talbot's pony was shot under him, and being pinned to the ground under the dead animal he was killed where he lay.

Young John Talbot, Lord Lisle, refused to leave his father, and fell by his side. The army was dispersed over Aquitaine, and the ancestral domains of seven generations of English kings passed from them for ever. By the irony of fate a Scottish soldier[1] was appointed to hold for the crown of France the French provinces that had clung with such attachment to England. Of all the great possessions of the English in France Calais now alone was left, to break in due time the heart of an English Queen.

At home the discontent over the national disgrace was profound. The people of course cast about to find a scapegoat, and after one or two changes finally fixed upon the blameless and unfortunate Henry the Sixth. Want of a strong central government was undoubtedly the disease from which England had suffered ever since the death of King Henry the Fifth, but for this the nation itself was principally responsible. It had chosen for its rulers the House of Lancaster because Henry of Bolingbroke had agreed to accept constitutional checks on the royal power before the country was ripe for self-government. It had thrown off the yoke of discipline which alone could enable it to tug the heavy load of English weal and English honour, and it paid the inevitable penalty. Numbers of republics have made the same mistake during the present century and have suffered or are suffering the same punishment. There is no surer sign of an undisciplined nation than civil war.

In the England of the fifteenth century the disease had been deeply aggravated by the interminable campaigns in France. All classes at home, from the highest to the lowest, were equally selfish and apathetic in respect of the national good: internal order was at an end, and riots and outrages which amounted to private war continued unceasingly and remained unrepressed. The system of indentures between king and subject for the supply of troops had been extended from subject to retainer and, as has been well said, the

[1] Robert Patillock.

clause "for the King's service" could easily be dropped out of the contract.[1] The red cross of St. George never appears in the English battlefields; red rose and white were indeed the emblems of contending factions, but we hear far more of the badges of great families, the ragged staff, the cresset and the like, and of the liveries, which, though forbidden by statute to any but the king, were conspicuous all through the Civil War. The loss of France furnished but too much material to the hands of violence and strife. England was full of unemployed soldiers, who had been trained in the undisciplined school of French faction to treachery and plunder and all that is lowest and most inhuman in war. Hundreds of men who had held comfortable posts in French garrisons, and had turned them to purposes of brigandage, were cast adrift upon England, barbarised, brutalised, demoralised, to recoup themselves in their own country. After the peace of Brétigny the disbanded soldiery had made France their chamber and swept down thence upon Italy; the like men[2] were now to be let loose upon England, and France was to be well avenged of her old enemy. Worst of all, the leaders of factions, in the madness of their animosity, were not ashamed to import foreign troops and set them at each other's throats.

I shall not dwell upon this miserable and disastrous period, marking as it does the wreck of our ancient military greatness. Such few military points as present themselves in the scanty chronicles of this time must be noted, and no more. Of the principal figures one only is to be remarked. Warwick the "King-maker" must be passed over as rather a statesman than a soldier; Margaret of Anjou—the pestilent, indomitable woman—must be remembered only for her importation of

[1] Oman's *Warwick*.

[2] Yet they were not all ruffians. In the *Paston Letters* some professional soldiers hired for private defence are described as gentlemanly comfortable fellows, and their employer is warned that they must not be put to sleep more than two in a bed (vol. ii. p. 327).

mercenaries; Edward the Fourth, full of the military genius of the Plantagenets, alone is deserving of lengthier mention. There was not an action at which he was present wherein he did not make that presence
1460. felt. It was he who at Northampton turned his treacherous admission to the left of the Lancastrian position to instant and decisive account. It was he who in the following year, still only a boy of twenty,
1461. crushed Owen Tudor at Mortimer's Cross; it was he who held supreme command at that more terrible Marston Moor of the fifteenth century, the battle of Towton.

1461, This action has a peculiar interest as an example of
March 28. English tactics and tenacity turned upon themselves. The Lancastrians, sixty thousand strong, were formed up on a plateau eight miles to the north of Ferrybridge, facing south—their right resting on a brook, called the Cock, their left on the Great North Road. It was a strong position, but too much cramped for their numbers, having a front of less than a mile in extent. They were probably drawn up according to the old fashion in three lines of great depth. The Yorkists numbered but five-and-thirty thousand, but they were expecting an additional thirteen thousand under the Duke of Norfolk, which, advancing from Ferrybridge, would come up on their own right and against the left flank of the enemy. Edward appears to have remedied his numerical inferiority after the pattern of his great ancestor at Creçy by forming his army in echelon of three lines, refusing his right. The foremost or left line of the echelon was commanded by Lord Falconbridge, the second by Warwick, and the third by Edward in person. The Yorkists advancing northward to the attack had just caught sight of the enemy on a height beyond a slight dip in the ground called Towton Dale, when there came on a blinding snowstorm, which so effectually veiled both armies that it was only by their shouts that they could know each other's position. Falconbridge with great readiness

seized the moment to push forward his archers to the edge of the plateau, whence he bade them shoot flight-arrows, specially adapted to fly over a long range, into the Lancastrian columns. This done he quickly withdrew his men. The Lancastrians thereupon poured in a tremendous shower of fighting arrows, all of which fell short of their supposed mark, and maintained it till their sheaves were well-nigh exhausted. Then Falconbridge again advanced and began to shoot in earnest; his men had not only their own stock of shafts but also those discharged by the enemy. The rain of missiles was too much for the Lancastrians: they broke from their position on the height and poured down across the dip to drive the Yorkists from the slope above it. Then the action became general and the whole line was soon hotly engaged.

What followed for the next few hours in the driving snow no one has told us, or, it is probable, could ever have told us. All that is certain is that the Lancastrians, though occasionally they could force the Yorkists back for a space, could never gain any permanent advantage, a fact that points to extremely judicious handling of the refused division by Edward. From five in the morning until noon the combat raged with unabated fury, and the pile of the dead rose so high that the living could hardly come to close quarters. At length at noon the Duke of Norfolk's column, timely as Blücher's, appeared in the Great North Road on the left flank of the Lancastrians, and began to roll them back from their position and from the line of their retreat. Slowly and sullenly the Lancastrians gave way; there was probably little attempt to alter their disposition to meet the attack in their flank; but for three long hours more they fought, disputing every inch of ground, till at last they were forced back from it upon the swollen waters of the Cock. Then the rout and the slaughter became general; thousands were drowned in the brook; and the pursuit, wherein we again see the hand of Edward, was carried to the very gates of York. Thirty-five thousand Lancastrians and eight thousand Yorkists perished in the fight, an appalling slaughter for so miserable a cause. But this was a contest not merely of faction against faction, but of North against South; and the North never spoke disrespectfully of the South again. This perhaps was the principal result of what must be reckoned the most terrible battle ever fought by the English.

1471, April 14. The decisive battle of Barnet furnishes a still more brilliant instance of Edward's skill, and of his quickness to seize the vital point in a campaign. All turned upon his forcing his enemies to action before they could gather their full strength about them. Edward marched his men up to Warwick's position actually after dusk had fallen, a rare accomplishment in those days, and drew up his men as best he could in the dark. When day broke with dense fog he discovered that his army far out-flanked Warwick's left, and was as far out-flanked by Warwick's on his own left. The result seems to have been that the two armies edged continually round each other until their respective positions were reversed,[1] for some of Warwick's cavalry, coming back from the pursuit of Edward's left, found itself on its return not, as it supposed, in rear of Edward's army, but of its own. The cry of treason, always common in the Wars of the Roses, was quickly raised, and in the general confusion the battle was lost to Warwick. None the less the victory was due to Edward's promptness; and indeed the rapidity alike of his decisions and of his marches stamp him as a soldier of no ordinary talent, and as in many respects far in advance of his time.

For the rest the Wars of the Roses show unmistakable signs of the changes that were coming over the art of war.[2] A most important point is the ever increasing

[1] The same thing has been seen at our autumn manœuvres.

[2] Allusion has already been made to the supplanting of the sheriff's authority by the barons in raising troops, and the consequent fashion of issuing liveries to the corps so formed. It is perhaps worth while to note and dismiss the minute point that the garrison of Calais, the only truly national force belonging at that moment

employment of artillery in the field and the greater value attached to it. Richard, Duke of York, is said to have had a great train of ordnance and so many as three thousand gunners with him at Dartmouth in 1452. Artillerymen were becoming far more common, and as a natural consequence bade fair to command a smaller price in the wage-market. From this time also it may be said that the duel of artillery tends to become the regular preliminary to a general action. Still more significant is the augmented prominence of the common foot-soldier, known from his peculiar weapon as the bill-man, who now begins to supplant the dismounted man-at-arms in the work of infantry, and as a natural consequence restores the latter to his proper station among the cavalry. New weapons again make their appearance in the hands of the foot-soldier. Both Edward and Warwick introduced hired bands of Burgundian hand-gun men, whereby the English became acquainted with the new arm that was to drive out the famous bow. Again, on the field of Stoke there were 1487.
seen two thousand tall Germans armed with halberd and pike, under the command of one Martin Schwartz, who fought on the losing side, but stood in their ranks till they were cut down to a man.[1] Lastly, the old order of battle in three lines was becoming rapidly obsolete. At Bosworth both armies were drawn up in a single line, with the cavalry on the wings; and the cavalry itself was beginning at the same time to forsake the formation in column for that in line, or as it was called, *en haye*.

All these changes were symptoms of a great movement that was passing over all Europe. The art of war, like all the other arts, was undergoing a transformation so fundamental that it has received the name of a renascence. England, cut off by her expulsion out of France from

to England, was clothed in scarlet jackets, and were the first English soldiers thus distinguished.

[1] Readers of *Kenilworth* will remember the ballad quoted by Giles Gosling—

"He was the flower of Stoke's red field
Where Martin Swart on ground lay slain."

her former contact with continental nations, exhausted by her civil wars, reduced to her true position as a naval power, and above all wedded to the peculiar system which had brought her such success, lagged behind other nations in the path of military reform. The century of the Tudors' reign is for the English army a century of learning, and to understand it aright we must first look abroad to the countries that were before her in the school, and glance at the innovations that were introduced by each of them in the course of the fifteenth and sixteenth centuries. Not without such study can we trace to their source innumerable points, great and small, that are observable in our army of to-day, nor grasp to the full the greatness of the English soldiers who, long before the renascence of the art of war, had divined its leading principles, had established for their country noble military traditions, and above all had made it a national principle that the English must always beat the French.

AUTHORITIES.—Monstrelet as before is the most important authority for the wars in France. The *Wars of the English in France* (Rolls Series) are valuable in elucidation. For the rise of the Scots in France M. Francisque Michel's *Les Ecossais en France*, and Forbes Leith's *Scots Men-at-Arms in France*. For the Wars of the Roses the sources of information are proverbially meagre, but the material has been worked up with admirable skill by Mr. Oman in his *Warwick*, to which I am greatly indebted. For the reorganisation of the French Army Daniel's *Ancien milice Française* may be consulted.

BOOK II

CHAPTER I

FIVE years after the battle of Agincourt the religious wars in Bohemia had given birth to one of the great soldiers of the world's history, John Zizka, the blind general of the Hussites. His military genius, quickened by fanaticism and spurred by the stern necessity of encountering an enemy always superior in numbers and equipment, had led him to ideas which were far in advance of his age. A master in organisation and discipline, he had evolved literally out of nothing the most famous army of its day in Europe, and by inexhaustible activity and resource had rendered it invincible. Beginning with such rude material of war as waggons and flails, and with no more skilful men than poor Bohemian peasants, he matured a system of tactics which defeated not only the chivalry of Europe but even the light irregular cavalry, soon to become famous as hussars, of Hungary. As victory supplied him with the means of procuring better arms, he rose rapidly to the occasion. Throwing all military pedantry to the winds he fought as his own genius dictated, and in the rapidity of his movements and unrelenting swiftness with which he followed up a victory he bears comparison with Napoleon. He was the first man to make artillery a manœuvrable arm, the first to execute complicated evolutions in the face of an enemy, and the first to handle cavalry, infantry, and artillery in efficient tactical combination. The employment of waggons for defence we have already seen copied by the English 1420.

at the battle of the Herrings, but Zizka's influence[1] spread far wider than this by breaking down the strength of European chivalry, and showing that drill, discipline, and mobility could make the poorest peasant more than a match for the armoured knight.

Zizka, however, had not been the first to deal a blow at the supremacy of feudal cavalry. The English archers and dismounted men-at-arms had been before him, and another power, which was destined to abolish that supremacy for ever, had been in some respects the predecessor even of the English. Allusion has already been made to the victory of the Swiss over the Austrian
1382. chivalry at Sempach; from that day it may be said that they began their advance to the highest military reputation of Europe. Appointed from the ruggedness of their country as well as by their own poverty to fight rather on their own feet than on horseback, cut off in great measure by the same causes from the feudalism that had overrun the rest of Europe, they were by nature destined to be infantry, and as infantry they developed their fighting system. Beginning like all primitive foot-men in all countries with the simple weapons of shield, spear, and axe, they improved upon them to meet their own peculiar wants. The problem before them was, how to defeat mounted men mailed from head to foot in the open field, how to keep the horses at a distance and cut through the iron shells that protected the men. The instinct of a Teutonic nation led them to give first attention to the cutting weapon. The English had turned their axes into broad-bladed bills; the Flemings had gone further and produced the *godendag*, a weapon good alike for cut and thrust; the Swiss, improving upon the *godendag*, invented the halberd, which combined a hook for pulling men out of the saddle, a point to thrust between the joints of their armour, and a broad heavy blade, the whole being set on the head of an eight-foot shaft.

[1] He has left us two words, howitzer and pistol, both of which are derived from the Czech.

The weight of the halberd made it, as an old chronicler[1] says, a terrific weapon, "cleaving men asunder like a wedge and cutting them into small pieces." Altogether it was calculated to surprise galloping gentlemen who thought themselves invulnerable in their armour.

But the halberd did not solve the problem of keeping horses at a distance. For this purpose the primitive spear was lengthened more and more till it finally issued in the long pike, the pike of the eighteen-foot shaft, which for nearly two centuries ruled the battlefields of Europe. The birthplace of the long pike is obscure,[2] but it was undoubtedly first brought into prominence by the Swiss, and that by a series of brilliant actions.
Arbedo attested the firmness of the new infantry in the 1422.
field; St. Jacob-en-Birs, where the Swiss detached 1444.
sixteen hundred men to fight against fifty thousand, its boundless confidence; and finally the three crushing
defeats of Charles the Bold at Granson, Morat, and 1476.
Nancy, established its reputation as invincible. For 1477.
action the Swiss were generally formed in three bodies, van, battle, and rear—the van and rear being each of half the strength of the battle or main body. These bodies were always of a very deep formation, and if not actually square were very solidly oblong. Occasionally the whole were massed into one gigantic battalion in order that the proportion of pikes to halberds, which was about one to three, might go further in securing immunity from the attack of cavalry. The van, from the desperate nature of its work, was called the *Verlorener Hauf*, from which is derived our own term, not yet wholly extinct, forlorn hope.[3] As regards discipline

[1] John of Winterthur. If the reader has ever plied a long bill-hook to cut down overhanging branches he will appreciate the power of the halberd.

[2] "The earliest mention of the long pike occurs in an order addressed to the burghers of Turin by Count Philip of Savoy in 1327; but whether Swiss borrowed it from Savoyards or Savoyards from Swiss is uncertain" (Köhler).

[3] Compare the French equivalent, *enfans perdus*. *Hauf* was the regular German word for any mass of soldiers, from a company to a

the Swiss appear to have been orderly and sober men until spoiled by the multitude of their successes, but at the last they became intolerably insubordinate. The cantons indeed were so deeply bitten with the military mania, that all great occasions, feasts, fairs, and even weddings, were made the occasion of some form of military display, while the very children turned out with drums, flags, and pikes, and marched with all the order and regularity of full-grown soldiers. In fact fighting became the regular trade of Switzerland, and as her people enjoyed for a time a practical monopoly of that trade they soon became grasping and avaricious, and would dictate to generals under threat of mutiny when and where they should fight, select their own position in the order of battle, and open the action at such time as they thought proper. Their officers lost control of them, and would plaintively say that if they could but enforce obedience in their men they would march through France from end to end. This insubordination was their ruin. The French, who were their chief employers, at last lost all patience with them, and gave
1515. them at Marignano a lesson which they did not speedily
forget. The suppression of this mutiny, which was in fact a two days' battle of the most desperate description, cost the Swiss twelve thousand men; and it speaks volumes for the fine qualities that were in them that the defeat attached them more closely than ever to the cause of France. But the spell of their invincibility was broken, and two more severe defeats at the hands
522, 1525. of a rival infantry at Bicocca and Pavia destroyed their
prestige for ever. Nevertheless they were superb soldiers, and as their good fortune delivered them from a meeting with the English archers, who would certainly have riddled their huge bristling battalion through and through, they became as they deserved the fathers of modern infantry. Let it be noted that they marched in step to the music of fife and drum, that they carried

battalion. The English word *hope* therefore is a corruption, *hauf* having more to do with heap than hope.

a colour in each company, and that several of the cantons carried a huge horn, whose sound was the signal for all to rally around it.

It was not to be expected that the Swiss should long enjoy their monopoly as the infantry of Europe without exciting competition. In the last quarter of the fifteenth century arose the rivals who were to wrest their supremacy from them, namely, the landsknechts of Swabia, or as the contemporary English called them, the lance-knights of Almain, who were the direct forerunners of the modern German infantry. The records that survive of them are very full, and as it was through them that the teaching of the Swiss was carried into England, with results that are visible to this day, a brief study of their history is essential to the right understanding of the history of our own army.

The Swabian infantry was called into existence by the imperative necessity for preventing any potentate who might be so fortunate as to enlist the Swiss, from dictating his will to Europe. Swabia being the province next adjoining Switzerland was not unnaturally the first to learn the methods of her neighbour; and though at first all fighting men who imitated the tactics and equipment of the mountaineers were known by the generic name of Swiss, yet the Swabians, as if from the first to point the distinction between them and their rivals, took the name of landsknechts, men of the plain, as opposed to men of the mountains. Maximilian the First, seeing how valuable such a force would be in the eternal contest of the House of Hapsburg against the House of Valois, more particularly since the Swiss were the firm allies of the French, gave them all possible countenance and encouragement; and very soon the landsknechts grew into one of the weightiest factors on the battlefields of Europe. Though mercenaries like the Swiss and the still earlier bands of Brabançons, and as such engaged on all sides and in all countries, they yet cherished not a little national sentiment; and the greatest of all their work was done in the service of the Empire.

When therefore the emperor needed infantry he issued a commission to some leader of repute to enlist for him a corps of landsknechts. The colonel[1] thus chosen thereupon selected a deputy or lieutenant-colonel and captains[2] according to the number of men required, and bade them help him to raise his regiment. Then the fifes and drums were sent into the district, with a copy of the Emperor's commission, to gather recruits. The recruits came, gave in their names and birthplaces to the muster-master, were informed of the time and place of assembly, and received a piece of money,[3] conduct-money as the English called it, to pay the expense of his journey thither and to bind the bargain. Here we draw a step closer to the Queen's shilling. At the assembly the men were formed in two ranks, facing inwards. An arch[4] was built by planting two halberds into the ground and laying a pike across them, and then every man passed singly beneath it under the eye of the muster-master and of his assistants, who watched every one sharply, rejecting all who were physically deficient or imperfectly armed, and above all taking care that no man should pass through twice, nor the same arms be shown by two different men. For captains were still unscrupulous, and were ever striving to show more men on their roll than they could produce in the flesh, and put the pay that they drew for them into their own pockets. So old was the trick and so deep-rooted the habit, that even in Hawkwood's bands the legitimate method of increasing a captain's pay was to allow him a certain number of fictitious men, called *mortes payes* (dead heads), and permit him to draw wages for them.

[1] *Feld obrist*, now *oberst*.

[2] *Hauptmann*. The Germans wisely cling to these old titles, and preserve them.

[3] *Laufgeld*.

[4] This seems to have been a reminiscence of the Roman *jugum*.

This practice in a legitimised form continued in our own army within the memory of living men.

Four hundred men was the usual number assigned to a company[1] of landsknechts, but there was as yet no certainty either in the strength of companies themselves or in the number of them that were comprised within a regiment. The muster[2] over, the men formed a ring round the colonel, who read aloud to them the conditions of service and the rate of pay, including under the former all the ordinary points of discipline. The men thereupon raised their hands, and with three fingers uplifted, swore by the Trinity that they would obey. The colonel then called into the ring the officers whom he had selected to be ensigns,[3] and delivered to each the colour of his company, exhorting him to defend it to the death. Nor must it be supposed that the ensign was then the beardless boy with which our own later experience has accustomed us to identify the title. He was rather a hardened, grizzled old warrior, who could be trusted at all critical times to rally the men around him. Pursuant to Oriental tradition, the fife and drum of each company were under the ensign's immediate orders, so that the position of the colour might always be known by sound if not by sight. The flag itself, which gave the officer his title, bore some colour or device chosen by the colonel, and among the landsknechts was always very large and voluminous, probably to contrast with the flags of the Swiss, which were the smallest in Europe. The landsknechts prided themselves on the grace and skill with which they handled these huge banners, and indeed all the dandyism (if the term may be allowed) observable in later years in the manipulation of the colour may be traced to them.

This ceremony over, the various companies separated

[1] *Fähnlein*, flag or ensign.

[2] Muster is a corruption of the French *monstre*, Latin *monstrare*. So to pass muster is to pass inspection.

[3] *Fähnlein*.

and formed each a distinct ring round its captain and ensign. The captain then selected his lieutenant,[1] and calling him under the colours bade the men obey him. He then chose also his chaplain and quartermaster, and having added to these a surgeon his patronage was exhausted. The men were then handed over to the senior non-commissioned officer,[2] a very important person, who was responsible for all drill and for the posting of all guards, and received his appointment directly from the colonel. Under his guidance the company elected a sergeant, who then in turn selected himself an assistant;[3] the assistant then chose a reconnoitrer,[4] and the reconnoitrer a quartermaster-sergeant. Finally, the company was distributed into files[5] of ten men apiece, which selected each of them a file-leader,[6] who, though he received no extra pay, enjoyed certain privileges within his file, such as the right to a bed to himself in quarters and the like. With his election, the file being the unit of the company, the hierarchy was complete.

It is unnecessary to trouble the reader with a list of the regimental staff, but a word must be said of the provost. His principal function was the maintenance of discipline, for which purpose he was provided with a staff of gaolers and an executioner, and his title is still attached to the same duties in the English army of to-day. But apart from this, it was his office to fix the tariff of prices of goods sold by the sutlers who accompanied the regiment. It was a most difficult and dangerous duty, for if he fixed the price too high the men became discontented and mutinous, and if too low the sutlers deserted the camp and left it to provide for itself, which was an alternative little less

[1] *Stellvertreter.* The Germans have since abandoned the word for "*leutnant.*"

[2] *Feldwebel.* We may call him the colour-sergeant.

[3] *Gemeinwebel.* [4] *Fourier.* [5] *Rot.*

[6] *Rottmeister.* Sir Walter Scott in the *Legend of Montrose* has inexplicably confounded the word with *Rittmeister,* which is a very different thing; a rare mistake with him.

formidable than the other. In consideration of the perils of his office the provost received certain perquisites in addition to his salary, such as the tongue of every beast slaughtered and an allowance for every cask broached, and even so was none too well paid.[1] It is hardly necessary to point out that in this commercial side of the provost's duties there lies the germ of our modern canteen, wherein the practice of taking perquisites, though strictly forbidden, still prevails among canteen-stewards.

The duties of another officer, whose name must be written down in the original, the *Hurenweibel*, show the early methods of coping with a difficulty which particularly besets our Indian army. Every regiment of landsknechts was accompanied by a number of followers on the march; and although by strict rule no woman was allowed to accompany a man except his lawful wife, yet we hear without surprise that there were many women following the colours whose status was not recognised by the rule above referred to. The poor creatures led a hard life. The washing, cooking, scavenging, and all manner of unpleasant duties, as well as the more congenial task of nursing the sick and wounded, was entrusted to them, and in case of a siege they were required to make the fascines and gabions. Their masters treated them very brutally, and as every colonel naturally wished to cut down their numbers as low as possible, no pains were spared to make their lives a burden to them. Over all this rabble the *Hurenweibel* was king, the sceptre of his office being a thick stick called a "straightener,"[2] which he used unmercifully. Yet these followers loved the life and tramped after their lords all over Europe, increasing their numbers as they went; the boys as they grew up being employed to carry

[1] It is a curious sign of the combination of his functions, that in every standing camp the Provost erected a gallows, which served to mark both the extent of his authority and the site of the market-place, or as we should call it, canteen.

[2] *Vergleicher*.

the men's weapons or harness on the march. Such boys, or rather fags, were called in French *goujats*, and are a curious feature in the armies of the time. The greatest of all *goujats*, if legend may be trusted, was Thomas Cromwell, the Hammer of the Monks.

For the trial of military offences a board of justices accompanied each distinct body, but there were some corps of landsknechts that enjoyed the privilege of the trial of the long pikes,[1] which gave the rank and file sole jurisdiction in respect of crimes that brought disgrace on the regiment. In such cases the provost laid his complaint; and the ensigns, thrusting their flags point downward into the ground, vowed that they would never fly them again until the blot on the fair name of the regiment was removed. The culprit was then tried according to a certain fixed procedure by his comrades alone, without the intervention of any officer. If he were found guilty, the men drew themselves up in two ranks, north and south, facing inwards; the ensigns, with colours flying, posted themselves at the east end of the lane thus formed, and the prisoner was brought to the west. The ensigns then exhorted him to play the man and make bravely for the colours, and the provost, clapping him thrice on the shoulder in the name of the Trinity, bade him run. Then the doomed man plunged into the lane, and every comrade plied pike and halberd and sword on him as he passed. The swifter he ran the sooner came the end, and as he lay hewn, mangled, and bleeding, gasping out his life, his comrades kneeled down together and prayed God to rest his soul. Then all rose and filed in silence three times round the corpse, and at the last the musketeers fired over it three volleys in the name of the Trinity.

The strength of a regiment of landsknechts varied very greatly. There might be thirty companies or there might be ten; the total force sometimes reached ten or twelve thousand men, and in such a case was frequently strengthened by a contingent of artillery. The weapons

[1] *Recht der langen Spiesse.*

were the pike, the halberd, and a proportion of firearms, which last tended constantly to increase. Every man found his own arms, and the dress of the landsknechts, being that which it pleased each man best to wear, was generally both fantastic and extravagant, for they had all the soldier's ambition to let their light shine before women. Maximilian's courtiers were so jealous of their gorgeous apparel that they begged him to forbid it, but the emperor was far too sensible to do anything so foolish. "Bah!" he said, "this is the cheese with which we bait our trap to catch such mice," a sentiment which English officers will still endorse. Not all the prejudices of dying feudalism could induce Maximilian to discourage his new infantry; on the contrary, meeting a regiment once on the march he dismounted, shouldered a pike, and marched with them for the rest of the day. It is worth noting that the drum-beat of the landsknechts, whereof they were extremely proud, probably the self-same beat as that to which Maximilian strode along that day, still preludes the marches of our own military bands.[1]

The drill of the landsknechts was probably crude enough. There was no exercise for pike or halberd, and there is no sign of the complicated manœuvres that were so common at the opening of the seventeenth century; but as they always fought, like the Swiss, in huge masses, there was probably little occasion for these. The men fell in by files, probably at sufficient distance and interval to allow every man to turn right or left about on his own ground; but for action they were closed up tight in vast battalions far too unwieldly for any evolution. Moreover, few of the officers knew anything of drill. They were selected for bravery and experience, no doubt, in some cases, but not for military knowledge; and it is the more probable that the colonels, according to custom, sold the position of officer to the highest bidder, since Maximilian could

[1] A roll on the two first beats of the bar, a single note on the third, and silence on the fourth.

rarely furnish them with money for their preliminary expenses. The one duty expected without fail of officers was that they should be foremost in the fight, and as a rule they one and all took their place in the front rank with the colonel for centre, and, armed like their men, showed the way into the enemy's battalion. Not one remained on a horse in action, though he might ride regularly on the march; and indeed the landsknechts disliked to see an officer mounted on anything larger than a pony at any time, admitting no reason for an infantry-man to ride a good horse except that he might run away the faster. The duties of officers being thus defined, it is easy to see why the colonel reserved to himself the appointment of the colour-sergeants, for they were practically the only men who knew anything of drill or manœuvre. The colonel might prescribe the formation of his battalion for action, but only the colour-sergeants could execute it; and hence arose the rule that sergeants should be armed with no weapon but a halberd, since any heavier weapon would impede them in the eternal running up and down the ranks which was imposed on them by their peculiar duty. The influence of these traditions was still visible in our army until quite recently. But a few years have passed since sergeants shouldered their rifles as though they carried a different weapon from the men, and officers have only lately ceased to depend on them greatly in matters of drill.

Such was the new infantry of Europe at the close of the fifteenth and the opening of the sixteenth centuries, not yet perfected, but advancing rapidly to an efficiency and importance such as had for many centuries been unknown in Europe. And now the nations poured down into the fair land of Italy to teach each other in that second birthplace of all arts the new-born art of war. France was the first that came; and few armies have caused greater wonder in Europe than that which marched with Charles the Eighth through Florence in 1496. The work begun for the expulsion of the

English from France had been steadily continued. Louis the Eleventh had hired Swiss sergeants to drill his infantry, and Picardie, the senior regiment of the old French line, was already in potential existence. But it was not these, but other men who set the Florentines at gaze. For there were to be seen the Scottish archers, the finest body-guard alike for valour and for stature in the world, the Swiss, marching by with stately step and incredible good order, the chivalrous gentlemen of France, mailed from top to toe and gorgeous in silken tabards, riding in all the pride of Agincourt avenged, mounted archers less heavy but more workmanlike as befitted light cavalry, and lastly a great train of brass artillery, cannons and culverins, and falcons, the largest weighing six thousand pounds and mounted on four wheels, the smallest made for shot no bigger than a doctor's pills and travelling on two wheels only. Already the quick-witted French had thought out the principle of the limber, and had made two wheels of their heavy guns removable. Already too they had trained the drivers of the lighter ordnance to move as swiftly as light cavalry.[1]

We cannot follow this army through the triumphs and the disasters of the next half century, but we must needs glance briefly at the rapid progress of French military organisation. Louis the Twelfth took the improvement of his foot-soldiers seriously in hand and increased the number of the companies, or bands as they were called, that had been begun by the bands of Picardy. The number of these bands, permanent and temporary, demanded the appointment of an officer who should be intermediary between the general and the captains of independent companies. About the year 1524 such an officer was established with the new title of colonel,[2] and

1 See the account in Paul Jove.

2 We need not enter into the controversy whether the word was derived from *columna* or *corona* or from neither. For a century or more it was written indifferently colonel or coronel, to which last the modern English pronunciation is doubtless to be traced.

the companies placed under his command were said, in French, to be under his regiment. The word soon grew to be used in a collective sense, and such and such companies under Colonel A.'s regiment became known simply as Colonel A.'s regiment. The colonel had a company of his own, but having no leisure to attend to it made it over to a captain, who was called the colonel's lieutenant or lieutenant-colonel. Another company was commanded by the sergeant-major, the word sergeant, which we met with first at the very beginning, having come into use in France with a new meaning in the year 1485. As already mentioned in speaking of the landsknechts, the name of sergeant became for some reason bound up with the functions of drill, and the sergeant-major was to the regiment what the sergeant was to the company. He was therefore the only officer who remained on his horse in action, his duties compelling him continually to gallop from company to company for the correction of bad formation, and for the ordering of ranks and files. It will be seen that the sergeant-major, or as we now call him major, originally did the work which is now performed in England by the adjutant.

Captain was of course an old title, and had been used for the chief of a band in France ever since 1355, having been borrowed possibly from the free companies. The captain's *locum tenens* or lieutenant had been instituted by the reforms of Charles the Seventh in 1444, and together with him his standard-bearer or ensign,[1] but there were other junior officers who came later even than the colonels to supplement the new military vocabulary. In 1534 we encounter for the first time *fouriers*, *caps d'escouade*, and *lancepessades*. The first of these, which existed for a time in the corrupted form *furrier*,

Brantôme writes always *couronnel;* Milton in his famous sonnet gives the word the dignity of the three syllables. Some say that it was borrowed from the landsknechts, but this is a palpable error. (See a paper by Mr. Julian Corbett, *American Hist. Review*, Oct. 1896, "The Colonel and his Command").

[1] *French* enseigne ; *Lat.* insigne, signum.

has passed from the English language.[1] The second is the French form of the Italian *capo de squadra*, head of the square, a reminiscence of the days when men were formed into square blocks, squads or squadrons, which passed into *caporal* and so into our English corporal. The third, again a French form of the Italian *lanz pesato*, signified originally a man-at-arms whose horse had been killed and who was therefore compelled to march with the foot. Being a superior person, he was not included among the common infantry-men but held this distinctive and superior rank, whence in due time was derived the prefix of lance to the titles of sergeant and corporal. Finally, in the year 1550 foot-soldiers in France began to be called by the collective name of *fanterie* or *infanterie*. This word, too, was a corruption from the Italian, for Italian commanders used to speak of their troops as their boys, *fanti*, and collectively as *fanteria*; and from them the term passed into all the languages of Europe. Nothing could better commemorate the situation of Italy in the sixteenth century as at once the cockpit of the nations and the school of the new art of war.

But before leaving France there is another aspect of her military institutions to be touched on. After the death of Francis the First, and particularly during the period of the religious wars, the discipline and tone of the French army underwent woeful deterioration. Captains from the first had been proprietors of their companies, which indeed were sometimes sold at auction by the colonel to the highest bidder; and, as they received a bounty in proportion to the numbers that they could show on their rolls, the rascality and corruption were appalling. The enforcement of strict discipline was bound to cause desertion, and every deserter meant a man the less on the captain's roll and a sum the less in the captain's pocket. No effort therefore was made to restrain the misbehaviour of soldiers when

1 But not until after the Seven Years' War, when Lord George Sackville applied for a "furrier."

off duty; they were allowed to rob and plunder at their own sweet will, and they had the more excuse since they were encouraged thus to indemnify themselves for the pay stolen from them by their officers. This recognised system of pillage was known as *picorée*,[1] a word which has passed through the English language in the form of pickeer. Yet another method there was among many of falsifying the muster-rolls, namely on the day of inspection to collect any yokels or men that could be found, thrust a pike into their hands, and present them as soldiers. They were duly passed by the muster-master, and as soon as his back was turned were dismissed, having served their purpose of securing their pay for the illicit gain of the captain till next muster. Such men were called *passe-volans*, a word which also was received into the military terminology of Europe, and like *mortes-payes* received at last official recognition. It must not be thought that such abuses were confined to France, but it is significant that she was the country to find names for them.[2] Nor must the reader be unduly impatient over the mention of these details in the military history of foreign nations. The English soldier for the next century and more is going to school, where like all pupils he will learn both good and evil; and it is impossible to follow his progress unless we know something of his schoolfellows as well as of his tutors.

Last of the nations let us glance at Spain, at the close of the fifteenth century just emerging triumphant from eight centuries of warfare against the Moors and girding herself for a great and magnificent career. Her training in war had been against an Oriental foe, swift, active, and cunning, and it is not surprising that when first she entered the field of Italy and met the
1495. massive columns of the Swiss at Seminara she should

[1] We even find the word incarnated by French writers as the strumpet Madame Picorée.

[2] As a matter of fact these abuses do seem to have been more flagrant in France than elsewhere, owing no doubt to the demoralisation caused by the religious wars. See for instance Brantôme, and the Memoirs of Sully.

have given way before them. But at the head of the Spanish troops was a man of genius, Gonsalvo of Cordova, who was quick to learn from his enemies. Confining himself for a time to the guerilla warfare which he understood the best, he mingled pikes among the short swords and bucklers which were the distinctive weapons of the Spanish infantry, and within a year had gained his first victory over the Swiss. His next Atella, campaign found him with a body of landsknechts in 1496. his pay, when he quickly perceived the possibilities that lay not only in the pikes but still more in the fire-arms which they brought with them. Before the year was 1503. past he had routed Swiss infantry and French cavalry in two brilliant actions at Cerignola and on the Garigliano, and fairly driven them out of Naples. He then set himself to remodel the Spanish foot by the experience which he had gathered in his later campaigns, and this with full appreciation of the moral and physical peculiarities of his countrymen. Thus though it was in the Spanish tongue that the pike was first named the queen of weapons, yet the value of the sword in the hand of a supple active people was never overlooked, and at Ravenna no less than Cerignola the rush of 1512. nimble stabbing Spaniards under the hedge of pikes had proved fatal to the lumbering unwieldy Teuton.

Still more remarkable was the rapid development 1522. of the power of musketry in Spanish hands. At Bicocca the Marquis Pescayra met the attack of a gigantic Swiss battalion by drawing up a number of small squares or squadrons of Spanish arquebusiers in front of his own battalion of pikes. His instructions were that not a shot should be fired without orders, a fact that points to early excellence in what is now called fire-discipline, but that each front rank should fire a volley by word of command and having done so should file away to the rear to reload, leaving the remaining ranks to do the like in succession. The results of this manœuvre were disastrous to the Swiss; and this ingenious method of maintaining a continuous fire of

musketry was the law in Europe for the next century and a half. In fact, if it were necessary to fix an arbitrary date for the first really effective use of small fire-arms in the battlefield the day of Bicocca might well be selected. But we must not fail to note concurrently the drill and discipline which made Pescayra's evolution possible. Three years later, at the famous
1525. battle of Pavia, this same skilful soldier attempted a still bolder innovation with his arquebusiers, and with astonishing success. Being threatened with a charge of French heavy cavalry (men-at-arms) he deployed fifteen hundred of his marksmen in skirmishing order before his front, who, taking advantage of every shelter and moving always with great nimbleness and activity, maintained a galling fire as the cavalry advanced, and finally, taking refuge under the pikes of the battalions which were drawn up in their support, smashed the unfortunate French as effectively as the English archers at Creçy. In truth, the effect of this daring experiment on military minds in Europe was hardly less than that of Creçy itself. Henry, Duke of Guise,[1] an excellent soldier, was so much struck by its success that he showed how the principle might be indefinitely extended and find ultimate shape, as many years later it did, in the formation of distinct corps of light-infantry. His own attempt to organise such a body in France was however a failure, and the Spanish arquebusiers long held their own as the first in Europe, a proud position which they had most worthily gained.

The remarkable prowess of the Spanish infantry soon made it popular with the nation. The cavalry, in the palmy days of chivalry the most gorgeous in Europe, lost its attraction for the young nobles, who enrolled themselves as private soldiers in the ranks of the foot, and carried pike and arquebus with the meanest of the people. Charles the Fifth himself once shouldered a piece, and marched, like Maximilian, in

[1] See the remarkable conversation in Brantôme, ed. Elzev. vol. i. pp. 376-382.

the ranks, until ordered by the commander-in-chief[1] of his own appointment not to expose himself to unnecessary danger, when like a good soldier he at once obeyed orders. And this leads us to another eminent feature of the Spaniards, the excellence of their discipline. English and French contemporary writers[2] agreed that they owed their victories to nothing else but obedience and good order, for that they were not in themselves remarkable as a fighting people. "I am persuaded," says Roger Williams, "that ten thousand of our nation would beat thirty thousand of theirs out of the field, excepting some three thousand [the choicest of the army] that are in the Low Countries." Gonsalvo was the man who had laid the foundation of this discipline, and it was worthily maintained by his successors. Charles the Fifth went so far in his respect for it as always to salute the gallows whenever he happened to pass them. And yet there are no signs of extraordinary brutality in the Spanish army, but on the contrary most remarkable tokens of good fellowship between officers and men, and of healthy *esprit de corps*. There was a system of comradeship which was the envy of all Europe. The two officers of each company, the captain and ensign,[3] would each take to themselves and entertain from three to six comrades from the young nobles who served in the ranks ; sergeants would also take one or two such comrades, and the privates formed little messes among themselves in like manner, with the result, unique in those days, that fighting and brawling were unknown in a Spanish camp. Quite as striking was the pride which the old soldiers took in themselves and their profession. It is recorded that a party of Spanish recruits, who had arrived at Naples, ragged, slovenly, and unkempt, and were staring about them in a clownish and unsoldierly fashion, were at once taken in hand by the old soldiers, who lent them

1 The Marquis del Vasto, of the same family as Pescayra.
2 For instance Roger Williams and Tavannes.
3 In Spanish called *alferez*.

good clothes, made them tidy, and taught them proper manners.[1]

For the rest the Spaniards originated a system which, though it now seems obvious enough, was in those days a new thing. It consisted simply in the maintenance of a nucleus, or as we should now call it a depôt, of trained men sufficiently numerous to teach recruits their duty. All recruits were trained in the garrisons at home, and from thence passed into the ranks of the regiment wherein they were needed; and every draft so disposed of was immediately replaced by an equal number of new recruits. When it is remembered that, according to the ideas of the time, seven thousand trained infantry and three thousand cavalry were judged sufficient to leaven an army of fifty thousand men, the strength which her system of recruiting gave to Spain is not easily exaggerated. The trained regiments of Spanish infantry were but four, and their united strength did not exceed seven thousand men, but their ranks were always full. The number of companies into which they were distributed was uncertain, and the strength of the companies themselves varied from one hundred and fifty to three hundred men, a curious defect in the most perfect organisation of the time. Lastly, the Spanish regiments were known by the name of tercios,[2] a term with which the reader must not quarrel, as he will encounter it on the battlefield of Naseby.

Not less remarkable than their forwardness in organisation and discipline was the ready quickness of the Spaniard in the improvement of fire-arms. The primitive hand-gun, as I have already said, differed little except in size from the smaller cannon of the time. It consisted simply of a barrel with a vent at the top, and though indeed attached to a wooden stock had no lock of any description. Hand-guns were often made so

[1] Brantôme.

[2] Tercio, like colonel, is a riddle which defies solution. It means a third, but a third of what is unknown (see Mr. Julian Corbett's paper, quoted above, p. 94).

short that they could be held even by a mounted man with one hand and fired with the other. Match-cord or tinder for purposes of firing the charge by the vent was 1475.
already in full use. The next step was to increase the length of the barrel and support it on a forked rest, a plan introduced by the Spaniards at Charles the Fifth's invasion of the Milanese in 1521. Ten years later a vast stride was made by the substitution of a pan at the side of the barrel for a vent at the top, and by the addition of a grip to the stock to hold the match-cord, which was brought in contact with the pan by pressing a trigger. In a word, the barrel was fitted with a lock. An extremely ingenious Italian in the French service, Filippo Strozzi, then took the improvement of fire-arms in hand, copying however, as always, from the Spanish model. The bore of the harquebus (for the primitive German *hakenbuchse* had by this time found its permanent corrupted form) was by him enlarged to bear a heavier charge and carry a larger bullet; and so perfect was the workmanship of the Milanese gunsmiths whom he employed that he succeeded in killing a man at four hundred and a horse at five hundred paces. The stock being long and the recoil very severe, men suffered not a little from bruises and contusions with this weapon; but its efficiency was proved. Strozzi also introduced another Spanish improvement, namely the practice of making all his arquebuses of one bore, which, though it now sounds obvious enough, waited for some years to find general acceptance in Europe. Hence the weapons were known as arquebuses of calibre, which phrase in England was soon shortened simply to calivers. These however were arms of small bore:[1] it was, as usual, the Spaniards who were the first to arm their infantry with muskets[2] of large calibre. Alva was the man who 1567.

[1] In a MS. treatise in the Record Office, of date 1570, the bore recommended is 28 bullets to the pound. This remained the standard bore in the French army all through the wars of Louis XIV.

[2] Musket is simply the word mosquito. Larger weapons were called drakes, falcons, and the like, and the smaller therefore after the lesser flying creatures.

introduced them, and the rebels of the Low Countries the first who felt their power. It needed but the substitution of a flint-lock for a match, and the abolition of the rest, to turn this weapon into Brown Bess, never so famous in English hands as in the battlefields of Alva's home. Bandoliers and cartridges had long been known to the Spaniards, and even to the French[1] before the middle of the sixteenth century, so that the general progress in arms and equipment was rapid.

But the weapons had hardly been improved for infantry before cavalry also began to crave for them. The simplest method of course was to place pike and arquebus in the hands of mounted men and turn them into mounted infantry, which was duly done in the French army by Piero Strozzi in 1543, and has earned him the title of the father of dragoons.[2] But still earlier in the century there had grown up in Germany a new kind of cavalry, called by the simple name of Reiters, which had perfected the smaller fire-arms, the petronel[3] and the pistol, and had finally adopted the latter for its principal weapon. The result was an important revolution in the whole tactics of cavalry.

Mention has already been made of the abandonment, at the close of the fifteenth century, of the dense column of mounted men-at-arms in favour of the less cumbrous formation in line, or as it was called *en haye*. The lance being still the principal arm of the cavalry, the freedom of movement gained by the change brought the attack of horse much nearer to the shock-action which is the rule at the present day. The new formation had, however, its disadvantages, for in the imperfect state of military discipline there was no certainty that the whole line would charge home. Retirement was so easy that cowards would drop back, feigning to bleed at the nose, to have lost a stirrup or cast a shoe,[4] while men of spirit,

1 Mem. de Vieilleville.
2 This again is a word which defies the skill of the etymologist.
3 *Poitrinal*, so called because it was held against the chest.
4 Mem. de La Noue.

and this was especially true of the impetuous French, would race to be the first into the enemy's squadron, and from premature increase of speed would arrive at the shock in loose order, and with horses blown and exhausted. So well was this defect realised that a shrewd French officer, Gaspard de Tavannes, at the battle of 1554. Renty deliberately reverted to the old dense column and overthrew every line that he met.

Yet another cause was contributing to restore the column as the favourite formation for the attack of cavalry. With the steady improvement in fire-arms, the bullet became more and more potent in velocity and penetration, and increasingly difficult to fend off by means of armour. It must never be forgotten that a bullet-wound, for a century and more after the introduction of fire-arms, generally meant death. The primitive surgery of the time, misled by the livid appearance of the edges of the wound, pronounced bullets to be in their nature venomous, and treated the hurt somewhat as a snake's bite, with such tortures of boiling oil and other descriptions of cautery as are sickening even to read of. Wise men took refuge in the virtues of cold water, and kept the surgeons at a safe distance. "Trust a doctor and he will kill you; mistrust him and he will insult you," wrote a Frenchman[1] who had suffered much from the profession. But above all, men relied on prevention rather than cure; so to keep bullets out of their bodies they made their armour heavier and heavier, covering themselves with stithies, to use the words of contemptuous critics,[2] till they could neither endure swift movements themselves nor find horses that could maintain any pace under the burden.[3] It was obvious therefore that if cavalry was to act by shock, the shock must be, as in former days, that of ponderous weight rather than of high speed.

1 Tavannes, ed. Petitot, vol. i. p. 304.

2 Tavannes, La Noue.

3 It is curious to compare the parallel contest of armoured ships and artillery at the present time.

Moreover, quite apart from all questions of formation there was much in the prevailing tactics of infantry to encourage cavalry to change the lance for the pistol. Huge square battalions, bristling with eighteen-foot pikes and garnished with musketeers, were not easily to be broken by a charge, but presented a large mark at a fairly safe range to the mounted pistolier. Thus all circumstances conspired to favour a great and radical reform in the tactics of cavalry, the change not only from line to column, but from shock to missile action. When once the pistol was recognised as the principal weapon of the horsemen, it was obvious that all other tactical considerations must give way to the maintenance of a continuous fire. To this end there was but one system known, namely the old method of Pescayra, that the front rank should fire first and file away to the rear to reload, leaving successive ranks to come up in its place, and go through the same performance in turn. Plainly, therefore, a reversion to the old dense column, as great in depth as in breadth of front, was imperative. It was accordingly re-introduced, and from its quadrate outline was called by the name of a squadron, which from this period tends to become a term applied exclusively to cavalry. Massed together in such squadrons men could move slowly and steadily, willingly sacrificing speed that they might take the better and surer aim.

Such was the new principle brought forward early in the sixteenth century by the mounted mercenary bands of Germany, and with ever-increasing success. Very soon the reiters become recognised as a valuable force, and received from Charles the Fifth something of the encouragement that the landsknechts had gained from Maximilian. The military aspirants of the Empire, forsaking the ranks of the once honoured infantry, hastened to enrol themselves among the new horse, and the landsknechts decayed that the reiters might flourish. That the new service was as honour-

able as the old may be doubted, for the reiters were proverbial for brutality, and their practice of blackening their faces betokens something of a ruffianly spirit; but, be that as it might, they forced their system, in spite of bitter opposition, upon the cavalry of Europe, and from the day of the battle of St. 1557.
Quentin may be said to have assured their evil supremacy.

It is therefore necessary to glance briefly at their organisation. The tactical unit was the squadron, which was of uncertain strength, varying from one hundred to three or even five hundred men. The officers were a captain,[1] lieutenant, ensign,[2] and quartermaster,[3] and the staff was completed by a chaplain, a sergeant[4] and a trumpeter. As every man brought his own equipment there was no precise uniformity, but it may be assumed as certain that all wore complete defensive armour to the waist, and some even to mid-thigh. For offensive purposes a pistol, or rather a brace of pistols, was indispensable. As in the case of the landsknechts, all matters of drill were the business of the sergeant, but it does not appear that the reiters ever attained great proficiency in manœuvre. Thus in action the successive ranks of the squadron seem to have been unable to file to the rear except to their left, so that it was impossible to post them on the right wing without bringing them into collision with the centre of their own line of battle. The trumpeters, it is worth noting, were required to be masters of but six calls,—Saddle, Mount, Mess, March, Alarm, Charge,—of which the French employed the first two and last two only. We shall presently make further acquaintance with these six calls, but it is sufficient meanwhile to call attention to their existence in the middle of the sixteenth century. The reiters however, should not be forgotten, for though not comparable to the landsknechts for quality as troops, they furnished

1 *Rittmeister.* 2 *Fähnrich.*
3 *Fourier.* 4 *Wachtmeister.*

the model for the first famous regiment of English cavalry.[1]

Lastly, let me close this necessarily brief and imperfect account of the renascence of the art of war by a remark which should perhaps have come first rather than last. Amid all the innovations which went forward during the sixteenth century in the province of armament, classical models reigned supreme in organisation and manœuvre. The whole story of the renascence resembles, if I may be allowed to use the metaphor, a long musical passage in pedal point, on the deep bass note of classical tradition. For this the revival of classical learning was doubtless responsible. When generals celebrated a triumph, as more than one general did, in the Roman manner after a victory, the pageant could hardly be complete without the presence of legions; and when Machiavelli declared that the Swiss tactics were those of the Macedonian phalanx, military students could be in no doubt where to seek out models for their own imitation. Francis the First adopted in 1534 both the name and organisation of the Roman legions for a time, while no military writer omitted to recommend the Roman ideal to aspirants of his profession. Every soldier steeped himself in ancient military lore, and quoted the Hipparchicus of Xenophon[2] and the Tactics of Ælian, the Commentaries of Cæsar and the expeditions of Alexander, Epaminondas' heavy infantry and Pompey's discipline. A Frenchman could not even praise the merits of the Englishman as a marine without calling him *epibates*. In a word Europe for two centuries, went forth to war with the newest pattern of musket in hand, and a brain stocked with maxims from Frontinus and Vegetius and Æneas Poliorceticus, and with examples from Plutarch and Livy and Arrian. She might well have found

[1] The particulars of the reiters' organisation are taken from the Kriegsbuch of Leonard Fronsberger, 1566.

[2] It is just possible that Xenophon's example may have favoured the abandonment of shock for missile tactics in cavalry.

worse instructors; but their lessons were for the most part imperfectly understood, and their broad principles seldom correctly deduced or intelligently applied. An opportunity was thus afforded for the demon of pedantry, which was eagerly and joyfully seized. Nevertheless, the present armies of Europe still double their ranks and files, by whatever name they may designate the evolution, after the manner prescribed by Ælian, and by him borrowed, it is likely, from the stern martinets of ancient Lacedæmon.

AUTHORITIES.—The chief authorities for Zizka's campaigns and organisation are Æneas Sylvius, Balbinus, *Miscellanea Rerum Bohem.* 1679; Dubravius, *Hist. Bohem.* 1602; Palacky, *Gesch. v. Böhmen.* His articles of war will be found in *Neuere Abhandlungen der königl. Böhm. Gesellschaft der Wissenschaft*, Band I. p. 375. For the Swiss, Simler, *de Repub. Helvet;* John of Winterthur, Pirckheimer, and the *Chronicle of Berne.* All the authorities for the battle of Sempach have been collected in Lickenau's memorial volume. A fantastic work, but not without useful information, is Karl Bürkli's *Der wahre Winkelried*, 1886. Köhler has handled both Bohemians and Swiss with his wonted thoroughness. For the landsknechts there are Adam Reissner's *Georg von Frundsberg* (1st ed. 1568, 3rd ed. 1620); Fronsperger's *Kriegsbuch;* Hortleder's *Der römischen Kaiser*, etc.; *Adelspiegel*, von Cyriack Spangenberg, 1594; the whole of which are more or less summarised in Barthold's *Georg von Frundsberg*, 1833, and in a still more compact form by Dr. Friedrich Blau, *Die deutschen Landsknechte*, 1882. The Spanish military reforms are more difficult to ascertain. I have relied principally on Roger Williams's brief account, sundry notices in Brantôme's *Vie des hommes illustres*; Paul Jove's *Vita Gonsalvi Magni*, and, perhaps most valuable of all, Reissner. For the French there are Daniel's *Ancien milice;* Susanne's *Hist. de l'ancienne infanterie française;* Paul Jove, and the *Memoires* of Vieilleville, Du Bellay, Villars, de Mergey, de la Noue, Tavannes, Onosandre, Brantôme, Monluc, and others. I have also consulted, among Italian writers, Julius Ferrettus, *De re militari*, 1575; Domenico Mora, *Il soldato*, 1570; Savorgnano's *Arte militare;* and of course Machiavelli. Lastly, I have not failed to study the classical authorities quoted in the text.

CHAPTER II

The accession of the Tudors to the throne of England marks an important period in our military history. The nation, after thirty years of furious internal war, during which it had lost all sense of national honour, began to settle down once more to a life of peace, and awoke to the fact that England was now no more than an insular power. France was lost to her except Calais, but Calais was something more than a mere sentimental possession. It was the bridge-head that secured to the English their passage of the Channel; and while it remained in the hands of an English garrison there was always the temptation to engage in Continental wars and to employ the army for purposes of aggression as well as of defence. Still the prospects of regaining the ancestral possessions of the Plantagenets in France seemed so hopeless that the English sovereigns might well doubt whether it were not now time to give the Navy the first and the Army the second place; and this question, already half decided by the keen good sense of King Henry the Eighth, was finally determined by the loss of Calais itself. There was, of course, always a frontier to be guarded on the Tweed, but with the cessation of expeditions to France, which had invariably called the Scotch armies across the border, there was no longer the same danger of Scottish invasion; and moreover, England and Scotland were now beginning to draw closer together. Thus it would seem that after the death of Queen Mary there should have been little reason for the existence of an English army, and indeed it will be seen that the national force became in many respects lamentably deficient. But meanwhile the wars of Europe changed from a contest between nation and nation to a death struggle between Catholic and Protestant. It was religion that drew the Scotch from their old alliance with the French to their former enemies the English; and it was religion which led the English to the battlefields of the Low Countries, where they learned the new art of war. The reign of the Tudor dynasty therefore falls for the purpose of this history into three periods, which are conveniently separated by the fall of Calais or the more familiar landmark of the accession of Elizabeth, and by the first departure of English volunteers to the Low Countries in 1572.

It is extremely difficult to discover the exact condition of England's military organisation when Henry the Seventh was fairly seated on the throne. The old feudal system, which had been turned by the nobles to such disastrous account for their own ends in the Civil War, seems to have been but half alive. Compositions, indents, and commissions of array had already weakened it in the past, and indents in themselves had been shown to be unsafe. The difficulties wherein Henry found himself are shown by two statutes imposing the obligation of military service on two new classes, namely holders of office, fees or annuities under the crown, or of honours and lands under the King's letters patent. It was stipulated that they should receive wages from the day of leaving their homes until the day of their return to them; but they were strictly forbidden to depart without leave, and their service was declared to be due both within the kingdom and without. But in fact the sovereign seems to have been driven back on the force which represented the old Saxon fyrd, and had its legal existence under the Statute of Winchester. Noblemen and gentlemen could of course still show a body of retainers, but many, indeed most, of the ancient magnates had perished, and recent experience had shown the danger of permitting their retinue to become too powerful. A curious complication, to which I shall presently return, in the collapse of the old feudal service was the extreme dearth of good horses. Altogether everything tended to compel resort to the national militia as the principal military force of England. Two allowances to the levies of the shire seem to have been finally established in this reign, namely coat-money and conduct-money. The first, as its name denotes, helped the soldier to provide himself with clothing and was a step further towards uniform; and indeed it is possible that it was deliberately designed to exclude the liveries of the nobility, already condemned by statute, in favour of the national white with the red cross of St. George. The conduct-money was simply the old allowance which was seen in the days of William Rufus, but which from henceforth apparently was refunded to the shire from the Exchequer. Both, however, though paid in advance to the soldier, were ultimately deducted from his pay, and are therefore of interest in the history of the British soldier's stoppages. Finally, we find indications of a stricter discipline in a statute that makes desertion while on service outside the kingdom into felony, and subjects captains who defraud men of their pay to forfeiture of goods and to imprisonment.

A few points remain to be mentioned before we
1485. pass to the reign of Henry the Eighth. The first was the establishment of that royal body-guard, which with its picturesque old dress and original title of Yeomen[1] of the Guard still survives among us. Though doubtless imitated from the Scottish Guard of the French kings, it is of greater interest as being composed not of aliens but of Englishmen, and as the first permanent corps of trained English soldiers in our history. Another smaller matter cannot be ignored without disrespect

[1] There were two kinds of soldiers, the gentleman soldier and the yeoman soldier. Hence the name points to the enlistment of men below the status of gentleman. The Navy still has "Yeomen of the Signals."

to military sentiment. After the victory of Bosworth Field Henry offered at the altar of St. Paul's Cathedral a banner charged with "a red fiery dragon" upon a field of white and green, the ensign of Cadwallader, the last of the British kings, from whom he was fond of tracing his descent. The scarlet of this red fiery dragon became from this time the royal livery, and was for the present reserved, together with purple, to the King's use alone.[1] But the green and white was more liberally distributed both to soldiers and mariners. A white jacket with the red cross of St. George had long been a common distinction of the English soldier, and the white as a colour of the Tudors now became so general that for a time "white coat" was used as a synonym for soldier.

Lastly must be noticed the definite establishment of the Office of Ordnance for the custody of military stores. The early history of the office is exceedingly obscure, and the existence of King Edward the Second's *artillator* hardly warrants us in assuming the permanent foundation of the department in the fourteenth century. The record of a Clerk of the Ordnance in 1418 sets the office on surer ground, and in 1483 the appointment of a Master-General advances it to a stage at which it becomes recognisable by us even at the present day; for the title of Master-General was held by John, Duke of Marlborough, and by Arthur, Duke of Wellington.

With Henry the Eighth we reach a new example in our history of an English soldier-king. Young, able, accomplished, and ambitious, he was strongly imbued with the military spirit, and possessed many qualities that must have made him a popular and might have made him a distinguished commander. He excelled

[1] I must confess that this should be put forward rather as a conjecture than an assertion; but it is remarkable that Henry VIII. should have permitted the use of any colours to the Artillery Company except purple and scarlet. Green and white were the favourite Tudor colours, being used even in ribbons for the attachment of the Great Seal.

in every exercise of arms; he was the finest archer in his kingdom; he had studied the art of war in the best authorities; he understood the conduct both of a siege and of a campaign; and lastly, he was no mean artillerist. This last attribute, however, he shared with several sovereigns of his time. Artillery was a favourite hobby with the crowned heads of Europe, possibly as a symbol of their military strength, for being unable to give themselves the pleasure of a great review owing to the inevitable confusion and expense, they were fain to console themselves with the several pieces, each one of them called by its pet name, that composed their park of ordnance. Altogether Henry was a prince who bade fair to restore the military prestige of England.

His first step was to increase his standing force by
1509. the creation of a second body-guard of men-at-arms,[1] composed of young men of noble blood; the reason given being that there were far too many such young men in the kingdom who were untrained in arms. The corps, as might have been expected with the best dressed sovereign in Europe, was so gorgeously arrayed that it perished after a few years under the weight of its
1511. own cost. His next act was more practical, a writ to the sheriffs for the better enforcement of the Statute of Winchester, which is interesting for its attempt to restore the command of the forces of the shore to their original holders.[2] Concurrently, however, we encounter a large number of the old-fashioned indents and commissions of array, all issued in prospect of English intervention in the eternal strife of the Hapsburgs and the Valois.[3] In 1512 an expedition was sent to the south of France, and there the defects of the army were lamentably seen. Although the importation of handguns and arquebuses shows that England was not blind to the progress of fire-arms in Europe, this force

[1] *Cal. S. P.* 20th November 1509.

[2] *Ibid.* 5th July 1511.

[3] *Ibid.* 3rd November 1509, 20th June, 1st July 1511, 8th April 1512. Rymer, vol. xiii. p. 329.

was armed principally if not exclusively with the old-fashioned bows and bills, and worse than all, these bows, which had been issued from the stores in the Tower, were found nearly all of them to be useless. Moreover, the victuals were "untruly served" to the men, their pay was withheld from them, and, acutest of all grievances, they could get no beer. The Council of War, in which the command was vested, could never agree as to a plan of operations, and though it kept the men thus inactive made no attempt to drill or exercise them. The natural result was a mutiny. One large band struck work for eightpence a day in lieu of the regular sixpence, several others swore that nothing should keep them from going home, and the disturbance was only quelled by the hanging of a ringleader.[1]

Henry seems to have had suspicions of the state of affairs, for in the same year Acts were passed to renew the existing statutes against desertion and fraud; though from the incessant re-enactment of these particular provisions it is clear that they were either easily evaded or negligently enforced. In the following year, however, Henry took the field in person in Normandy, where 1513.
his presence appears materially to have altered the complexion of affairs. His force was designed to have consisted of thirty thousand men, but was reduced by impending trouble with Scotland to less than half that number. The details of its organisation are still extant, and it is curious to find that, after but two generations of severance from France, the French terms vanguard, battle, and rearguard have given place to fore-ward, mid-ward, and rear-ward. Another novelty is the addition of wings, which had formerly been attached to the vanguard only, to the midward also; which was clearly a new departure.[2] There is again a strong tendency, which after a year becomes a rule, to make the tactical units of uniform strength, one hundred men being the common establishment for a company.

[1] *Cal. S. P.* 5th August 1512.
[2] Stow.

1513. Every captain too has an officer under him called his petty captain, a name which appears in the statutes of the previous reign, and was not yet displaced by the title, as yet reserved to the King's deputies only,[1] of lieutenant. The ensign[2] does not yet make his appearance, for the grouping of companies is strictly territorial, and one standard apparently alone is allowed to each shire. Every company, however, has the distinctive badge of its captain, and the archers of the King's Guard are dressed in uniform of white gaberdines. Lastly, there are in the army fifteen hundred Almains, the landsknechts of whom account was given in a previous section, eight hundred of whom, "all in a plump," marched immediately before the King. Possibly this place of honour was granted to them to kindle the emulation of the English, but more probably because Henry, following the evil example of the French, trusted more to trained mercenaries than to his own subjects. We shall constantly meet with such contingents of aliens among the English during the next forty years, until at last England awakes, like every other nation in Europe, to the truth that her own children, as carefully trained, are worth just double of the foreigners.

The most remarkable of the mounted men in this army were the Northern Horsemen, who, called into being at some uncertain period by the eternal forays on the Scottish border, now appear regularly on the strength of every expedition as perfectly indispensable. They were light cavalry, the first deserving the name

[1] Such at least is my impression. The commander-in-chief of a force not commanded by the King in person is styled the lieutenant or King's lieutenant. So also the commander of the body-guard is styled lieutenant, the King himself being captain. Compare the title, which we shall presently see introduced, of lord-lieutenant. But we meet also with the phrase lieutenant (*i.e.* commanding officer) of the rearguard or other of the three divisions in the army. The word is always used of a high office.

[2] In 1542, however, Wallop constantly speaks of ensigns (see *State Papers*, Henry VIII. (ed. 1830, 1849), vol. ix. *anno.* 1542).

ever seen in our army, and probably the very best in Europe. They wore defensive armour of back and breast and iron cap, carried lance and buckler or sometimes a bow, and were mounted on "nags" which were probably nearer thirteen than fourteen hands high. For duties of reconnaissance they were perfect, and they must be reckoned the first regular English horse that were the eyes and ears of the army. We shall see them at a later stage merged in a mounted body much resembling them, namely the demi-lances, which were destined, during the period of transition that is before us, to fill the place already almost vacated by the men-at-arms. 1513.

There is no need to dwell on the incidents of a not very eventful campaign. The panic flight of the French at the Battle of the Spurs upheld the old belief that they could not stand before the English; and the siege and capture of Terouenne under the personal direction of Henry helped to confirm it. A fruitless attack on an English convoy, curiously resembling the Battle of the Herrings in its main features, also helped to maintain the ancient reputation of the English archers. Lastly, the siege of Tournay gave Henry an opportunity of showing off some of his new artillery. There were twelve huge pieces, called the twelve apostles, of which he was particularly proud; but as St. John stuck in the mud and was unfortunately captured, it is well not to say too much of them. But the French were by no means impressed with the appearance of their old enemies in the field. "The English," wrote Fleuranges in a patronising way, "are good men and fight well when parked in a strong position, but otherwise I make no great account of them."

1513, September. But while Henry was plying his apostles against Tournay, some still older enemies of the nation had formed a very different opinion of the English. For in September, Thomas, Earl of Surrey, met the Scots at Flodden Field, and dealt them a blow from which they never wholly recovered. The odds against the

1513, September 9. English were heavy, for they could bring but twenty-six thousand men against forty thousand or, as some say, eighty thousand Scots, and the position taken up by James the Fourth was so strong that Surrey could not venture to attack it. With ready intelligence he made a detour from south to north of the Scottish host, and James, who had not attempted to molest him during the movement, hurried down, fearful of being cut off from his base, to meet him in the open field. The sequel is an example of the helplessness of pedantry, even of the newest pattern, in the face of genuine military instinct. The Scotch had studied the methods of the landsknechts; they were armed principally with pikes; they were drawn up in five huge battalions, after the Swiss model, and they advanced to the attack in silence "after the Almain manner." Lastly, they had with them some of the finest artillery hitherto seen.[1] Yet all this availed them nothing. The English too were formed, after a method which had lately come into fashion, in two divisions, fore-ward and rear-ward, each with two wings; but Surrey boldly wheeled both into one grand line,[2] holding but one small body of horse in reserve, and appears to have overlapped the cumbrous masses of the enemy. There is no need to give details of the battle; it began between four and five in the evening and was over in an hour. The English leaders seem to have shown not only bravery but skill. The English archers as usual wrought havoc against unarmoured men; the English bills got the better of the Scottish pikes, and the English light cavalry, admirably handled, twice saved the infantry from defeat. Ten thousand Scots were slain, and James himself, with the head and heir of almost every noble house in Scotland around him, lay covered with ghastly wounds among the dead. He had, from some whimsical return to an obsolete practice, dismounted his men-at-arms, who, in obedience to the new fashion which coun-

[1] *Cal. S. P.* 1513. 4460.

[2] *Ibid.* 4441.

selled protection against the new-fangled bullets, were clad in the heaviest armour. Arrows fell harmlessly from them, and even bills could not cut them down with less than half a dozen strokes; but they could not fly, and the bill-men did not weary of killing. And so on Flodden Field was shown a forecast of what was to be seen later in Italy, when infantry, finding men-at-arms prostrate on the ground, hammered them to death like lobsters within their shells before they could break through their armour.

Still the lesson of Flodden to the English was mainly that bows and bills were still irresistible; and to a conservative people none could have been more welcome. Henry, who was an enthusiastic archer, had already renewed a statute of his father's prohibiting the use of the cross-bow without a licence, and he now withdrew all licences and extended the prohibition to hand-guns.[1] The long-bow, on the other hand, received all the encouragement that enactments and sentiment could afford it. Henry dressed himself and his body-guard in green, which was the archer's peculiar colour; and the Venetian ambassador Giustiniani writing in 1519 described, with but slight exaggeration, the English military forces as consisting of one hundred and fifty thousand men, whose peculiar though not exclusive weapon was the long-bow. Men-at-arms were extinct, light cavalry insignificant in number. Giustiniani, however, did not add that the archers were now more efficiently equipped than at any previous period, being provided with two stakes instead of one, and further protected by a breastplate.[2] Nor did he notice a new weapon, the Moorish or Morris pike, which had lately come into use among the English, and had brought them a little closer to the famous infantry of the Continent.

It is, however, almost with a smile that we see Henry with undiminished satisfaction flaunting his

1 *Cal. S. P.* vol ii. part i., 6 Henry VIII. caps. 2, 11, 13.

2 *Ibid.* vol. iii. part i. p. 402.

1520. archers in the face of Francis at the Field of Cloth of Gold. Francis on his side produced his Swiss, and gave the English an opportunity of studying the first infantry in Europe. Fleuranges was at their head, and as his eye wandered from the scarlet and gold of the body-guard to the white and green of the other English troops, he probably felt justified in his opinion that they could not meet his own men in the open field. Henry, however, was unchangeable,[1] and the only sign of novelty that we see at this famous pageant is a horn-shaped flag borne in the retinue of Cardinal Wolsey, the *cornette*, which was in due time to give its name to the standard-bearers of the English cavalry.[2]

1522. Peace never endured long in those days, and in 1522 Henry was again at war with Francis, in alliance with Charles the Fifth. Again the English deficiencies became patent. In his expedition to France, which led to little result, Henry was forced to rely principally on Charles for cavalry;[3] and when it was evident that France would require to be fought on the Scottish border also, the Earl of Surrey, who held command in the north, begged for a reinforcement of four thousand landsknechts. The French, he said, would certainly bring pikes with them, and the English were not accustomed to pikes, though they would soon learn from the Almains.[4] In plain words, the English soldiers with their existing equipment were unfit to meet the French in the field. Fortunately the Duke of Albany, who was opposed to Surrey, was a coward, and little came of the alarm in the north. But the danger seems for the moment to have aroused Henry
1523. to a sense of his backwardness, for we find in 1523 a scheme for the purchase of ten thousand eighteen-

[1] At the meeting with Francis and Charles V. Henry took for his device an English archer in a green coat drawing an arrow to the head (Camden).

[2] *Cal. S. P.*, Henry VIII., vol. iii. part i. 869.

[3] *Ibid.* vol. iii. part ii. 2012, 2013.

[4] *Ibid.* 2995.

foot pikes and corselets, five thousand halberds, and ten thousand hand-culverins with matches,[1] bullet-moulds and powder-flasks complete. This is the first indication of a design to equip the army according to the best rules of the age, and, if it had been adopted, little change would have been needed for a century and a half. It is difficult to say why it was not, for at this time there are signs of an intention to take the improvement of the army seriously in hand.[2] But Henry changed his policy. Peace was made, and 1525.
was immediately followed by a proclamation to enforce the statute for the encouragement of the long-bow and the discountenance of cross-bows and hand-guns.[3] We must come down to the prolonged rejection of breech-loading artillery by the country in our own day before we can find a parallel to such perversity.

Nevertheless, in spite of all Henry's efforts fire-arms seem to have taken some hold on England, and particularly on London. In the general alarm that followed the insurrection known as the Pilgrimage of Grace, the King relied principally on London; and in 1537 he granted a Charter of Incorporation to the Artillery Company of the city, an association formed for the improved training of the citizens in weapons of volley, which term included hand-guns and cross-bows as well as the long-bow. This association survives as the Honourable Artillery Company. Again, at the great review of the London trained-bands two years later 1539.
we find like symptoms of a change. The old account of this pageant is of singular interest for the sight which it gives us of the most efficient soldiers in England. The force consisted of fifteen thousand picked men, all able-bodied and properly equipped, and all, except the officers, clothed in white even to their shoes.

1 In the original *lontes*. Lunt was the Scotch name for a musket-match to the end (*Cal. S. P.*, Henry VIII., vol. iii. part i. 3494).

2 See the armed strength of England in 1524. *Ibid.* vol. iv. part i. 972.

3 *Ibid.* 2086.

White was at once the old colour of England, the colour of the city, and the colour of the Tudors. The men paraded at Mile End, the famous drill-ground which was later to pass into a proverb, at six o'clock in the morning, and at eight moved off on their march to Westminster, in the three orthodox divisions of fore-ward, mid-ward, and rear-ward. First came the artillery, thirteen field-pieces, with their ammunition and "gun-stones," for shot was not yet always made of metal, in carts behind them. Then came the banners of the city, and then the musketeers, five in rank, with five feet of distance between ranks; after them came the bowmen in open order, every man a bow's length[1] from his neighbour; then followed the pikemen with their morris-pikes, "after the Almain manner," and lastly came the bills. Every one of the five divisions in each ward had its own band, its own colours, and its officers riding at its head; and it is worthy of note that the hand-guns and pikes took precedence of the bows and bills. So they marched on in their spotless white to Westminster, where the King awaited them on a platform. As the musketeers passed him they fired volleys, for a volley was of old the salute to the living as well as to the dead, the great guns were manœuvred and "shot off very terribly," doubtless to an accompaniment of female screams, and the force marched back through St. James' Park to the city. The review was intended as a demonstration against the menaces of foreign powers, and it had its due effect.

The danger passed away; but within four years
1544. Henry was again in the field fighting with Charles the Fifth against the French. There is little that is worth remarking in the campaigns that followed. The English as usual took with them their bows and bills, and the archers still came off with credit. A contingent of landsknechts was with them, who behaved so ill

[1] Six feet. A horse's length was reckoned at the same figure a hundred years later.

as to draw upon themselves more than ordinary dislike; and indeed the palmy days of the landsknechts were over. One portion of the English army alone provoked the warm admiration of Charles, namely, the Northern Horsemen. Wallop, the English commander, took justifiable pride in them, and detached them to clear the country before the Emperor on his departure. Away started the sturdy border-men on their tough little ponies, while Charles watched with all his eyes; and when he saw them breast an ascent before them and "hurl" up the hill, he cried out with honest delight.[1]

Nevertheless it must be confessed that Henry, though the eight and thirty years of his reign were perhaps the most eventful in the history of the modern art of war, did singularly little for the army. The passion for the bow, which evinced itself in repeated enactments and proclamations to the very close of his reign, and the false system of hiring mercenaries, led to a neglect of the infantry which might easily have proved disastrous. For the cavalry, though here again he was inclined to use mercenaries, he showed more care. He was much exercised by the decay of the English breed of horses, and passed three several Acts for its remedy. The wording of these throws a flood of light on our ancient troop-horse. To improve the breed it was enacted that every owner of a park should keep from two to four brood-mares not less than thirteen hands high, and that no stallions under fourteen hands should be employed for breeding; the hand to be reckoned as four inches and the measurement to be made to the withers. From the operation of this Act the counties of Northumberland, Cumberland, and Westmoreland, the home of the Northern Horsemen, were excluded. By a subsequent Act it was ordained that all chases, forests, and commons should be driven once a year, the unlikely mares and foals slaughtered, and no stallions allowed to run free that were under fifteen hands in height. What effect these measures may have wrought

[1] *State Papers* (ed. 1830-1849), vol. ix. pp. 523, 524.

I am unable to say; but the knowledge of the small stature of brood-mares can help us to a better understanding of the difficulties which beset the maintenance of an efficient cavalry.[1]

But the arm wherein Henry worked most improvement was undoubtedly the artillery. We find him at first purchasing all his guns abroad, for the most part in Flanders, and procuring his gunners also from foreign parts; but it is clear, from the number of Englishmen whose appointment to the post of gunner remains on record, that the English were rapidly learning their business from their instructors, while as early as 1514 we find Lord Darcy pleading for the employment of native gunners.[2] There is evidence too that the artilleryman's art was by no means so rare as it had been, gunners receiving no more than the ordinary soldier's pay of sixpence a day.[3] The casting of ordnance in England was less common, though there are scattered notices of English gun-founders from the beginning of the reign. Finally, in the year 1535 John Owen began to make even the largest guns, and obviated the necessity of depending on foreign makers for artillery. In 1543, moreover, Henry induced two foreigners to settle in England, Peter Bawd and Peter van Collen, who among other improvements devised mortar-pieces[4] of large calibre and shells to fire from them. Shell, indeed, was frequently used in the campaign of 1544, and Henry was early in appreciating its advantages. There was, however, still the difficulty of finding horses to draw the field-guns, which he seems to have attempted to overcome as early

[1] Henry in 1519 tried to procure horses from Italy, but was informed by Alfonso of Ferrara that there, too, the breed was decayed (*Cal. S. P.* vol. iii. part i. 171). Henry gave as much as £35, a great sum, for his own horses.

[2] *Cal. S. P.* 1514. 4902.

[3] *Ibid.* 1513. 4375.

[4] *Stow.* Mortar is the German *meerthier*, sea-beast. So other pieces were called after reptiles and monsters and birds,—serpentines, dragons, basilisks, falcons, culverins (couleuvrines), etc.

as in the third year of his reign by some kind of registration of waggoners and teams. The drivers were to wear the white coat and red cross, and to be mustered and paid every month; and for their protection it was ordered that their paymaster should take no bribes from them beyond one penny a month from each man, a curious commentary on the financial morality of the army. Be that as it may, however, there exists no doubt that Henry the Eighth created the British gunner who, as his proud motto tells, has since worked his guns all over the world. 1513.

His zeal as an artillerist led Henry also, perhaps almost insensibly, towards the peculiar organisation for defence which was copied at a later period by the colonies, and for a short time was expanded even into an imperial system. The mounting of valuable guns entailed the necessity of maintaining a small body of trained men to keep them in order; and thus grew up the practice of stationing small independent garrisons in all the principal fortresses, which garrisons were immovably attached to their particular posts and constituted what was really a permanent force. Thus almost at a stroke the military resources of England fell into three divisions—the standing garrisons just mentioned, the militia which could be called out in case of invasion, and the levies, nominally feudal but in reality mercenary, which were brought together for foreign service and disbanded as soon as the war was over. The attention devoted by Henry to the defence of the coast identifies his name peculiarly with certain modern strongholds, which stand on the same site and bear the same appellation as he gave them three centuries ago. Nor must it be forgotten that, though he did comparatively little for the army, Henry did very much for the navy, and perceived that the true defence of England was the maintenance of her power on the sea.

Two small points remain to be mentioned before we dismiss the most popular of English kings. A dear lover of music he took an interest in his military bands,

1542. and we find him sending all the way to Vienna to procure kettle-drums that could be played on horseback "after the Hungarian (that is to say the Hussars') manner," together with men that could make and play them skilfully. Ten good drums and as many fifers were ordered at the same time, with advantage, as may be hoped, to the English minstrels. Lastly, Henry was the first man of whom we may authentically say that he brought the English red-coats into the field for active service. Red garded with yellow was the uniform
1544. worn by his body-guard at the siege of Boulogne; and perhaps it was right that the scarlet should have made its first appearance in the presence of such old and gallant enemies as the French.

Under the rule of his boy successor we find little change in the old order of things. There was the usual fight with the Scotch on the border, and yet
1547. another crushing defeat, at Pinkie, of the old inveterate enemy. But hired Italian musketeers contributed not a little to the victory; and the state of the forces of the shires was most unsatisfactory. Fraudulent enlistment and desertion, doubly expensive since the payment of coat- and conduct-money had been instituted, were as common as ever, and the dishonesty of officers was never more flagrant. A stringent Act was passed to check these irregularities, with apparently the usual infinitesimal measure of success. Foreign troops were never so much employed in England, though even
1549. they complained of unjust dealing. The insurrection in the west was suppressed principally by landsknechts and Italian harquebusiers, not however before they had suffered one repulse from the men of Devon, beyond doubt to the secret joy of all true Englishmen. Nevertheless the reign saw the rise of the Gentlemen Pensioners and, more important still, the appointment of a lord-lieutenant in every county, to be responsible for the forces of the shire. The latter was no doubt a stroke in the right direction, but it did not touch the heart of the matter. The worn-out machinery which

had been patched and tinkered for five centuries was not so easily to be repaired; and a new fly-wheel, though it might turn magnificently on its own axis, could not keep the other broken-down wheels in motion.

The reign of Queen Mary brought the most 1553. important change in the military system of the country that had occurred for two centuries. The Statute of Winchester was superseded and a new Act enacted in its place. The reform, however, was in reality quite inadequate to the occasion. It provided for the supply of more modern weapons and for a new distribution, according to a new assessment, of the burdens entailed by the maintenance of a national force; but in substance the new statute was drafted on the lines of the old, and the variations were very superficial. The extinction of men-at-arms hinted at by Guistiniani is sufficiently proved by the mention of two different kinds of cavalry, "demi-lances" or "medium" horse and the light horse with which we are already acquainted; and progress in the equipment of the infantry is shown by the mention of long pikes and corselets and of harquebuses. But alongside of these improved weapons are the familiar bows and bills; and a clause which, considering that Mary had married the heir of Spain is truly marvellous, provides that a bow shall in all cases be accepted as an efficient substitute for an arequebus. These details, however, are comparatively unimportant. The difficulty was one, not of arms, but of men; and Mary knew it. She would have formed a standing army if she had dared, but as she designed it principally for the coercion of her own subjects she ventured neither to ask for the money to establish it nor to brave the indignation that would have followed on its establishment.

Her unpopularity at the close of her reign, so strikingly in contrast with the devoted loyalty which she had enjoyed on first mounting the throne, told heavily against the efficiency, always largely dependent

on sentiment, of the forces of the shire. Never children crept more unwillingly to school than the English contingent which joined the Spaniards after the battle
1557. of St. Quentin. Never half-witted woman looked on with more helpless, impotent distraction at the robbery of her jewels than the once iron-willed Mary, when
1558. Guise marched up to Calais. The English garrison made all the resistance that could be expected of brave men, but they were outnumbered, and the commanders asked in vain for reinforcements. The Government awoke to the danger too late; and, yet more sadly significant, the forces of the shires came unwillingly to the musters and came unarmed. Yet Mary's name is bound up with two material benefits conferred on the British soldier. The men who went to St. Quentin received eightpence a day, the sum for which her father's men had mutinied forty years before; and from this time, for two full centuries, eightpence replaces sixpence as the soldier's daily stipend. More thoughtful too than any of the kings that came before her, she left directions in her will for the provision of a house in London, with a clear endowment of four hundred marks a year, "for the relief and help of poor, impotent and aged soldiers" who had suffered loss or wounds in the service of their country. For all her man's voice and masculine will, she had a woman's heart which warmed to the deserving old soldier, and whatever her demerits in the eyes of those who wear the gown, her memory may at least be cherished by those who wear the red coat.

CHAPTER III

We enter now on the fateful reign of Queen Elizabeth. 1558.
The condition of England at its opening after the previous years of misgovernment was most unpromising. Wrenched from its moorings by the Reformation, the country had been tossed about by a hurricane of religious fanaticism, which, working round through all points of the compass, had left her helpless and bewildered, uncertain by which course to steer or for what port to make head. Elizabeth was by political exigency rather than religious conviction a Protestant, but her great object in life was to sail, if she could, clear of the circular storm and lie outside it. The design was an impossible one, and her obstinate persistence therein went near to bring England to utter ruin, but in the extremely difficult position wherein she found herself on her accession to the throne she had much excuse for a tortuous policy. The finance was in hopeless disorder, and the realm through long neglect virtually defenceless. There was no discipline in such forces as the country could raise; and the military stores, which her father had taken such pains to collect, appear to have perished. The French were in Scotland in considerable force, and, as the Council pointed out, France was a state military, while England was established for peace. There in reality lay the kernel of the whole matter. England was behind all Europe in military efficiency, and all Europe was keenly alive to the fact.

The situation was so desperate that heroic measures,

however distasteful to the Queen from their expense, were inevitable. Arms were purchased hastily in vast quantities in Flanders, the forces of the shire were called out, and Elizabeth exercised in St. James' Park with fourteen hundred men of the trained-bands, who had been equipped by the city with caliver, pike, and halberd. But up in the north, the loyalty of the troops was doubtful, and their discipline more doubtful still. Fraud again was rife among the officers. The landsknechts during their stay had set the fashion of extravagance in clothing, and some captains, as it was quaintly said, carried twenty to forty soldiers in their hose. Thus, though the muster-rolls of the army in Scotland showed eight thousand men for whom the Queen paid wages, but five thousand were actually with the colours, and the pay of the remaining three thousand went of course into the captains' pockets. This state of things was put down with a strong hand by special Commissioners, and the little army round Leith became orderly and efficient; but corruption had sunk so deep that it had eaten its way even among the officials of the ordnance at the Tower of London.

The French, however, were in due time compelled
1560. to evacuate Scotland, and the danger in the north ceased to be pressing. There was, however, constant trouble in Ireland; and to provide the necessary troops to keep it in order, resort was made to an instrument of which we shall hear much in the years that follow, namely, the press-gang. None the less the revelations discovered by the war in Scotland prompted Cecil to require a report from the magistrates all over England as to the condition of the population and the working of the statutes enacted for national defence. The answer was by no means complimentary to the influence of the Reformation, nor encouraging in respect of military efficiency. The people, reported the magistrates, were no longer trained to the use of arms, because the gentlemen no longer set them the example. In plain words the old system of the fyrd, a people in arms,

was obsolete. Not one but many causes had conspired to make it so. The country was passing through a social as well as a religious revolution ; old landmarks were vanishing, old customs dying out; and the loss of the old faith had become to many an excuse for disburdening themselves of every irksome duty. Again, Calais was lost, and though there were still vague hopes that it might yet be regained, England was now strictly insular and France was closed as a field of national adventure. The people had awaked to the fact that their heritage was the sea ; and the life of the corsair, free, stirring, lucrative, and dangerous, appealed powerfully to a race at once adventurous and grasping, energetic and casual, bold and born gamblers.

Moreover, the national weapon, the long-bow, and the tactics that went with it, were things of the past, while the new arms were at once distasteful and costly, and in the unsettled state of the country not to be trusted in every man's hand. The whole business of war, too, was becoming difficult and elaborate, and was passing through transitions too rapid to permit it to be learned once for all. Military training no longer consisted in friendly matches at the archery butts, but in precise movements of drill and manœuvre, unwelcome alike because their advantages were unrecognised, and because they could no longer be learned from the old masters. The acknowledged leaders in hundred and parish and shire gave place to experts trained in foreign schools, men who swaggered about in plumed hats and velvet doublets and extravagant hose, swearing strange oaths of mingled blasphemy taught by Spanish Catholics and Lutheran landsknechts, and prating of besonios and alferez, of camp-masters and rote-masters, of furriers and huren-weibels, of false brays, mines and countermines, in one long insolent crow of military superiority. Such instructors were not likely to soften the painful lesson that war had become a profession, and could no longer be tacked on as a mere appendage to the everyday life of the citizen.

Now, therefore, if ever, was the time for the establishment of a standing army in England. She was menaced by foreign enemies on all sides, and in perpetual peril of intestine insurrection. There was unceasing trouble in Ireland, and eternal anxiety on the Scottish border. The forces of the shires had been proved to be worthless, and the service was not only inefficient but unpopular; the people came unwillingly to the muster, and would gladly have paid to be relieved of the burden. Great results would have followed from the institution of a standing force; order would have been maintained at home; interposition in foreign affairs would have had redoubled weight; untold expense through unreadiness, knavery, and inefficiency would have been spared; and finally, the British Army would have grown up to be honoured as a great national possession, called into existence to stave off a great national peril, instead of to be abused as an instrument of tyranny, and to be condemned to a blighting heritage of jealousy and suspicion.

But Elizabeth would have none of such things. She refused, to her credit, to employ foreign mercenaries, and by breaking off that evil tradition did lasting good. But she was incapable of living except from hand to mouth. She hated straight dealing for its simplicity; she hated conviction for its certainty; above all she hated war for its expense. She loved her money as herself, and to these twain she would sacrifice alike the most faithful servant and the most friendly State. She was so mean and dishonest in defrauding even such troops as she employed of their due, that no one seems to have dared even to hint to her the expediency of keeping a standing army. It may be urged that this was well for the liberties of England, but, on the other hand, it went near to destroy them altogether; and, after all, a standing army did not save either James the Second of England or Louis the Sixteenth of France. The people of England, however, saw more clearly than their tricky inconstant Queen, and made good her delinquencies in their own way.

The French had not long evacuated Scotland when the desperate condition of the Protestants in France 1562.
forced the Prince of Condé to offer Elizabeth Havre and Dieppe as pledges for the restoration of the lost Calais, if she would send him money and men. Elizabeth consented; and seven or eight thousand men were despatched to garrison these two ports. Five hundred of them, English and Scots, at once volunteered to cut their way into Rouen, which was closely besieged by Guise, and fell at the capture of the town, fighting desperately till they were cut down almost to a man. These volunteers should be remembered, for they cleared the ground for the foundation-stone of the British Army, English and Scots fighting side by side for the Protestant cause in a foreign land. The remaining troops were, as was inevitable under the parsimonious rule of Elizabeth, ill-equipped and ill-provided, a miserable contrast to the armies of the Plantagenets, and a shameful example which has been followed only too faithfully since. War between France and England at once broke out in earnest, and the garrison of Havre required reinforcement. No troops of course were ready, and it was necessary to raise recruits in a hurry. The prison doors were opened; the gaols were swept clean; robbers, highwaymen, and cutpurses, the sweepings of the nation, were driven into the ranks; and a second evil precedent, companion to the press-gang, was set for the misleading of England the Unready. None the less these poor men fought gallantly enough against the besieging French, until the plague suddenly broke out among them; and then they went down like flies. Between the 7th and 30th of June the effective strength of the garrison of Havre sank from seven thousand to three thousand men. More men were hurried across the channel to perish with them, but the waste was greater than the repair, and in another fortnight but fifteen hundred of the

whole force were left. Further requests for men and arms were met by the despatch of raw boys and of all the worn-out ordnance in the Tower—"The worst of everything is thought good enough for this place," wrote the General, Lord Warwick, in the bitterness of his soul—and finally after a grand defence Havre was surrendered.

Nevertheless, little or nothing was done to make good defects in the years that followed. The dishonesty of the officers and the indiscipline of the men in Ireland was past all belief; but it was only with extreme difficulty that Elizabeth was induced to remedy the evil, which brought untold misery and oppression upon the forlorn Irish, by the simple process of paying her soldiers their wages. It was not until 1567, when the movements of Philip the Second gave the alarm of invasion, that a corps of arquebusiers, four thousand strong, was formed for the defence of the coast towns from Newcastle to Plymouth, and prizes were given for the encouragement of marksmanship with the new weapon. Even so, practice with the bow was still enjoined upon the villagers, as though no better arm could be discovered for them.[1]

1569. Then came the rebellion, which but narrowly missed a most serious character, of the Catholic nobility in the North. Disloyalty was widespread in Yorkshire, and it was proverbial that the Yorkshire levies would not move without pay; but Elizabeth was too economical to send the train-bands from London to nip the insurrection in the bud, and only at the last moment consented to provide money for the payment of the troops on the spot. The difficulties of the commanders were frightful. The numbers that came to muster were far short of the true complement; horsemen were hardly to be obtained by any shift, and the footmen that presented themselves came with bows and bills only, there being but sixty firearms, and not a single pike, among two thousand five hundred infantry. The

[1] See *Cal. S. P.*, Dom., Addenda (1561-1579), pp. 78-84.

rebels, on the other hand, were very well equipped, and had a force of cavalry armed after the newest pattern of the Reiters. "If we had but a thousand horse with pistols and lances, five hundred pikes and as many arquebuses," wrote Elizabeth's commanders, "we should soon despatch the matter"; but even so trifling a contingent as this could not be produced except after infinite difficulty and delay.[1]

For all this Elizabeth was responsible; but the peril was so great that it stirred even her avaricious soul. From this year bows and bills began slowly to make way for pikes and firearms; and a manuscript treatise in the State Papers shows that the reform was brought under the immediate notice of the Royal Council.[2]

An alarm of invasion by the French in the follow- 1570.
ing year led also to a general stirring of the sluggish forces of the shire. The French ambassador reported that one hundred and twenty thousand men could take the field in different parts of the country; and the muster-rolls showed the incredible total of close on six hundred thousand men. Yet when we look into these muster-rolls we find simply a list of able-bodied men and of serviceable arms in each shire without attempt at organisation. In truth, throughout the long reign of Elizabeth we feel that in military matters one effort and one only is at work, namely, in Carlyle's words, to stretch the old formula to cover the new fact, to botch and patch and strain the antiquated web woven by the Statute of Winchester and newly dyed by the Statute of Philip and Mary to some semblance of the pattern given by the armies of France and Spain.

But when we turn from the Queen to the people we perceive the energy of a very different force. The

[1] *Cal. S. P.*, Dom., Addenda (1566-1579), pp. 111-113, 115-116, 121-123, 126-127, 129.

[2] One sentence gives a clue to Henry VIII.'s long discouragement of firearms. "Is not the safety of the country worth more than the saving of a few wild-fowl?"

English army indeed was not created by a sovereign or a minister; it created itself in despite of them. The superior equipment of the northern rebels over that of the forces of the Queen was typical of the whole course of English military progress in the sixteenth and seventeenth centuries. The army was conceived in rebellion, born in rebellion, nurtured in rebellion. Protestantism all over Europe went hand in hand with rebellion; and Elizabeth, always irresolute and incapable of conviction, was distracted between a political preference for Protestantism and a natural abhorrence of disloyalty. For years she struggled by the most contemptible trickery to be true to both these opposing principles, and for a time, by the help of extraordinary good fortune, she attained the success which only a false woman could compass. But long before she could make up her mind, the people had taken matters into their own hands, and thereby begun the creation of our present army. It was on May Day 1572, four years later than the first rising of the Low Countries against Spain, that the army took its birth from a review of Londoners before the Queen at Greenwich. In the ranks that day were many captains and soldiers who had served in Scotland, Ireland, and France, and were now adrift without employment on the world. Subscriptions were raised by sympathetic Protestants in the city, and three hundred of them were organised into a company and sent to fight for the Dutch under Captain Thomas Morgan. From this beginning we must presently trace the history of the English regiments in the Low Countries to the eve of the Civil War; and for the next seventy years therefore our story must flow in two distinct streams—the slender thread that runs through England itself, and the broader flood which glides on with ever-increasing volume in the Low Countries, on the Neckar, and even in distant Pomerania. And since at every great national crisis the two streams for a time unite, the lesser tributary may be dismissed forthwith by a brief review of the progress of the military art in England to the close of the sixteenth century.

London as usual led the van of military improvement. In the year following the departure of Morgan's 1573.
company, three thousand men of the train-bands were formed into a special corps, which was mustered three times a week for exercise, and having been armed with weapons of the newest pattern was regularly drilled by experienced officers on the once famous ground at Mile End. William Shakespeare, it is evident, was one of the spectators that went from time to time to see them, and no doubt laughed his fill at the failings of the recruits. These were sometimes not a little serious. Thus one caliverman left his scouring-stick in the barrel, and accidentally shot it into the side of a comrade, whereof the comrade died; so that the whole body of calivermen gained the enjoyment of a military funeral in St. Paul's Churchyard, whither they followed the corpse with trailing pikes and solemn countenances, and at the close of the ceremony fired their pieces over the grave.[1]

Something therefore had at least been learned from the landsknechts, and other changes were coming fast. The old white coat and red cross seems to have disappeared abruptly at the beginning of the reign, and coats, or, as they were called, cassocks,[2] generally red or blue, were provided by shires and boroughs in their stead. Once, indeed, these bright hues are found condemned as too conspicuous for active service in Ireland, and some dark or sad colour, such as russet, is recommended in its stead,—a curious anticipation of our modern *khaki*.[3] Again, to turn to smaller changes, the word petty captain had dropped out of use since 1563, to yield place to the title of lieutenant, and the word ensign seems to have been accepted generally at about the same time. Sergeant had been the title of the expert at drill since 1528, but

[1] Stow.

[2] The word was borrowed from the French *casaque*, the regular term for a livery-coat. Facings were soon added. *Cal. S. P.*, Dom. (1595), p. 22.

[3] *Cal. S. P.*, Dom. (1581-1590), p. 16.

in 1585 there is a distinct order that the men appointed to instruct the bands of the shires shall be called cor-
1587. porals.[1] Two years later we find officers of higher rank asking for a new denomination, and proposing that they may bear the title of colonel and the officers next below them that of sergeant-major, or, as we now call it, major. It was indeed time, for the word regiment came likewise into use at the same period, and a regiment without a colonel is naught. Before the end of the century the term infantry had also passed into the language, while the flags of the infantry, from their diversity of hues, had gained the name of colours.[2]

But far more striking than these superficial changes is the sudden deluge of military pamphlets which burst over England from the year 1587 onwards. The earliest military treatise, so far as I have been able to discover, that was delivered to the English in the vulgar tongue is *The Ordering of Souldiours in battelray*, by Peter Whitehorn, which was published in 1560. This book produced, no doubt, some effect in its time, but it is of small import compared with those that follow. The earliest written by an Englishman, though not published until four years after his death, was the work of one William Garrard, gentleman, who had served with the King of Spain for fourteen years and died in 1587. It is a remorseless criticism of the existing English military system. The author sweeps away bows and bills in a single contemptuous sentence, and lays it down for a dogma that there are but two weapons, for the tall man the pike and for the little nimble man the arquebus. But in the matter of equipment, he notes that the English are lamentably deficient. As good an arquebus could be made in England as in any country, but the armourers had already learned to make cheap and nasty weapons for common sale to the poor men of the shire.

[1] *Cal. S. P.*, Dom. (1581-1590), p. 255.

[2] One bitter critic avers that the expression was due to the number of low-born captains, who, having no arms to bear on their ensigns, were obliged to trust to distinctions of colour only.

Again, other nations carried their powder in flasks or metal cartridges, but the English actually carried theirs loose in their pockets, ready to be kindled by the first spark or spoiled by the first shower, and in any case certain to suffer from waste. Such slovenliness, says the indignant Garrard, is fit only "for wanton skirmish before ladies"; it is impossible for such arquebusiers to attain to the desirable consummation of "a violent, speedy, and thundering discharge." The pikemen, again, instead of a light poniard carried "monstrous daggers like a cutler's shop," fitter for ornament than use. Moreover, the dress of both was open to objection. Colour was a matter of indifference, though some fine hue such as scarlet was preferable for the honour of the military profession, but all military garments should be profitable and commodious, whereas nothing could hamper the limbs more than the great bolstered and bombasted hose that were then in fashion. I cannot resist the temptation of transcribing Garrard's picture of the march of the ideal soldier, and the delicate appeal to the soldier's vanity.

"Let the pikeman march with a good grace, holding up his head gallantly, his face full of gravity and state and such as is fit for his person; and let his body be straight and as much upright as possible; and that which most important is that they have their eyes always upon their companions which are in rank with them and before them, going just one with another, and keeping perfect distance without committing the least error in pace or step. And every pace and motion with one accord and consent they ought to make at one instant of time. And in this sort all the ranks ought to go sometimes softly, sometimes fast, according to the stroke of the drum. . . . So shall they go just and even with a gallant and sumptuous pace; for by doing so they shall be esteemed, honoured and commended of the lookers on, who shall take wonderful delight to behold them."

Earlier in appearance though not earlier composed

than Garrard's was a shorter work by one Barnaby Rich, which appeared in 1587, and wherein the writer had the courage to condemn the practice of emptying the gaols into the ranks; but the great military book of the year was a translation from the French of La Noue, one of the noblest and ablest of the Huguenot commanders. Though written of course for Frenchmen, the soundness of doctrine in respect of discipline and equipment and the commendations of the Spanish system were of value to all; while of still greater import to England was the impassioned advocacy of the missile tactics of the Reiters for cavalry. But perhaps most striking of all in the light of later events is the deep note of Puritanism to which every page of the treatise is attuned. In La Noue's Huguenot regiments there were no cards, no dice, no swearing, no women, no leaving the colours for plunder or even for forage, but stern discipline at all times and public prayers morning and evening. It is difficult to suppress the conjecture that this book had been read and digested by Oliver Cromwell.

The strong opinions expressed in these books of course provoked controversy. Sir John Smyth, knight, an officer of some repute, boldly took up the cudgels on the other side, and undertook to prove even in 1591 that the archer was more formidable than the arquebusier and the arrow than the bullet, which was an argument only too welcome to old-fashioned insular Englishmen. On the other hand, he enters minutely and intelligently into points of drill and manœuvre, condemns the bombasted hose as vehemently as Garrard himself, and prescribes a more serviceable dress for the soldier. From him we learn our first knowledge of the manual exercise of the pike, how it should be advanced and how shouldered with comely and soldierlike grace, and how men should always step off with the right foot. From him also we obtain sound instruction for the shock attack of cavalry, and some mention of the Hungarian light horsemen, called "ussarons"; and from him finally we gather information of the extraordinary inefficiency even at the close of the reign of the shire-levies of England, of the neglect of the arms and the corruption of the muster-masters.

Roger Williams, whom I have already quoted, also entered the lists at this time with an account of the Spanish organisation, and combated warmly for the superiority of the lance over the pistol as the weapon of cavalry; and a translation by Sir Edward Hoby from the Spanish of Mendoza (1597) also upheld the cause of shock-action. Hard upon these followed a version of the striking work of Martin du Bellay, with its complete scheme for what we now call the short-service system; and in the same year (1598) appeared a dialogue by one Barret, which sought to close the whole controversy. A conservative gentleman who upholds bows and bills is utterly demolished by a captain who pleads for pike and musket, would abolish the shire-levies bodily as useless, and would substitute a reorganised force on the favourite model, already once adopted in France, of the Roman legion. But Barret knew his countrymen and expected little. "Such as have followed the wars," he says, "are despised of every man until a very pinch of need doth come"; and military reform then as now could not be pushed forward except under pressure of a scare of war.

So matters drifted on to the close of the sixteenth century and beyond it. The military spirit was abroad, and the military pen busy beyond precedent. The character of the old soldier became a favourite with beggars and vagabonds, and was rewarded so freely at the hands of the charitable that it was necessary to suppress the imposture by special statute. Yet in spite of all this simmering and seething nothing was done in England for the English army. Soldiers who wished to learn their profession sought service elsewhere than with the Queen; even in Ireland the value of a company sank to fifty pounds;[1] and the most conspicuous type of warrior that was to be found at home was the worst.

[1] Collins.

Shakespeare, who saw everything and into the heart of everything, marked these impostors and reproduced them with such genial satire, such incomparable humour, that in our delight in the dramatist we overlook the military historian. Yet he is as truly the painter of the English army in his own day as was Marryat of the navy in later years. Falstaff the fraudulent captain, Pistol the swaggering ensign, Bardolph the rascally corporal, Nym the impostor who affects military brevity, Parolles, "the damnable both sides rogue," nay, even Fluellen, a brave and honest man but a pedant, soaked in classical affectations and seeking his model for everything in Pompey's camp—all these had their counterparts in every shire of England and were probably to be seen daily on the drill ground at the Mile End. Not in these poor pages but in Shakespeare's must the military student read the history of the Elizabethan soldier.

CHAPTER IV

THE arrival of the first English volunteers, under Thomas Morgan, in the Low Countries was, as fate willed it, most happily timed to synchronise with the movement that laid the foundation of Dutch Independence. In April 1572 an audacious enterprise of the fleet of Dutch privateers under the Count de la Marck had led to the surprise and capture of the town of Brill, a success which at once fired the train of revolt in the seven provinces north of the Waal and shook the hand of Spain from town after town first in Holland and Zealand, and later in Friesland, Gelderland, Utrecht, and Overyssel. The incident, which time was to prove so far reaching in its results, was a curious commentary on the latest phase of Elizabeth's policy. She had just reconciled herself with Alva and forbidden De la Marck's privateers to enter English ports: the sea-rover's reply was to beard Alva in his own stronghold and deal Elizabeth's friend a blow from which he never recovered. The whole island of Walcheren, excepting Middelburg, fell into the hands of the insurgents, and Alva, who was a splendid soldier, whatever his other failings, lost no time in attempting to recover the port of Flushing. By the irony of fate Morgan's volunteers arrived in the very nick of time to save it, and in the sally which brought them first face to face with the dreaded troops of Spain they made a brilliant beginning for the new British Army. Of the three hundred, fifty were killed outright in this action, the first of fifty thousand or twice fifty thousand who were to lay their bones in Holland during the next seventy years.

Morgan, having rescued Flushing, at once wrote letters to England to point out the importance of the town which he held and to beg for reinforcements. In the autumn accordingly appeared Colonel Sir Humphrey Gilbert, with a regiment, the first of many English regiments that were to enter the Dutch service, of ten companies and fourteen hundred men, raw troops under a raw leader. Morgan would have been the better commander, but he was a modest unambitious man; Gilbert, on the other hand, suffered from fatal ignorance of his own incapacity. Sir Humphrey at once launched out boldly into complicated operations which he was utterly incompetent to direct, was outwitted and out-manœuvred, fell back on swearing when things went wrong, and not only lost his own head but completely broke the spirit of his men. The new regiment in fact behaved very far from well. "I am to blame to judge their minds," wrote Roger Williams, the ablest of Morgan's officers, after Gilbert's first defeat, "but let me speak truth. I believe they were afraid." He adds elsewhere a gentle but telling criticism, that lays the blame on the right shoulders. "A commander that enters the enemy's countries ought to know the places that he doth attempt: if not he ought to be furnished with guides." So ignorant were even educated Englishmen of the alphabet of war. Gilbert, however, did not learn his lesson quickly. A slight success, wherein the English displayed conspicuous gallantry, heated his ambition once more to boiling-point; he essayed another adventure in the grand manner, failed utterly, and sailed home with the scanty remnant of his regiment, a sadder and wiser man.

Morgan meanwhile had gone home and raised ten more companies, with which however he could do very little. The men were not paid on their disembarkation in Holland, as William of Nassau had promised them, and they became discontented and insubordinate. Morgan naturally took their part, and the result was, that after some few petty engagements against the

Spaniards, he took his departure in dudgeon and sailed with the seven hundred men that were left to him to England. He had done good work, and his name deserves to be remembered; for he was the first man who made perfect arquebusiers of the English, and the first who taught them to love the musket. Fifty years had flown since the Spaniards had shown the way, and the English were only just beginning to follow. Roger Williams on Morgan's retirement took service with the Spaniards for a time, in order to learn his duty the better, and presently returned, without reproach, to wield the knowledge that he had gained against themselves. To such shifts were British officers reduced who wished to master their profession.

To follow the actions of sundry other corps of volunteers during the succeeding years would be tedious. I pass at once to the landing in July 1577 of a company of three hundred Englishmen under the command of John Norris, one of the first and most eminent of the new school of officers who were the fathers of our Army. He had learned his work first in Ireland, and later in France under a great disciplinarian, the Admiral Coligny. He too arrived at a critical time. A few months after his disembarkation, while he was still in garrison at Antwerp, Don John of Austria surprised the Army of the States at Gemblours, and not only defeated 1578,
it but shattered it to fragments. Six months later Don January 29.
John attempted to repeat the blow against a second August.
Army of the States, a heterogeneous force of English, Scotch, and Flemings, under the command of the veteran Huguenot, De la Noue. Having but fourteen thousand men against thirty thousand of the finest troops in Europe, De la Noue took up a strong position at Rymenant, near Malines, and stood on the defensive. After trying in vain to draw him from his entrenchments Don John finally launched a desperate attack on the quarter held by the English and Scotch under Norris. Four companies of Scots bore the first brunt of the assault, but were presently reinforced, just

in time, by Norris's eleven companies of English; and then the struggle became as desperate as ever was fought by British soldiers. The Spanish troops were the flower of the army, the Old Regiment,[1] which had not its peer in Europe; but with all their magnificent training and discipline they could not carry the position. Three times they forced the British back, and three times when success seemed assured they were met by a resistance that would not be broken, and were hurled back in their turn. The day was intensely hot, and the British, scorning all armour, fought in their shirt-sleeves, but they fought hard, and not only hard but, thanks to John Norris, in good order. Norris himself, always in the thickest of the fight, had three horses killed under him in succession, but never lost hold of his men; and at last the famous infantry of Spain drew back, beaten, and Don John abandoned the attack. It was a great day for old "Bras de fer" De la Noue, but a still greater for John Norris and his British. They had, by general admission, not only saved the day, but they had repulsed the most formidable troops in the world.

During the years that follow Norris and his companies were incessantly engaged, generally victorious, though once at least defeated with heavy loss; their gallant leader, though frequently wounded, reappearing always whenever work was to be done. Their highest trial was when they encountered the greatest General of the day, Alexander of Parma, and the whole Spanish army with him, in a rearguard action, and beat them off with such persistent bravery that the French volunteers after the engagement crowded to their colours and begged to be allowed to serve under them. Norris indeed was the Moore of the sixteenth century, alike as a teacher in the camp and as a General in the field.

Nevertheless, brilliant as his service was, he could not stay the victorious advance of the Spaniards. After ten years of fighting the Dutch States had lost almost the whole of Spanish Flanders except a few large towns and

[1] *Tercio Viejo.*

the sea-coast from Dunkirk to Ostend, and still Elizabeth would not move to help the Dutch insurgents in a task, no less vital to England than to them, which lay beyond their strength. At last the assassination of William the Silent forced her to make up her uncertain mind to the inevitable rupture with Spain. The United Provinces were in the utmost need; the strong hand of Alexander of Parma was at the throat of Antwerp, and unless its grip could be relaxed the city must inevitably fall. The States threw themselves upon the English Queen, entreating her even to make them a part of her realm, and at last, after much paltry haggling, Elizabeth consented to send them four or five thousand men, taking over the towns of Brill, Flushing, Rammekins, and Ostend as security for their obligations towards her. Elizabeth was always careful to look after the money. 1584, July 10.

This agreement being at last concluded the press-gang[1] was at once set to work in England; four thousand men were raised and dressed in red coats, and within a fortnight after the signing of the Treaty they had crossed the North Sea, only to find that Antwerp was already in Parma's hands and that they had come too late. Norris, however, at once took the force in hand, and was carrying on active operations with brilliant success when he was stopped by a peremptory rebuke from the Queen; the troops had been transported for the relief of Antwerp, and she would not have them employed on any other service. The States, naturally exasperated by this contemptible double-dealing, received the troops reluctantly into the cautionary towns and left them with no very good grace to take care of themselves. Elizabeth, as her nature was, had refused to send a penny of money or an ounce of supplies, and the soldiers, ill-fed, ill-clothed, and ill-lodged, began to melt away by hundreds through death and desertion. 1585.

[1] The press-gangs were not very scrupulous. On one occasion they took advantage of Easter Sunday to close all the church doors in London and take a thousand men from the various congregations.—Stow.

1585. In December, however, Robert, Earl of Leicester, was sent out as Commander-in-Chief of the forces in the Low Countries, and as he brought with him a reinforcement of cavalry, and also money sufficient to pay the arrears of the soldiers' wages, it was hoped that matters would be placed on a better footing. But it was not to be. Elizabeth was not yet in earnest in breaking with Spain, and Leicester, gathering an inkling of her intentions from her refusal to provide him with additional funds, went very unwillingly to take up his command. On arriving in Holland he found things even worse than he had anticipated. The men were in a shocking state, dying fast of cold and hunger; they had not a penny wherewith to supply themselves; and their clothing was so deficient that for very nakedness they were ashamed to appear in public. Leicester with all his faults had evidently a genuine tenderness for his unfortunate soldiers; he wrote letter after letter pressing vehemently for money, but Elizabeth would not give a farthing.
1586. The natural consequences followed. By February half the men were dead, and the half that remained alive were in a state of suppressed mutiny. No good officer would accept a command in the army on such terms, and the companies fell into the hands of unscrupulous swindlers who sent their men out to plunder and did not omit to take their own share, rejoicing over every soldier who died or deserted for the money that would pass into their pockets when the long-deferred pay-day should come. There have been many sovereigns and many ministers in England who have neglected and betrayed their soldiers, but none more wantonly, wilfully, and scandalously than Elizabeth.

Nevertheless, as the spring of 1586 approached, it behoved Leicester to open a campaign of some kind. Parma was advancing along the line of the Maas, evidently bent on taking every fortified town on the river, and it was necessary if possible to check him. The Generals, however, were ill-matched; Parma easily brushed aside Leicester's feeble opposition, and having secured the line of the Maas turned next to that of the Rhine. Meanwhile a large reinforcement of men, unarmed and untrained, had been sent from England ; and Leicester concentrated his forces, summoning all the garrisons of the cautionary towns to join him at Arnheim. Philip Sidney came from his government at Flushing, Lord Willoughby came from Bergen-op-zoom, John Norris and his brother Henry hurried up likewise, the veteran Roger Williams joined them, and lastly, in the retinue of Lord Willoughby, came a young man of greater promise than any, named Francis Vere. The plan of operations was soon determined ; since Parma could not be checked on the Rhine, he must be called away from it by a diversion in the north on the Yssel, where the Spaniards still held the towns of Doesburg and Zutphen. 1586. July.

All turned out as had been expected. Doesburg was easily captured, and Parma no sooner heard that Leicester was before Zutphen than he abandoned his operations on the Rhine and marched north to relieve it. Halting on the evening of the 21st of September at some distance from the town, he sent forward a convoy of supplies towards it, protected by an escort of three thousand men under the command of the Marquis of Pescayra.[1] The convoy was to start at midnight, and it was reckoned that it would be within a mile and a half of Zutphen by daybreak. Pescayra was then to halt at an appointed place, send a messenger into the town and concert arrangements with the Governor for a sortie to facilitate the entrance of the convoy.

Intelligence of Parma's design was duly brought to Leicester, who, calling John Norris, ordered him to take two hundred horse and three hundred foot and lie with them in ambuscade by the road by which the convoy was expected to arrive. Norris readily picked out two hundred horse, ordered Sir William Stanley to follow them with three hundred pikemen, and before dawn of the 22nd had successfully taken up the position

[1] The grandson of the victor of Pavia.

1586, Sept. 22 / Oct. 2. assigned to him. No force appears to have been detailed by Leicester to support the ambushed party, and no scouts to have been sent forward by Norris to give warning of the enemy's approach. The morning broke with dense impenetrable fog, amid which the English could hear a distant sound of rumbling waggons and tramping men. Presently Norris was joined by all the adventurous gentlemen—Lord Essex, Lord Audley, Lord North, and many others—who were to be found in Leicester's camp: they had not been able to resist the temptation of an action, and came galloping up with their retinue at their heels to see the sport. The sounds of the approaching convoy became more distinct, but nothing could be seen till the fog suddenly rolled away and revealed straight before them the three thousand Spaniards, horse and foot, marching by their waggons in beautiful order.

The English gentlemen threw all discipline to the winds at the sight: they never dreamed of anything but a direct attack, and one and all went at once, each in his own way, to work. Young Lord Essex called on his squadron of troopers to follow him, and couching his lance flew straight upon the enemy's cavalry, overthrew the foremost man and horse, flung away his broken lance for his curtel-axe, and with his handful of men hard after him burst into a heavy Spanish column and shivered it to pieces. The routed Spaniards fled in disorder to the shelter of their musketeers, with Essex still spurring at their heels; and then Spanish discipline told. The musketeers fired a volley which brought down many of the English horses and compelled the rest to wheel about. Then the action became simply a series of furious personal combats. Sir Philip Sidney's horse was killed under him at the first charge, but he mounted another and plunged into the hottest of the fight. Lord North, unable owing to a recent wound to draw on more than one boot, dashed in half-booted as he was and fought as busily as any. Sir William Russell swung his curtel-axe so murderously that the

Spaniards vowed he was a devil and no man. Lord Willoughby was so beset with enemies that only great good fortune and immense personal strength served to pluck him out. Sir William Stanley's horse was struck by seven bullets but found strength to carry him safe out of action. And meanwhile the drivers of the waggons had fled, and English and Spanish soldiers were tugging the heads of the teams this way and that with oaths and yells and curses; but still Spanish discipline told, and still the convoy moved slowly forward. Again and again the Spanish horsemen shrank before the English cavaliers, but the firm ranks of the musketeers always gave them shelter, and, charge as the English might, the waggons crept on and on till they fairly entered the town. Nothing was gained by the action. The attack, if supported, might have been fatal to Pescayra, but no support could be looked for from Leicester, and there was so little intelligence in the onslaught that no one seems to have attempted even to hamstring the waggon-horses. Zutphen therefore remains no more than one of the maddest of the many mad exploits performed by English officers of cavalry, and is remembered chiefly through the death of one of the noblest of them. Before the action, Philip Sidney had given the thigh-pieces of his armour to the Lord Marshal, Sir William Pelham; at its close he was seen riding painfully back, with the unprotected thigh shattered by a musket bullet. He lingered in agony for some days and then died. His body was brought back to England to be followed to St. Paul's Churchyard by the London train-bands and laid to rest, as befitted a good and gallant soldier, under the smoke of their volleys.[1]

1586, Sept. 22 / Oct. 2.

Yet another scene of desperate valour was witnessed at Zutphen before the campaign came to an end. One principal protection of the town was an external

[1] Stow says that they fired two volleys only, which I hope is incorrect. The passage, however, shows that the reason for the three volleys was already unknown to many.

1586. sconce,[1] which on a former occasion had resisted the troops of the States for a whole year, and was now carried by the English by assault. The breach was barely practicable, the footing on the treacherous sandy soil being so uncertain that the storming party could hardly mount it. Their leader, Edward Stanley, however, was not to be turned back. Dashing alone into the breach he caught the head of a Spanish soldier's pike that was thrust out against him and tried to wrench the weapon from his grasp. Both men struggled hard for a time, while a dozen pikes were broken against Stanley's cuirass and a score of bullets whistled about his ears. At last Stanley, without quitting his hold, allowed the Spaniard to raise the pike, used the purchase so gained to help him up the wall, scrambled over the parapet and leaped down alone into the press of the enemy with his sword. His men, redoubling their efforts, hoisted each other up the breach after him and the sconce was won. Stanley, marvellous to say, escaped unhurt, and received not only warm commendation in Leicester's despatches, but a pension for life from Leicester's own pocket, for the most daring act that is recorded of the whole of that long war.

1587. The plot of the Spanish Armada now began to thicken, and the scene must be shifted for a moment to England. In the Low Countries Parma was looking about for a port of embarkation from which to ship his men across the North Sea. He fixed upon Sluys, and in spite of a desperate resistance from a handful of gallant Englishmen, led by Roger Williams, he succeeded in capturing it after a siege of three months. At the end of 1587 Leicester resigned his command and returned to England; and in the following year all the best officers, and many of the English companies, were gathered together in the camp at Tilbury. Leicester was in chief command, with John

[1] That is to say a fort or intrenchment. German *schanze*. It seems a pity that we should have allowed so useful a term to become obsolete.

Norris for his second, and Roger Williams among 1588. others for assistant, but these officers were not on very friendly terms with each other; and, indeed, the less said of Tilbury Camp as a whole the better. Contemporary writers indeed aver that it was a pleasant sight to see the soldiers march in from the various shires, "with cheerful countenances, courageous words and gestures, leaping and dancing";[1] but such a display was a better indication of loyalty than of discipline, and sadly different from the pace, full of gravity and state, which had been enjoined by the best authorities. There was, moreover, great disorder and deformity of apparel; most of the men wore their armour very uncomely, and the whole army refused point-blank to use the headpieces issued from the Tower. Ammunition again was short, provisions were scanty, organisation was extremely defective, and the general confusion incredible. Four thousand men who had marched, pursuant to orders, twenty miles into Tilbury, found that they must go that distance from the camp again before they could find a loaf of bread or a barrel of beer. A thousand Londoners who were likewise in the march were ordered to halt unless they could bring their own provisions with them. Leicester might safely remark that "great dilatory wants are found upon all sudden hurly-burlies,"[2] but there was no excuse for such chaos after the incessant warnings of the past thirty years. Elizabeth must bear the chief share of the blame. The woman who in her imbecile parsimony starved the fleet that went forth to fight the Armada could not be expected to show better feeling towards the army. It was no thanks to the Queen that the Spanish invasion was repelled.

I shall not follow the veterans John Norris and Lord Willoughby on their expeditions to Corunna 1589. and Brittany in the following year. Far more important to us is the rise of a great leader, and the opening of a new era in the war of the Low Countries.

[1] Stow. [2] *Cal. S. P.*, Dom. (1588), p. 513.

1590. On Leicester's resignation of the chief command, there was appointed to succeed him a man whose name must ever be venerated in the British Army, Prince Maurice of Nassau,[1] second son of William the Silent. Though but twenty years of age when selected as Governor and Commander-in-Chief of the United Provinces, he had already made up his mind that if the War of Independence were to end in victory it must be fought not, as heretofore, with a mob of irregular levies, but with a trained, disciplined, and organised army. His own natural bent lay chiefly towards mathematics, which he cultivated as a means to the mastery of military engineering, and eventually reduced to practice by so sedulous a use of the spade in all military operations as to provoke many a sneer from soldiers of a more primitive type. But Maurice knew his own mind, and was not to be deterred by sneers. His principal assistant was his cousin, Louis William, Stadtholder of Friesland, an industrious student of classical antiquity with the rare faculty of adapting old systems to modern requirements. To his diligence was due the instruction of the army in drill and discipline, and to his influence must be ascribed Maurice's admiration for the *Tactics of Ælian*.[2] His new and elaborate manœuvres also elicited the scorn of the old school of officers,[3] but he too was not easily discouraged; and the two cousins worked hand-in-hand, the one at the broader principles, the other at the hardly less important details, of their profession, until they raised up an army which supplanted the Spanish as the model for Europe. Not the least

[1] Born 14th November 1567.

[2] See the English translation of the *Tactics*, by Captain John Bingham, 1619.

[3] Hear, for instance, Tavannes, whom his writings prove to have been in many respects an excellent soldier: "Cette grande invention d'exercice pratiquée en Flandre avec leurs demi-tours à gauche et à droit—les anciens qui n'en usaient pas (!) ne laissaient de combattre aussi bien ou mieux que maintenant" (*Memoires*). Tavannes began to write in 1599-1600, and died in 1629.

weighty of Maurice's reforms was the regular payment 1590. of the men, and the stern repression of fraudulent practices among the officers. In a word, he appreciated the value of sound administration no less that of pure military skill and training in the conduct of a war.

The tactical organisation of the new army was not so perfect as, with the Spanish model before us, we might with reason have expected. The tactical unit of infantry was the company, and the regiment still consisted of an uncertain number of companies temporarily united under the command of a colonel. The composition of the companies again was uncertain. The normal strength was one hundred and thirteen men, which was later reduced to eighty, but colonels had double companies—some even double regiments—and there appears to have been no very great exactitude, probably because men could only be persuaded to serve under the captain of their choice. The officers of a company were of course captain, lieutenant, and ensign; the non-commissioned officers included two sergeants and three corporals, as well as a "gentleman of the arms," who was responsible for the condition of the weapons. Lastly, there were two drummers, who, it should be noted, like the trumpeters in the cavalry, were not the mere signal-makers that they now are, but the men regularly employed in all communications with the enemy, and as such expected to possess not only discretion but some skill in languages. They received far higher pay than the common soldier, and if they did a tithe of that which was expected of them they were worth every penny of it.

Every company was divided into three corporalships, each of which was the peculiar care of one of the three corporals and of one of the three officers. In equipment there were at first three descriptions of arms—halberds, pikes, and muskets—of which however the halberds soon disappeared, leaving pikes and shot in equal numbers, but with an ever-growing tendency towards preponderance of shot. The normal formation of a company was

1590. in ten ranks; and the men were never less than three feet apart from each other, such open order being essential to the execution of the prescribed evolutions. To increase the front, the ranks were doubled by moving the even ranks into the intervals of the odd; to diminish the front, the files were doubled by the converse process.[1] To take ground to flank or rear every man turned to right or left or about on his own ground, and it is worth remarking that the best men were always stationed in the front rank and the next best in the tenth, and that while the captain was posted in front of his company, the lieutenant, except in a charge, remained always in the rear.

The musketeers were usually drawn up in two divisions, one on either flank of the pikes; and the problem that eternally confronted the captain was how to handle the two elements in effective combination and yet contrive never to confuse them. In action the musketeers generally moved in advance of the pikes, firing by ranks in succession, according to Pescayra's method, and filing to the rear to reload. Sometimes they were extended across the front of the pikes, but more often they kept their place on the flanks. Meanwhile the pikemen, heavily weighted by helmet, corselet, and tassets (thigh-pieces), moved stolidly on: as they drew nearer the enemy the musketeers fell back until they were first aligned with them, and then abreast of the fifth or sixth rank. If neither side gave way, matters came to push of pike and a general charge, wherein the musketeers ceased firing and fell in with the butt, a method of fighting which was peculiarly favoured by the English. To resist cavalry the musketeers fled for shelter under the pikes, generally in considerable disorder, and the

[1] Perhaps the following explanation will make this clearer:—Where an English officer would now give the word "Form fours" (to convert two ranks into four), the Dutch officer would have given, "To the right hand double your files." Where the Englishman would give the word "Front" (to reconvert four ranks into two), the Dutchman would have said, "To the left hand double your ranks."

outer ranks of pikemen, lunging forward, stayed the butts of their pikes against the hollow of the left foot. 1590.

The cavalry was divided at first into lancers and carbineers, the former being fully covered with armour to the knee; but the lance, in deference to the fashion of the Reiters, was soon[1] discarded for the pistol. The carbineers carried a carbine[2] with a wheel-lock, and were trained to shoot from the saddle, the ranks firing in succession according to Pescayra's system. The tactical unit was the troop or cornet, which, after many changes, was finally fixed at a strength of one hundred and twenty men, and divided, like the company, into three corporalships. Captain, lieutenant and cornet, three corporals, a trumpeter, a farrier, and a quartermaster made up the higher ranks of the troop, no such title as a sergeant appearing in the cavalry. Of artillery I shall say nothing, since the Dutch organisation was in this respect peculiar, and could not serve like that of the infantry and cavalry as a model for the English.

Concurrently with the rise of Maurice as Commander-in-Chief must be noted that of a new English General, whose name is bound up for ever with the actions of his countrymen in the Low Countries. Francis Vere came of the old fighting stock of the Earls of Oxford. The seventh Earl had fought with the Black Prince at Crecy and Poitiers, the twelfth with King Harry at Agincourt, and succeeding holders of the title had distinguished themselves on the Lancastrian side in the Wars of the Roses. Francis, grandson of the fifteenth Earl, was born about 1560, came to Holland with Leicester in 1585, and after brilliant service at the defence of Sluys and elsewhere rose to be sergeant-major of infantry, a sure proof that he was not only a gallant man but an adept in his profession. Finally, in August 1589 he was appointed sergeant-major-general of the Queen's forces

[1] 1599.
[2] Its bore was of thirty bullets to the pound.

1589. in the Low Countries, where he was joined by two gallant brothers, Horace and Robert, who worthily upheld the honour of the name.

His task, as that of every officer who had to do with such a woman as Elizabeth, was at first no easy one. His force being very small required constant reinforcement, and was accordingly strengthened by five hundred of the "very scum of the world," such being the description of recruit that Elizabeth preferred to supply. He took care, however, to procure for himself better
1591. material, and at the opening of 1591 had no fewer than eight thousand men under his command. But as fast as he trained them into soldiers Elizabeth required their services for her own purposes, and frittered them away in petty meaningless operations in France, filling their place with some more of the very scum of the world, which could be swept out of the gaols and taverns at a moment's notice. The system was in fact that of drafting, in its most vicious form. Vere for a time bore it in silence, but at last he protested, and like all of Elizabeth's best men was soundly abused for his pains. Still the Queen knew his value well enough to withdraw not only his troops but himself from the expedi-
1595. tion to Cadiz, and the disastrous island-voyage to the
1596. Azores.

A far more serious difficulty was the corruption of departments and contractors at home and the vicious system of paying the men. The wages of a private at eightpence a day were reckoned for the year at £12 : 13 : 4, of which £4 : 2 : 6 was deducted for two suits of summer and winter clothing,[1] £6 : 18 : 6 paid in imprests at the rate of 2s. 8d. a week, and the balance, £1 : 2 : 6, alone made over in money. Even in theory the allowance does not sound liberal, but in practice it was ruinous. The men drew their pay and clothing from their captains, and the captains received the money in uncertain instalments, the balance due to

[1] These stoppages were known even then by the name of "off-reckonings."

them being made good at the close of every six months. This in itself was wasteful, since it enabled the captain to put in his own pocket the wages of soldiers who had died or had been discharged in the interval. But apart from this the captains frequently withheld the clothing altogether, or served out material of uncertain quality, charging the men treble the just price for the same; or again they would make their own contract for victualling the men, of course to their own profit, in lieu of paying to them the weekly 2s. 8d. which was due to them for subsistence. How widely the practice may have obtained among officers it is difficult to say, but the system was presently altered to the advantage alike of the State and the soldier by the officials in London. The officers also had their complaints, not a whit less sweeping, against those officials, and they preferred them in uncompromising terms. Such representations were not likely to meet with encouragement. Elizabeth was not friendly to soldiers, and hated to be troubled with obligations towards men who had faithfully served her. An Act had been passed in 1593 throwing the relief of crippled or destitute soldiers on their parishes, and she could not see what more they could want. Bloody Mary had shown them compassion; not so would Good Queen Bess; she would not be pestered with the sight of the "miserable creatures." As to the complaints of officers, she had heard enough of their ways, and would take the word of the Treasurer of the Forces against theirs. Still Vere and his captains persisted, and at last the shameful truth was revealed that the Treasurer himself was the culprit, and had for years been cheating alike his Queen, her officers, and her men.

It is easy therefore to understand the relief with which the English commanders in the Low Countries must have welcomed a new treaty made in 1598, whereby Elizabeth was quitted of her engagement to furnish the United Provinces with auxiliary troops, and all English soldiers were ordered henceforth to take their pay from the States and their orders from the Dutch Generals. The troops in the Low Countries were now comparatively freed from the caprices of the Queen and could work in harmony with their masters. From this point therefore the English fairly enter the school of the new art of war.

CHAPTER V

So far I have abstained from any attempt to describe 1600.
the military operations of the States, or even the brilliant little enterprises of Vere himself, since his assumption of the command: but at this point, when we enter upon the palmy days of the English in Holland, it is worth while to be more precise. So far Maurice had occupied himself principally with the task of recovering the towns occupied by the Spaniards within the seven provinces;[1] the States-General in the year 1600 resolved upon the bold step of carrying the war into the enemy's country. Ostend, which was held by the Queen of England, was to be the base of operations, and the design was to land a force on the Flemish coast and besiege first Nieuport, to the west of Ostend, and afterwards Dunkirk. Maurice and Vere both thought the enterprise hazardous in the extreme, but they were overruled by the civilians. A force of twelve thousand infantry, sixteen hundred cavalry and ten guns was assembled at Flushing, and a fleet was collected to transport it to its destination. The army was organised in the three familiar divisions, vanguard, battle, and rearguard, of which the rearguard under Sir Francis Vere consisted of sixteen hundred English veterans, two thousand five hundred Frisians, two hundred and fifty of Prince Maurice's body-guard, and ten cornets of horse, making in all four thousand five hundred men. With Vere were men whose names through themselves

[1] Holland, Zealand, Utrecht, Gelderland, Overyssel, Frieland, Groningen.

1600. or through their successors were to become famous—Sir Edward Cecil, Sir Charles Fairfax, Captain Holles, and others. In another division of the army was a regiment of Scots under Sir William Edmunds, which had recently been recruited to the high strength of one hundred and fifty men to each company. English and Scots already loved to fight side by side.

The force embarked on the 21st of June, but being delayed by calms landed short of Nieuport, marched overland, capturing the fort of Oudenburg on the way, and on the 1st of July was before Nieuport. The Spanish commander, the Archduke Albert, no sooner heard what was going forward than he at once concentrated his army at Ghent for an immediate advance; and Maurice, who was busily preparing for the siege of Nieuport, was surprised by the sudden intelligence that his little garrison at Oudenburg had been overwhelmed, and that the Spanish forces were in full march for his camp. The situation in which he found himself was now very critical. Expecting no such movement Maurice had divided his forces round Nieuport into two parts, which were cut off from each other by the haven that runs through the town. Though dry at low water this haven was unfordable at high tide, and the bridge which was constructing across it was still unfinished. Worst of all, it was the weakest division of the force, three thousand five hundred men under Ernest of Nassau, that stood on the side of the haven nearest to the enemy; and a battle within twenty-four hours was inevitable.

The question therefore arose whether the action should be fought in dispute of the enemy's passage over a stream called the Yser leet which barred the line of his advance, or on the sandy dunes by the sea-shore, where the Spaniards would certainly seek it if the passage were successfully accomplished. Vere was for the former course, and Maurice, thinking the advice good, ordered Count Ernest's division to march straight for the bridge on the Yser leet, saying that he would shortly follow with the rest of the army. Vere protested in vain that this was a perversion of his counsel: either the whole army must march with Count Ernest, or no part of it must move at all; for to send forward a weak division in the hope of delaying the Spanish advance was simply to court defeat. Maurice, however, stuck to his opinion, and at midnight Count Ernest marched off with his division unsupported to the bridge. He arrived too late, for the Spaniards had already secured the passage, and he therefore took up the best position that he could find, behind a dyke, to defend himself as well as he could. The first shot had hardly been fired when his men began to run. It was such a panic as has rarely been matched in the annals of war. Cavalry and infantry, Dutchmen and Scots, threw down their arms, took to their heels and fled like swine possessed of devils into the sea. The Scotch officers of Sir William Edmunds' regiment strove to rally the fugitives, but in vain: they were cut down one after another, and the men that escaped death by lead or steel were swallowed up literally in the waves. Two thousand five hundred men, including a thousand massacred at Oudenburg, were thus lost, and Maurice had now to face his enemy with a weakened army and with his retreat barred by the haven behind him. Defeat would mean not only annihilation but the undoing of all the work of the rebellion. With superb courage he ordered his fleet of transports to sea, and staked all on the hazard of the coming battle.

1600.

July 2.

Meanwhile Vere, whose division had this day the place of the vanguard, had moved at daybreak down to the bank of the haven and was waiting for the ebb-tide to cross it, when the news came that the Archduke's army was in full march along the sea-shore. As soon as the tide permitted he forded the haven with all haste, not allowing the men to strip, for, as he said, by nightfall they would have dry clothes or want none. Presently he came in sight of the enemy, ten thousand foot, sixteen hundred horse and six guns, moving along the

1600, July 2. flat sands of the sea-shore. The space between the sea and the enclosed country was broken up into three descriptions of ground running parallel one to another; next the sea was the narrow plain of the strand between high- and low-water mark, next the strand were the broken hillocks of the sand-dunes, and between the dunes and the enclosed land ran a margin of unbroken green, called by Vere the Greenway. Vere lost no time in taking up a position at the narrowest point that he could find, distributing his division skilfully among the hillocks to repel an advance through the dunes, and posting two guns, by Maurice's order, to command the Greenway. To his right rear stood the battle or second division, one thousand strong, and in rear of the battle the third division of rather more than two thousand men. The army was thus formed in echelon of three lines with the right refused, its left resting in the sea, its right on the enclosed land.

Weak in cavalry, the Spaniards halted till the rising tide had covered all but thirty yards of the strand, and then moved the whole of their horse to the Greenway and of their infantry into the dunes. Maurice likewise withdrew his cavalry from the shore and massed it in columns on the Greenway, leaving but two troops, both of them English, still standing on the beach. For two whole hours of a beautiful summer's afternoon the two armies waited each for the other to advance, and at last, at half-past two, the Spaniards began to move. Vere, taking every possible advantage of the sandhills to protect and conceal his men, had thrust forward small parties to contest every inch of ground; and it was against the foremost of these, two and fifty English and fifty Frisians, that the first attack of five hundred of the flower of the Spanish infantry was directed. Meanwhile the Spanish cavalry moved forward along the Greenway. This cavalry, disordered by the fire of Vere's two guns and galled in flank by a detachment of his musketeers, soon gave way before the cavalry of the States; but the struggle of the infantry in the van was very severe.

The first attack of the Spanish vanguard was repulsed, 1600,
but being quickly reinforced it moved forward again July 2.
and the fight then became desperate. For a time the battle seems to have resolved itself into a furious contest for the possession of a single sandhill, round which, as round the two-gun-battery at Inkermann, both sides fought madly hand to hand, each alternately repelling and repelled, till at last this "bloody morsel," as Vere called it, was finally carried by the English.

The Archduke without delay brought up his centre in line with his vanguard, and essayed to force his way through Vere's right. The columns were met by a murderous fire from a party of musketeers which had been posted by Vere to check any such movement, and were driven back; and then the whole strength of the Spanish attack was concentrated once more upon Vere's main position. Husbanding his strength to the utmost, Vere gradually drew the whole of his English into action and fought on. So far, owing to the skill of his dispositions, little more than half of his force had been engaged, but seeing that they were likely to be overwhelmed by numbers, he sent messengers to summon his reserve of two thousand Frisian infantry, and to beg Maurice to help him with cavalry from his right. Messenger after messenger was despatched without result. Vere went down among his few remaining men, and the little force, cheered by his presence, fought gallantly on and still held the enemy at bay. He was struck by a musket ball in the thigh and by a second in the leg, but he concealed the wounds and held his men together. Yet the expected reinforcements came not, and the English were slowly forced back, still in good order and still showing their teeth, from the dunes on to the beach, the Spaniards following after them, but afraid to press the pursuit. As the English retired, Vere's horse was shot under him and fell, pinning him helpless to the ground. Three of his officers ran up and freed him; and mounted on the

1600, July 2. crupper behind one of them, he continued calmly to direct the retreat.

Arrived on the sands he found his reserve of Frisians still halted in their original position, having never received orders to move, and with them the two troops of English horse. A charge of the cavalry, supported by two hundred infantry under Horace Vere, soon swept the Spaniards back into the dunes, and then at last Sir Francis made himself over to the surgeon, while Maurice came forward, cool and unmoved, to save the day. The Spaniards now massed two thousand infantry together for a further advance, while the English officers, weary with fighting and parched with heat and sand, exerted themselves to rally their men. The English were quickly reformed, so quickly that the Spaniards, who had sent forward a party to disperse them, promptly withdrew it at the sight of Horace Vere returning with his two hundred men from the beach. Maurice saw the movement and exclaimed joyfully, "Voyez les Anglais qui tournent à la charge." He at once ordered up the cavalry from the right under Sir Edward Cecil; and meanwhile Horace Vere and his brother officers hastily decided that their only chance was at once to charge the two thousand Spaniards with their handful of men. They rushed desperately down upon them; the Spaniards, worn out by a long march and hard fighting, gave way, and Maurice catching the supreme moment launched Cecil's troopers into the thick of them. A second charge disposed of the Spanish horse; Maurice ordered a general advance, and the battle was won. Three thousand Spaniards were killed outright; six hundred more with all their guns and one hundred and twenty colours were captured. On the side of the States the loss fell almost wholly on the English. Of their captains eight were killed, and but two came out of the field unhurt; of the sixteen hundred men eight hundred were killed and wounded. They with the Frisians had borne the brunt of the action, and Maurice gave them credit for it. So ended the fight of

Nieuport,[1] the dying struggle of the once famous 1600.
Spanish soldier, and the first great day of the new English infantry.

Next year the Archduke Albert sought revenge for 1601.
his defeat by the investment of the one stronghold of the United Provinces in Flanders, the little fortified fishing-town of Ostend. The garrison had made itself so obnoxious to the surrounding country that the States of Flanders petitioned the Archduke to stamp out the pestilent little fortress once for all; and hence it was that in the following years the principal operations grouped themselves around the siege. The Archduke's army consisted of twenty thousand men with fifty siege-guns; the garrison of barely six thousand men, half English and half Dutch, of which fifteen hundred English, all dressed in red cassocks, were a reinforcement just imported from across the sea. Francis Vere was in supreme July 9.
command, and his brother Horace commanded a regiment under him.

I shall not weary the reader with details of Vere's skill and resource in improving the defences of the town, or of the incessant encounters that took place during the first weeks of the siege. The Spanish fire was so hot and the losses of the besieged so heavy that the garrison was fairly worn out with the work. Vere was dangerously wounded in the head within the first three weeks and compelled to throw up the command until restored to health, and at the close of the first month hardly a red cassock of the fifteen hundred was to be seen, every man being wounded or dead. Nevertheless, the sea being always open to the besieged, fresh men and supplies could always be poured into the town to repair the waste. Two thousand English, for a wonder well equipped and apparelled, were the first to arrive, and were followed by a contingent of French and Scots.

[1] I have followed the narrative of Sir Clements Markham (*The Fighting Veres*) in preference to that of Motley in the description of the battle, being satisfied after careful consultation of the authorities that his account is the more accurate.

1601. They too went down with terrible rapidity. The town was but five hundred yards across, and the Spanish batteries were built within musket-shot of the defences. Hardly a house was left standing, and the garrison was compelled to burrow underground as the only refuge from the incessant rain of missiles. The winter set in with exceptional rigour, the defenders dwindled to a bare nine hundred effective men, and at Christmas Vere, in the face of foul winds and failing supplies, was compelled to resort to a feigned parley to gain time. By a fortunate change of wind four hundred men were able to enter the harbour and recruit the exhausted garrison.

1602. So far the Spaniards had fired one hundred and sixty-three thousand cannon-shot into the town, and they now decided on a general assault. On the 7th of January Vere received intelligence of the coming attack, and, though his force was far too weak to defend the full extent of his works, made every preparation to repel it. Firkins of ashes, barrels bristling with tenterhooks, stones, hoops, brickbats, clubs, what not, were stored on the ramparts, and at high tide the water was dammed up into the ditch. At nightfall the Spanish columns fell on the devoted town at all points. They were met by a shower of every description of missile; flaming hoops were cast round their necks, ashes flung in their eyes, brickbats hurled in their faces; and storm as they might they could gain no footing. Thrice they returned to the assault, and thrice they were beaten back, and at last they retired, sullen and furious, for the tide was rising, and on one side they could advance to the town only by a passage which was not fordable at high water. Vere opened the sluices of the ditch as they retreated, and the rush of water swept scores if not hundreds of them out to sea. The Spanish loss was two thousand men; that of the garrison did not exceed one hundred and thirty.

I shall not further follow this memorable siege. Vere and his brother Horace left the town worn almost to death in March 1602, but still the defence was maintained. Reinforcements from England came in by hundreds and by thousands. Rogues, vagabonds, idle, dissolute, and masterless persons were impressed impartially together with men of honesty and reputation, clapped into red or blue cassocks and shipped across to Ostend. Volunteers of noble and of humble birth, some in search of instruction, some with a thirst for excitement, hurried likewise to the siege, and Ostend became one of the sights of Europe. Governor after governor, gallant Dutchmen all of them, came to take command. Three of them were killed outright, but still the defence continued, until at last on the 13th of September 1604 the heap of ruins which marked the site of Ostend was surrendered into the generous hands of Spinola. The siege had lasted three years and ten weeks, and had cost the lives of one hundred and twenty thousand men. 1603-1604.

Before the town fell the campaigns of Francis Vere were ended. In 1602 he accompanied Maurice to the siege of Grave, where he was once more dangerously wounded, and in the summer of 1604 he retired from the service of the States, from whom he deservedly received a pension for his life. In the very same year King James the First made a treaty with the Archdukes of the Spanish Netherlands, which left the Dutch patriots henceforth to fight their battles by themselves; but nations like the English and Scotch are not bound by the decisions of such a creature as James. The British troops not only remained in the service of the State but grew and multiplied exceedingly, and Francis Vere, who had made their service honourable and given their efforts distinction, could feel that his work was well done. A few short years of rest closed a life that was shortened by hardship and wounds; and on the 28th of August, 1609, within four months of the signing of the truce which gave breathing time to the exhausted combatants of the Dutch war, the old soldier died peacefully in his house in London. His tomb in Westminster Abbey is admired by thousands who know not one of his actions, but surely it is no derogation to art to remember that the recumbent marble effigy, and the four noble figures that kneel around it are those not of conventional heroes, but of honest English fighting men, typical of many thousands who perished in the cause of Dutch freedom and lie buried and forgotten in the blood-stained soil of the Netherlands.

The twelve years' truce gave the English regiments a rest which, though not wholly unbroken, left some of the more daring spirits free for other adventure. The cause of the Elector Frederick, a prince less interesting to the English as the Winter King than as the husband
1619 of their favourite Princess Elizabeth, called Horace Vere and many another gallant gentleman with four thousand good soldiers into the Palatinate, where however their bravery could not avail to save them from inevitable failure. King James of course had no part in the venture; so far from moving a finger in aid of the Protestant cause in Germany, he even conspired secretly with Spain for a partition of the Netherlands, which was to be effected by the English troops in the Dutch service, the very men who had made the cause of the United Provinces their own and had carried it through the perils of Nieuport and Ostend. It is hardly surprising that such a man should, not indeed without searching of heart but without stirring a hand, have suffered Germany to drift into the Thirty Years War.

1621. The lapse of the twelve years' truce found a large contingent of English under the command of Sir Edward Cecil attached to the army of Prince Maurice; and
1624. three years later the final breach of England with Spain increased its number from six to twelve thousand, and in 1625 even to seventeen thousand men. It would be tedious to follow them through the operations of the ensuing campaigns; it must suffice to call attention to the rise of men who were to become famous in later days and thus bridge over by a few stepping-stones the connection of the British army with the old Dutch schools of war. The first names are those of Philip Skippon, whom we find wounded before Breda in 1625,

and of Captain John Cromwell, a kinsman of the great 1625-1637. Oliver, who was also wounded in the same action. Coming next to the siege of Bois le Duc in 1629 we find the list far longer—Lord Doncaster, Lord Fielding, who trailed a pike in Cecil's regiment, Lord Craven, a Luttrell, a Bridgeman, a Basset, a Throgmorton, a Fleetwood, a Lambert, a second Cromwell, Thomas Fairfax, Philip Skippon, Jacob Astley, Thomas Culpeper, the veterans Balfour and Sandilands from north of the Tweed, and many more. Lastly, at the siege of Breda in 1637 we see Prince Rupert and Prince Maurice, sons of the Winter King, as forward in the trenches as any needy cadet could be, working side by side with Philip Skippon, Lord Warwick, and George Goring. Of these Skippon and Goring divided the honours of the siege. Skippon at a post of extreme danger drove off two hundred Spaniards at push of pike with thirty English; he was struck by five bullets on helmet and corselet and at last shot through the neck, but he merely sat down for ten minutes and returned to his work until recalled by the Prince of Orange. Goring in the extreme advanced sap paid extra wages from his own pocket to any who would work with him, and remained there while two-and-twenty men were shot down round him, until at last he was compelled to retire by a bullet in the ankle. Meanwhile fresh volunteers kept pouring in—Herbert, son of Lord Herbert of Cherbury, Sir Faithful Fortescue of the King's cavalry in Ireland, Sir Charles Slingsby, with many more, and lastly Captain George Monk of Potheridge in Devon, one day to be the first colonel of the Coldstream Guards, and even now distinguished by peculiar bravery.

There they were, brave English gentlemen, all wearing the scarf of orange aud blue, fighting side by side with the pupils of Francis Vere, learning their work for the days when they should be divided into Cavaliers and Roundheads and flying at each other's throats. It was a merry life enough, though with plenty of grim earnest. Before each relief marched off for the night

1625-1637. to the trenches it drew off in *parado*[1] to the quarters of the colonel in command, heard prayers, sang a psalm and so went to its work; but though there was a preacher to every regiment and a sermon in the colonel's tent, there was no compulsion to attend, and there were few listeners except a handful of well-disposed persons.[2] It was to be a very different matter with some of them ten years later, but that they could not foresee; and in truth we find among the gentlemen volunteers some very familiar types. One of them arrived with eighteen suits of clothes, got drunk immediately on landing and remained drunk, hiccuping "thy pot or mine," for the rest of his stay. It is not difficult to understand why this gentleman was sent to the wars. Another, Ensign Duncombe, came for a different reason; he had fallen in love with a girl, who though worthy of him was not approved of by his parents. So he too was sent out to forget her, as such foolish boys must be; and he became a great favourite and did well. But unluckily he could not forget; so one day he sat down and wrote two letters, one full of passion to his beloved, and another full of duty to his father, and having done so, addressed the passionate epistle, as is the way of such poor blundering boys, to his father and the dutiful one to the lady. And so it came about that some weeks later the regiment was horrified to hear that young Duncombe had shot himself; and there was an ensign the less in the Low Countries and a broken heart the more in England, sad silence at the officers' table and much morbid discussion of the incident in the ranks. It is such trifles as these that recall to us that these soldiers of old times were really living creatures of flesh and blood.

The men too were learning their business with all the elaborate exercise of musket and of pike, and familiarising themselves with the innumerable words of

[1] Hexham. This is the first instance that I have encountered of the word parade, which is evidently of Spanish origin.

[2] Hexham.

command and with the refinements in the execution of the same. The pikeman learned by interminable directions to handle his weapon with the better grace, and listened to such cautions as the following. "Now at the word *Order your pikes*, you place the butt end of your pike by the outside of your right foot, your right hand holding it even with your eye and your thumb right up; then your left arm being set akimbo by your side you shall stand with a full body in a comely posture." The musketeer too grasped that the minutest motion must be executed by word of command. Stray grains of powder spilled around the pan disappeared at the word *Blow off your loose corns*, sometimes by a puff or two sometimes by a "sudden strong blast," but always in accordance with regulation. At the word *Give fire* again he learned the supreme importance of "gently pressing the trigger without starting or winking," and soon revived the old English reputation, first won by the archers, for fine marksmanship. An eye-witness records with delight that after each shot they would lean on their rests and look for the result as coolly as though they had been so many fowlers watching for the fall of their bird. Lastly, they learned a new feat, untaught in any drill-book, with which this section may fitly be closed. Pikemen and musketeers were drawn up in line, every pike with a wisp of straw at its head, and every musket loaded with powder only; and at the word every wisp was kindled and every musket fired in rapid succession. The volley met with a stop at first, to use the words of our authority, as was perhaps natural at a first attempt, but eventually it ran well; and thus was fired before Bois le Duc in the year 1629 the first *feu de joie* that is recorded of the British Army.[1] 1625-1637.

AUTHORITIES.—The chief sources of information for the actions of the British in the Low Countries are the histories of Meteren,

[1] The capture of Wesel was the occasion of rejoicing; and the details of the description leads me to infer that the *feu de joie* was a novelty.

Grimeston and Commelyn; Roger Williams's *Actions of the Low Countries;* Hexham; Vere's *Commentaries;* the *Leicester Correspondence* (Camden Society); the *Calendars of State Papers, Domestic and Foreign Series;* and the *Holland Papers* in the Record Office. These last, consisting of several scores of portfolios of manuscript documents, I cannot pretend to have studied exhaustively. Sir Clements Markham's *Fighting Veres* and Mr. Dalton's *Life of Lord Wimbledon* are the best modern books on the subject, and I wish to acknowledge to the full my obligation to them. Hexham's *Principles of the Art Military* is the best authority for the Dutch system of drill. The *Tactics of Ælian*, translated with commentary by Captain John Bingham, 1616, is also valuable. Last, but not least, the reader will supply for himself the familiar name of Motley.

CHAPTER VI

It is now needful to turn to the second and perhaps more important school of the British Army. As in the Low Countries we found English and Scots fighting side by side, but gave to the English, as their numerical preponderance demanded, the greater share of attention, so now in the German battlefields of the Thirty Years' War we shall see them again ranked together, but must devote ourselves for the same reason to the actions of the Scots.

The North Britons seem to have found their way very quickly to the banners of Gustavus Adolphus, and to have fought with him in his earlier campaigns long before he had established himself as the champion of Protestantism. To mention but two memorable names, Sir John Hepburn and Sir Alexander Leslie had risen to high rank in his service many years before he crossed the Baltic for his marvellous campaigns in Germany. But to trace the history of the famous Scottish regiments aright, they must be briefly followed from their first departure from Scotland to take service under King Christian the Fourth of Denmark, who curiously enough forms the link that connects the two schools of Maurice of Nassau and Gustavus Adolphus.

It was in reliance on promises of subsidy from the English King Charles the First that Christian first levied an army and took the field for the Protestant cause. His plan was for a defensive campaign, but this was impossible unless his soldiers were regularly paid, which they would be, as he hoped, with English money.

Needless to say, Charles when the moment came was unable to fulfil his promise; Christian was driven to take the offensive and was completely defeated by Tilly at Lutter. The unhappy king appealed indignantly to Charles for help, but Charles could send nothing but four English regiments which had been raised for service in the Low Countries two years before, and were now, through the prevailing maladministration in every department of English affairs, weak, disorganised and useless. Their numbers were however supplemented by the press-gang, and a body of some five thousand men, unpaid and ill-found, ripe for disease and disorder, were shipped off to the Elbe.

626, August 17.

A little earlier than the defeat at Lutter one of the many gentlemen-adventurers in Scotland, Sir Donald Mackay, had obtained leave from King Charles to raise and transport five thousand men for King Christian's ally, the famous free lance, Count Ernest Mansfeld. It does not appear that he succeeded in recruiting even half of that number, for heavy drafts had already been made upon the centre and south of Scotland for levies. Still some two thousand men were collected by fair means or foul, and even if some of them were taken from the Tolbooth at Edinburgh, it was fitting that in a corps so famous there should be representatives from the Heart of Midlothian. But it is certain that a goodly proportion were taken from the northern counties and in particular from the district of the Clan Mackay, and that these took the field in their national costume and so were the first organised body deserving the name of a kilted regiment. The officers, from their names and still more from their subsequent behaviour, seem to have been without exception gentlemen of birth and standing, worthy to represent their nation. Some of them probably had already experience of war; one at least, Robert Munro, the historian of the regiment, had served in the Scottish body-guard of the King of France, and had learned from sad experience the meaning of the word discipline.[1]

[1] "I was once made to stand at the Louvre Gate in Paris, being

The regiment sailed in divisions from Cromarty and Aberdeen and arrived at Glückstadt on the Elbe in October 1626. The winter was spent in training the men, but not without riot and brawling. The officers were constantly quarrelling, and there was so little discipline among the men that a sergeant actually fell out of the ranks when at drill to cudgel a foreign officer who had maltreated one of his comrades. Meanwhile Count Mansfeld, who had originally hired the regiment, was dead, and in March 1627 Sir Donald Mackay offered its services to the King of Denmark. Christian accordingly reviewed it, and having first inspected the ranks on parade, "drums beating, colours flying, horses neighing," saw it march past and paid it a handsome compliment. The men were then drawn, after the fashion of the landsknechts, into a ring, where they took the oath and listened to a rehearsal of the articles of war; and so their services began. Half of them were despatched with the English regiments to Bremen, and the remainder were stationed at Lauenburg to guard the passage of the Elbe.

1626.

1627.

After a vast deal of marching and counter-marching four companies, under Major Dunbar, were left at Boitzenburg, at the junction of the Boitze and the Elbe, while Mackay with the remaining seven was moved to Ruppin. Three days after Mackay's departure, Tilly's army, ten thousand strong, marched up to Boitzenburg and prepared to push forward into Holstein. Dunbar knowing his own weakness had strengthened his defences, but eight hundred men was a small garrison against an army. On the very first night he made a successful sortie; and on the next day the Imperialist army assaulted his works at all

July.

then in the King's regiment of guards passing my prenticeship, for sleeping in the morning when I ought to have been at my exercise. For punishment I was made to stand from eleven before noon to eight o'clock of the night sentry, with corselet, headpiece, braselets, being iron to the teeth, in a hot summer's day, till I was weary of my life."—Munro's *Expedition*, p. 45.

1627. points. The first attack was repulsed with loss of over five hundred men to the assailants. Reinforcements were brought up; the attack was renewed and again beaten off, and finally a third and furious onslaught was made on the little band of Scots. In the midst of the fighting the ammunition of the garrison failed and its fire ceased. The Imperialists, guessing the cause, made a general rush for the walls. The Scots met them at first with showers of sand torn from the ramparts, and presently falling in with pike and butt of musket fought the Imperialists hand to hand, and after a desperate struggle drove them out with the loss of another five hundred men. Tilly then drew off and crossed the Elbe higher up, and Dunbar by Christian's order marched proudly out of Boitzenburg. This was the first engagement of Mackay's regiment, a fitting prelude to work that was to come.[1]

The headquarters of the regiment was presently moved from Ruppin to Oldenburg to hold the pass against October. Tilly's advance, and here they too came into action. They were ill supported by their foreign comrades, for the Danes gave way, the Germans of Christian's army took to their heels, and the brunt of the engagement fell upon half the regiment of Scots. After two hours of heavy fighting they were relieved by the other half, and so the two divisions, taking turn and turn, maintained the struggle against vastly superior numbers from seven in the morning till four in the afternoon, when the enemy at last drew off owing to the darkness. The spirit shown by the Scots was superb. Ensign David Ross received a bullet in the chest; he retired for a few minutes to get the wound dressed, and returned to the fight; nor did he afterwards miss an hour's duty on the plea that he was wounded. Hector Munro of Coull, being shot through the foot, refused

[1] But poor Dunbar and his four companies were to have little part in it. Shortly after he again defied the whole of Tilly's army, and after a desperate resistance the eight hundred men were annihilated, seven or eight alone escaping to tell the tale.

to retire till he had fired away all his ammunition, and 1627.
before he could do so was shot in the other foot also. Yet another, Hugh Murray, being ordered to bring away his brother's corpse under a heavy fire, swore that he would first empty his brother's bandoliers against the enemy, and was shot in the eye, though not fatally, while fulfilling his oath. Yet these were young soldiers, of so little experience that they left their reserve of ammunition exposed, and suffered heavily from the explosion of a barrel of powder. They lost sixteen officers and four hundred men that day.

That night the Danish army retreated to Heiligenhaven, but some German Reiters that were attached to it were so unsteady that they speedily turned the retreat into a flight; and when the harbour was reached the cavalry crowded on to the mole to seize all the transport-vessels for themselves. Sir Donald Mackay, who was himself wounded, was not the man to suffer his regiment to be sacrificed; he calmly ordered his pikemen to advance, swept the whole of the Reiters into the sea, seized the nearest ship, brought others out of the roadstead and proceeded to the work of embarkation. The last boat's load shoved off surrounded by the enemy's horse, and the last of the Scots, a gallant boy named Murchison, though wounded in the head and shot through the arm, swam off to the boat under a heavy fire, only to die two days later of his injuries. The rest of the Danish army, thirty-five troops of horse and forty companies of foot, surrendered without a blow. Hence it is hardly surprising that, when next the Scots found themselves in quarters alongside Danish horse, there was a furious riot which cost the lives of seven or eight men before it could be suppressed. But in truth Mackay's regiment was so much weakened by its losses that both colonel and lieutenant-colonel returned perforce to Scotland to raise recruits.

I shall not follow the various small actions of the 1628.

1628. earlier part of the campaign of 1628 in Holstein, though many of them were brilliant enough. It must suffice that Scotch and English fought constantly side by side not only against the enemy, but once riotously against the Danes themselves, whom they considered to be unduly favoured in the matter of rations. In May the Imperialists moved up in force to occupy Stralsund; and the burghers having appealed to Christian for assistance received from him the seven companies, now reduced to eight hundred men, of Mackay's regiment.

On arrival their commanding officer at once selected the most dangerous post in the defences, as in honour bound, and for six weeks the regiment was harassed to death by exhausting duty. The men took their very meals at their posts, and Monro, who was now a major, mentions that he never once took off his clothes. They suffered heavily too from the enemy's fire, a single cannon shot strewing the walls with the brains of no fewer than fourteen men; but still they held out. At

June 26. last Wallenstein came up in person, impatient at the delay, and vowed that he would take the town in three nights though it hung by a chain between heaven and earth. His first assault was hurled back by the Scots with the loss of a thousand men. But the Highlanders also had been severely punished; three officers and two hundred men had been killed outright, and seven more officers were wounded. On the following night the attack was renewed and again repulsed, but the garrison was now compelled to open a parley in order to gain time; and the negotiations were prolonged until the arrival of a second Scottish regiment under Lord Spynie enabled the defenders to renew their defiance.

Shortly after the King of Sweden charged himself with the defence of Stralsund. Alexander Leslie, whom we shall meet again, was appointed to take the command, and Mackay's and Spynie's regiments after a final sortie were withdrawn to Copenhagen. Of Mackay's, five hundred had been killed outright in the siege, and a bare hundred only remained unwounded; in fact the regiment required virtually to be reconstructed. The 1630. work of recruiting and reorganisation occupied the winter months, at the close of which the corps, now raised to ten companies and fifteen hundred men, was honourably discharged from the service of Denmark, and free to join itself, as it presently did, to Gustavus February. Adolphus.

Its first duty was to learn the new drill and discipline introduced by the King of Sweden; and as his system was destined to be accepted later by all the armies of Europe, no better place can be found than this, when it was just brought to perfection and first taught to British soldiers, to give some brief account of it.

The infantry of Gustavus Adolphus, as of all other civilised armies at that period, was made up of pikemen and musketeers, and beyond all doubt had originally been trained and organised on the models of the Spanish and the Dutch. Enough has already been said of these to enable the reader to follow the reforms introduced by the Swedish king. First as regards weapons: the old long pike was cut down from a length of fifteen or eighteen feet to the more modest dimension of eleven feet, and the old clumsy musket with its heavy rest was replaced by a lighter weapon which could be fired from the shoulder without further support. The defensive armour of the pikeman was also reduced to back, breast, and tassets; and thus both divisions of the infantry, carrying less weight than heretofore, were enabled to move more rapidly and to accomplish longer marches without fatigue. This was a first step towards the mobility which the great soldier designed to oppose to the old-fashioned forces of mass and weight.

Next as to the tactics of infantry: Gustavus's first improvement was to reduce the old formation from ten ranks to six; his second and more important was to withdraw the musketeers from their old station in the flanks of companies, and to mass pikes and shot into separate bodies. It is abundantly evident that he looked upon the development of the fire of musketry as of the first importance in war, and to this end he sought to render the musketeers independent of the protection of the pikes. This idea led him to a curious revival of old methods, nothing less than a modification of the stakes which were seen in the hands of the English at Hastings and Agincourt, and which now took the name of hog's bristles or Swedish feathers. This, however, was a small matter compared to his improvement in the method of maintaining a continuous fire. Pescayra's system was one which, on the face of it, was not suited to young or unsteady troops. In theory it was a very simple matter that the ranks should fire and file off to the rear in succession, but in practice the temptation to men to get the firing done as quickly as possible and to seek shelter behind the ranks of their comrades was a great deal too strong. The retirement was apt to be executed with an unseemly haste which was demoralising to the whole company, and there was no certainty that the retiring ranks, instead of resuming their place in rear, would not disappear from the field altogether. Gustavus therefore made the ranks that had fired retire through[1] instead of outside their companies, where, through judicious posting of officers and non-commissioned officers, any disposition to hurry could be checked by the blow of a halberd across the shins or by such other expedients as the reader's imagination may suggest. In an advance, again, he made the rear ranks move up successively through the front ranks, and in a retreat caused the front ranks to retire through the rear.

This reform was as much moral as tactical; but the next made a great stride towards modern practice. Not content with reducing ten ranks to six Gustavus on occasions would double those six into three, and by making the front rank kneel enabled the fire of all

[1] There were only two "orders" in the Swedish army: *Open order* for parade, which meant six feet from man to man, outstretched hand to outstretched hand; and *Battle order*, three feet from man to man, elbow to elbow.

three to be delivered simultaneously. Here is seen the advantage of abolishing the old musket-rest, with which such a concentration of fire would have been impossible. Still following out his leading principle, he encouraged the use of cartridges to hasten the process of loading; and finally to perfect his work he introduced a new tactical unit, the *peloton*, called by Munro *plotton* and later naturalised among us as the platoon of musketeers, which consisted of forty-eight men, eight in rank and six in file, all of course carefully trained to the new tactics. Yet with all these changes the drill was of the simplest; if men could turn right, left, and about, and double their ranks and files, that was sufficient.

In the matter of pure organisation Gustavus again improved upon all existing systems. First he made the companies of uniform strength, one hundred and twenty-six men, distributed into twenty-one *rots* or files, and six corporalships. A corporalship of pikes consisted of three files, and of musketeers of four files;[1] and to every file was appointed a *rottmeister*[2] or leader, who stood in the front, and an *unter-rottmeister* or sub-leader, who stood in the rear rank. Both of these received higher pay than the private soldier. Two sergeants, four under-sergeants and a quartermaster-sergeant completed the strength of non-commissioned officers, while three pipers and as many drums made music for all. Moreover each company carried a kind of reserve with it in the shape of eighteen supernumerary men who bore the name of *passe-volans*, the old slang term for fictitious soldiers since the days of Hawkwood, and were allowed to the captain as free men, unmustered. The officers of course were as usual captain, lieutenant, and ensign.

1 A file in those days consisted, of course, of six men, not as now of two. So a corporalship of pikes would be eighteen, and of musketeers twenty-four men.

2 The *rottmeisters* were fifteen in number, the six corporals bringing up the total to the necessary twenty-one.

Eight such companies constituted a regiment, which was thus one thousand and eight men strong, with a colonel, lieutenant-colonel, and major over all. The regimental staff included many officials borrowed from the landsknechts' model for the trial and punishment of offenders, and for a complete novelty, four surgeons. The provision of medical aid had formerly been left to the captains, and it is to Gustavus that we owe the first example of a sounder medical organisation.

Four companies or half of such a regiment were called either a squadron or by the Italian name *battaglia*, to which must be traced our modern word battalion. Two such regiments were called a brigade, which marks the latest advance in organisation made by Gustavus. Maurice of Nassau had been before him in the formation of brigades but had not reduced them to uniform strength. The Swedish brigades had a stereotyped formation for battle, and were called after the colour of their standards, the white, the blue, the yellow, and finally the green, better known as the Scots Brigade, which is that wherein we are chiefly interested.

Passing next to the cavalry, the marks of Gustavus's reforming hand are not less evident. The force at large was divided into cuirassiers and dragoons. Of these the latter, who were armed with muskets and were simply mounted infantry, may be dismissed without further observation. The cuirassiers, except outwardly, bore a strong resemblance to the Reiters, for, though stripped of all defensive armour except cuirass and helmet, they still carried two pistols as well as the sword. Gustavus, however, here as with the infantry, took a line of his own. He began by reducing the depth of the ranks from the bottomless profundity of the Reiters to three or at most four; and though he still opened his attack with the pistol and so far adhered to missile tactics he to a considerable extent combined with them the action by shock. As in the infantry, it was Pescayra's system that he wished to supersede. The Reiters, as we know by the testimony of many eye-witnesses, were often so anxious to go to the rear and reload that they fired their pistols at absurd ranges, sometimes indeed hardly waiting to fire before they turned about. Unable to apply to cavalry the system which he had adopted for the infantry, and failing in common with all his contemporaries to grasp the principle that, since a horse has four legs and a man two, the evolutions of horse and foot must be fundamentally different, Gustavus none the less determined that his cuirassiers should at all events come to close quarters with their enemy. He therefore trained them not to fire till they could see the white of their opponents' eyes, and having fired to strike in with the sword.

Hence he has the credit, which is not wholly undeserved, of having restored shock-action, and is said to have made his cavalry charge at the gallop; but the first statement is misleading, and the second in the face of contemporary accounts incredible. In the first place, the sword is a singularly ineffective weapon against mailed men, and a true restorer of shock-action would almost certainly have reverted to the lance. In the second place, mounted men who open their attack with pistols will infallibly check their horses at the moment of firing in order to ensure greater accuracy of aim. Lastly, Gustavus's favourite plan for the attack of cavalry was to intersperse his squadrons with platoons of musketeers, which advanced with them within close range [1] and fired a volley into the enemy's horse. This preliminary over, the cuirassiers advanced, fired their pistols, fell in with the sword, and retired; by which time the musketeers had reloaded and were ready with another volley. Close range of the musket of those days would not have allowed space for a body of horse to gather way for a shock-attack in the modern sense, and it is therefore more than doubtful whether the Swedish squadrons charged at higher speed than the trot. Gustavus's system was in fact simply a revival of Edward the First's at Falkirk, which had already been

[1] See Monro, vol. ii. p. 65.

developed with great success by Pescayra at Pavia. Nevertheless, by reducing the depth of squadrons and insisting that his men should come to close quarters, Gustavus unquestionably did very much for the improvement of cavalry.[1]

Most remarkable of all were his reforms in the matter of artillery. Profoundly impressed by the power of field-guns he spared no effort to make them lighter and more mobile, so as to be at once easily manœuvred and capable of transport in larger numbers. Here again Maurice had been before him, not without success, but Gustavus possessed in the person of a Scotch gentleman, Sir Alexander Hamilton, an artillerist of wider views than lay to the hand of the great Dutch soldier. Hamilton's first experiment was to make leathern guns,[2] strengthened by hoops of metal and with apparently a core of tin, which could easily be carried on a pony's back or stacked away by the dozen in a waggon. Gustavus used them frequently in his earlier campaigns but discarded them at latest after the battle of Breitenfeld, finding that their life did not extend beyond ten or a dozen rounds. He then fell back on light two-pounders and four-pounders, which required few horses for draught, and could be loaded and fired by a skilful crew more rapidly even than a musket. A few such guns were attached to each regiment and called regimental pieces ; and very effective they were presently found to be.

Further, Gustavus was a consummate engineer, as fond of the spade as Maurice himself, and a past master of field-fortification. On stepping ashore in Germany he first fell on his knees and prayed, and then picking up a spade began to dig with his own hands. This, it

[1] Stress has been laid upon the fact that Gustavus always led the cavalry in person. Doubtless he was fond of his Horse, but since at that period cavalry was always stationed in the wings, and the right wing was the post of honour, this does not count for very much.

[2] They were called after their inventor by the name of "Sandy's stoups," and were used by the Scots at the battle of Newburn in 1640.

may here be mentioned once for all, was the one point in his system which the Scots could not endure; they always grumbled when called upon to use the spade, and in spite of the King's occasional reproaches, always made less progress with field-works in a given time than any other corps in the army.

Lastly, to turn to broader principles, the great innovation of Gustavus, visible in all his reforms, was to match mobility against the old system of weight. He never massed his troops in unwieldy bodies, but distributed them in smaller and more flexible divisions, allowing plenty of space for facility of manœuvre. His order of battle was that which was customary in his time, consisting of two lines with infantry in the centre and cavalry on the flanks; but he always allowed three hundred yards of distance between the first and second line, and erected the practice of keeping a reserve, which had been intermittently observed for centuries, into an established principle. Again, he carefully studied the effective combination of the three arms with a thoroughness unknown since the days of Zizca, supplying artillery to his infantry, and supporting impartially horse with foot and foot with horse. Finally, as the backbone of all, he enforced with a strictness that had never been seen before him the observance of discipline.

Such was the Army and such the General to which 1630.
Mackay's regiment now joined itself. In June 1630 it embarked for Germany as part of the thirteen thousand men which formed the Swedish army, half of the companies at Elfsknaben, the remainder under Munro at Pillau. The latter detachment was wrecked off Rügenwalde, which was held by the Imperialists, and lost everything; but having made shift to obtain arms calmly attacked the Imperial garrison and captured the town—as daring a feat of arms as ever was done by Scotsmen. After several small engagements Monro rejoined his headquarters at Stettin, and in January 1631 Gustavus, who boasted with justice that his army was as effective for a winter's as for a summer's campaign, invaded

1631. Brandenburg and marched for the Oder. The Scotch were organised into the famous Scots Brigade, consisting of four picked regiments—Hepburn's, Mackay's, Stargate's, and Lumsden's, the whole under the command of Sir John Hepburn.

We must pass over the operations in Brandenburg, where the Scots Brigade distinguished itself repeatedly, and come forthwith to Saxony, whither Gustavus had been called from the Oder by Tilly's advance upon May. Magdeburg. Arriving too late to save the unhappy city he entrenched himself at Werben, at the junction of the Elbe and the Havel, and gave the world a first notable example of his skill as an engineer. Tilly, having lost six thousand men in the vain attempt to storm the entrenchments, invaded Saxony, whither Gustavus at once followed him and offered him battle on the plain of Leipsic.

On the 7th of September Tilly took up his position facing north, on a low line of heights running from the village of Breitenfeld on the west to that of Seehausen on the east. His army was drawn up in a single line. On each wing as usual was posted the cavalry, seven regiments under Pappenheim on the left, seven more under Furstenburg on the right, all drawn up in the dense columns beloved of Charles the Fifth. In the centre was Tilly himself, with eighteen regiments of infantry, his famous Walloons among them, massed together in the old heavy Spanish formation. On the heights above him were his guns. The whole force numbered forty thousand men, and their General was a man who, though seventy years of age, had never lost a battle.

On the other side the armies of Gustavus and of his allies the Saxons were drawn up in two lines. On the left were the Saxons, fourteen thousand strong, and on the right, with which alone we need concern ourselves, the Swedes. In touch with the Saxon right, the Swedish left under Field-Marshal Horn was made up, both in the first and second lines, of six regiments of horse, with four platoons of musketeers between each regiment.

The right wing under Gustavus himself was similarly 1631,
composed. In the centre the first line was made up of Sept. 7.
four half brigades of foot, supported by a regiment of cavalry and eight platoons of Scots; and the second line of three brigades, of which Hepburn's was one. In rear of both lines was a reserve of cavalry, and in the extreme rear a further reserve, the first ever seen, of artillery.

The battle opened as usual with a duel of artillery, which was continued from noon till half-past two, the Swedish guns, more numerous and better served than Tilly's, firing three shots to the enemy's one. Then Pappenheim, on Tilly's left, lost patience, and setting his cavalry in motion without orders came down upon the Swedish right. He was met by biting volleys from the platoons of musketeers and charges from the cuirassiers at their side; his men shrank from the fire, and edging leftward across the front of Gustavus's wing swept down towards its rear. General Bauer, in command of the reserve cavalry of the first line, at once moved out and broke into them; and the whole Swedish right coming into action drove back Pappenheim's horse, after a hard struggle, in disorder. Gustavus checked the pursuit, for Tilly had pushed forward a regiment of infantry in support of Pappenheim, and turning all his force on this unhappy corps annihilated it.

On the Imperialists' left Furstenburg, following Pappenheim's example, had also charged, and had driven the entire Saxon army before him like chaff before the wind.[1] He followed them in hot pursuit; and had Tilly at once advanced with his centre against Field-Marshal Horn, the situation of the Swedes would have been critical, for their left was now completely uncovered. But owing to the faulty disposition of his artillery Tilly could not advance directly without putting his guns out of action, and he therefore followed in the track of Furstenburg to turn Horn's left flank. The delay gave Horn time to make dispositions to meet the attack.

[1] Tallard fatally repeated this independent formation of two armies at Blenheim.

1631, Sept. 7. Hepburn's brigade came quickly up with another brigade in support, and the Scots after one volley charged the hostile infantry with the pike and routed it completely. Gustavus meanwhile had again advanced with his cavalry on the right, and sweeping down on the flank of Tilly's battery captured all his guns and turned them against himself. The battle was virtually over, but four splendid old Walloon regiments stood firm to the last, and though reduced to but six hundred men retreated at nightfall in good order.

The victory was crushing ; and yet of all the Swedish infantry two brigades alone had been engaged, and of these the Scots had done the greater share of the work. The battle marks the death-day of the old dense formations and the triumph of mobility over weight, and is therefore of particular interest to a nation whose strength is to fight in line.

From Leipsic Gustavus marched for the Main, where the Scots were as usual put forward for every desperate service, and held his winter court at Mainz. 1632. In the spring of the following year he marched down to the line of the Danube with forty thousand men, forced the passage of the Lech in the teeth of Tilly's army, entered Bavaria and by May was at Munich. Then hearing that the towns on the Danube in his rear were threatened he turned back to Donauwörth, whence he was called away by the movements of Wallenstein in Saxony to Nürnberg. Such marching had not been since the days of Zizca. He now turned Nürnberg, as he had turned Werben in the previous year, into a vast entrenched camp ; for he had now but eighteen thousand men against Wallenstein's seventy thousand, and it behoved him to make the most of his position. Wallenstein, however, without risking an engagement, took the simpler course of making also an entrenched camp, cutting off Gustavus's supplies from the Rhine and Danube, and reducing him by starvation. Reinforcements came to the Swedes, which raised their army to five-and-thirty thousand men ; Wallenstein allowed them to pass in unmolested to consume the provisions the quicker. The pinch of hunger began to make itself felt in the Swedish camp, pestilence raged among the unhappy troops, and at last Gustavus in desperation launched his army in a vain assault upon Wallenstein's entrenchments. For twelve hours his men swarmed up the rugged and broken hill with desperate courage, three times obtaining a momentary footing and as often beaten back. The cannonade was kept up all night, and it was not till ten o'clock on the following morning that the Swedes retreated, leaving four thousand dead behind them. The Scots Brigade suffered terribly. Monro, out of a detachment of five hundred men, lost two hundred killed alone, besides wounded and missing. His lieutenant-colonel who relieved him at night brought back but thirty men next morning. Other corps had lost hardly less heavily, and Gustavus, foiled for once, retreated to Neustadt, leaving one-third of his force dead around Nürnberg. 1632.

Sir John Hepburn, in consequence of a quarrel with the Swedish king, now took leave of him and entered the service of France; and the Scots Brigade, weakened to a mere shadow, was left behind at Dunkerswald to await reinforcements, while Gustavus marched away to his last battlefield at Lützen. We need follow the fortunes of the Brigade little further. The famous regiments, together with the other Scots and English in the Swedish service, now some thirteen thousand men, did abundance of hard and gallant work before the close of the war. The ranks of Mackay's regiment were again swelled to twelve companies and fifteen hundred men, but at Nördlingen it was almost annihilated, and emerged with the strength of a single company only. Times had changed, and discipline had decayed since the death of Gustavus; and in 1635, on the alliance of France with Sweden, and the outbreak of war between France and Spain, the fragments of all the Scotch regiments were merged together, and passed into the service of France under the command of the veteran Sir John Hepburn as the Regiment d'Hebron. 1634, August 26.

There for a short period let us leave it, wrangling with Regiment Picardie for precedence, claiming, on the ground that some officers of the Scottish Guard had joined it, to be the oldest regiment in the world,[1] and earning the nickname of Pontius Pilate's guards. Hepburn commanded it for but one year, for he fell at
1636. its head at the siege of Saverne, but it fought through many actions and many sieges, the battle of Rocroi not the least of them, before it returned to the British Isles. We shall meet with it again before that day under a new name, and under yet a third name shall grow to know it well.

Authorities.—Munro's *Expedition* is far the most valuable; it has been abridged and supplemented by Mr. John Mackay in his *Old Scots Brigade*. Harte's *Life of Gustavus* wrestles manfully with the military details, which are very clearly summed up in Mr. Fletcher's *Gustavus* in the Heroes of the Nations Series. Some few details will be found also in Fieffé's *Histoire des troupes Etrangéres*.

[1] As I believe that this pretension is still advanced by patriotic North Britons, it is as well to say that it is preposterous. The true Scottish Guard enjoyed an independent existence till the Revolution, and to claim its privileges for Hepburn's regiment is as absurd as though a corps raised to-morrow, and officered by half a dozen gentlemen of the Grenadier Guards, should claim precedence of all British infantry.

CHAPTER VII

Once more we return to England and take up the thread of the army's history within the kingdom. Of the reign of James the First there is little to be recorded except that at its very outset the Statute of Philip and Mary for the regulation of the Militia was repealed, and the military organisation of the country based once more on the Statute of Winchester. James was not fond of soldiers, and military progress was not to be expected of such a man. Enough has already been seen of his methods through his dealings with the Low Countries, and there is no occasion to dwell longer on the first British king of the House of Stuart.

Charles the First was more ambitious, and sufficiently proud of the English soldier to preserve the ancient English drum-march.[1] Soon after the final breach with 1625.
Spain he imbibed from Buckingham the idea of a raid on the Spanish coast after the Elizabethan model, which eventually took shape in the expedition to Cadiz. Of all the countless mismanaged enterprises in our history this seems on the whole to have been the very worst. There was abundance of trained soldiers in England who had learned their duty in the Low Countries; and Edward Cecil, he whom we saw some few years back in command of the cavalry at Nieuport, begged that liberal offers might be made to induce them to serve. Officers again could be procured from the Low Countries, and therefore there should have been no difficulty in organising an excellent body of men. In

[1] Dalton, vol. i. p. 234.

1625. the matter of arms, however, though English cannon was highly esteemed, Charles was forced to purchase what he needed from Holland, which was a sad reflection on our national enterprise. Accordingly over a hundred officers were recalled from Holland; and two thousand recruits were collected, to be sent in exchange for the same number of veterans from the Dutch service. Eight thousand men were then pressed for service in various parts of England, and the whole of them poured, without the least preparation to receive them, into Plymouth, where they gained for themselves the name of the plagues of England. Sir John Ogle, a veteran who had served for years with Francis Vere, eyed these recruits narrowly for a time, old, lame, sick and destitute men for the most part, and reflected how without stores, clothes, or money he could possibly convert them into soldiers. Then taking his resolution he threw up his command and took refuge in the Church. Very soon another difficulty arose. The States-General firmly refused to accept two thousand raw men in exchange for veterans, and shipped the unhappy recruits back to England. They too were turned into Plymouth and made confusion worse confounded. Then the arms arrived from Holland, and there was no money to pay men to unload them. The port became a chaos. Buckingham had already shuffled out of the chief command and saddled it on Cecil, and the unfortunate man, good soldier though he was, was driven to his wit's end to cope with his task. His tried officers from Holland were displaced to make room for Buckingham's favourites, who were absolutely useless; and yet he was expected to clothe, arm, train, discipline, and organise ten thousand raw, naked men, work out every detail of a difficult and complicated expedition, and make every provision for it, all without help, without encouragement, and without money. Cash indeed was so scarce that the king could not afford to pay the expenses of his own journey to Plymouth.

Under such conditions it is hardly surprising that

the enterprise was a disastrous failure. A few butts of 1625. liquor left by the Spaniards outside Cadiz sufficed to set the whole force fighting with its own officers, and after weary weeks at sea, aggravated by heavy weather and by pestilence, the result of bad stores, Cecil and the remains of his ten regiments returned home in misery and shame.[1]

A similar enterprise under Lord Willoughby in the 1626. following year failed in the same way for precisely the same reasons; but Buckingham, still unshaken in his confidence, led a third and a fourth expedition to Rochelle with equal disaster and equal disgrace. The captains had no more control over their men than over a herd of deer.[2] At last, at the outset of a fifth expedition, which promised similar failure, the dagger of Lieutenant Felton, a melancholy man embittered by deprivation of his pay, put an end to Buckingham and to all his follies. On the whole he had not treated the soldiers worse than Elizabeth, but a man of Elizabeth's stamp was more than could be borne with.

Nevertheless, amid all these failures there were still plenty of men in England who had the welfare of the military profession at heart. Foremost among them was the veteran Edward Cecil, now Lord Wimbledon, who strove hard to do something for the defence of the principal ports, for the training of the nation at large, and in particular for the encouragement of cavalry. The mounted service had become strangely unpopular with the English at this time, whether because the eternal sieges of the Dutch war afforded it less opportunity of distinction, or because missile tactics had lowered it from its former proud station, it is difficult to say. Certain it is that officers of infantry, and notably Monro, never lost an opportunity of girding at horsemen as fitted only to run away, and as preferring to be mounted only that they might run away the faster. But Cecil, though in this respect unique, was by no means the only man who

[1] Mr. Dalton has told the story very fully in his *Life of Cecil.*

[2] Ward, *Animadversions of Warre.*

made his voice heard. Veteran after veteran took pen in hand and wrote of the discipline of Maurice of Nassau and, as time went on, of the system of Gustavus Adolphus; while on the other hand one ingenious gentleman, still jealous of the old national weapon, invented what he called a "double-arm," which combined the pike and the bow, the bow-staff being attached to the shaft of the pike by a vice which could be traversed on a hinge. Strange to say this belated weapon was not ill-received in military circles and found commendation even among Scotsmen.[1] On one important point, however, there was a general consensus of opinion, namely that the condition of the English militia was disgraceful, its system hopelessly inefficient and the corruption of its administration a scandal. The trained bands were hardly called out once in five years for exercise; few men knew how even to load their muskets, and the majority were afraid to fire a shot except in salute of the colours, not daring to fire a bullet from want of practice.[2] The Londoners, as usual, alone made a favourable exception to the general rule.

1639. The real root of the evil was presently to be laid bare. The disputes between Charles the First and his subjects were assuming daily an acuter form, until at last they came to a head in the Scotch rebellion of 1639. It was imperative to raise an English force forthwith and move it up to the Border. Charles, as usual in the last stage of impecuniosity, thought to save money by an exercise of old feudal rights, and summoned every peer with his retinue to attend him in person as his principal force of cavalry. It was a piece of tactless folly whereof none but a Stuart would have been guilty: the peers came in some numbers as they were bid, but they did not conceal their resentment against such pro-

[1] See *Pallas Armata*, by Sir T. Kellie, 1627. This writer deserves mention as the first who introduced the system of drilling by numbers. He talks as glibly of odd and even numbers as a modern drill sergeant.

[2] Barriffe and Ward.

ceedings. The foot were levied as usual by writ to the 1639. lord-lieutenant with the help of the press-gang, they behaved abominably on their march to the rendezvous, and on arrival were found to be utterly inefficient. Their arms were of all sorts, sizes, and calibres, and the men were so careless in the handling of them that hardly a tent in the camp, not even the king's, escaped perforation by stray bullets. In other respects the organisation was equally deficient; no provision had been made for the supply of victuals and forage; and altogether it was fortunate that the force escaped, through the pacification of Berwick, an engagement with the veterans from the Swedish service under old Alexander Leslie that composed a large portion of the Scottish army.

The following year saw the war renewed. This 1640. time the farce of calling out a feudal body of horse was not repeated, but unexpected difficulties were encountered in raising the levies of foot. In 1639 the infantry had been drawn chiefly from the northern counties, where the tradition of eternal feuds with the Scots made men not altogether averse to a march to the Border. But in 1640 the trained bands of the southern counties were called upon, and they had no such feeling. It is possible that unusual rigour was employed in the process of impressment, for the authorities had been warned, after experience of the previous year, to allow no captains to play the Falstaff with their recruits. Be that as it may, the recalcitrance of the new levies was startling. From county after county came complaints of riot and disorder. The Wiltshire men seized the opportunity to live by robbery and plunder; the Dorsetshire men murdered an officer who had corrected a drummer for flagrant insubordination; in Suffolk the recruits threatened to murder the deputy-lieutenant; in London, Kent, Surrey, and half a dozen more counties the resistance to service was equally determined; and when finally in July four thousand men reached the rendezvous at Selby, old Sir Jacob Astley could only designate them as the arch-knaves of the country. Money being of course

1640. very scarce, the men were ill-clothed and ill-found, and their numbers were soon thinned by systematic desertion. A new difficulty cropped up in the matter of discipline. Lord Conway, who commanded the horse, had executed a man for mutiny; he now found that his action was illegal and that he required the royal pardon. If, he wrote, the lawyers are right and martial law is impossible in England, it would be best to break up the army forthwith: to hand men over to the civil power is to deliver them to the lawyers, and experience of the ship-money has shown what support could be expected from them.

There, in fact, lay the kernel of the whole matter; indiscipline was not only rife in the ranks but widespread throughout the nation. From long carelessness and neglect the organisation of the country for defence by land and sea had become not only obsolete but impossible and absurd. For centuries the old vessel had been patched and tinkered and filed and riveted, occasionally by statute, more often by royal authority only, but chiefly by mere habit and custom. But now that the reaction which had established the new monarchy was over, and men, stirred by a counter-reaction, subjected the military system to the fierce heat of constitutional tests, the whole fabric fell asunder in an instant, and brought the new monarchy down headlong in its fall. The story is so instructive to a nation which has not yet given its standing army a permanent statutory existence, that it is worth while very briefly to trace the progress of the catastrophe.

According to ancient practice, the various shires were called upon to provide their levies for the Scotch war with coat-money and with conduct-money to pay their expenses till they had passed the borders of the county, from which moment they passed into the king's pay. The writs to the lord-lieutenants distinctly stated that these charges would be refunded from the Royal Exchequer, and though the chronic emptiness of the Royal Exchequer might diminish the value of the pledge, the form of the writ was distinctly consonant with custom and precedent. Many of the county gentlemen, however, refused to pay this coat- and conduct-money; they had been encouraged by the attacks made on military charges in the Short Parliament; and the Crown, aware of the general opposition to all its doings, did not venture to prosecute. Another incident raised the general question of military obligations in an acuter form. In August 1640, Charles, sadly hampered by the general objections to military service on any terms, fell back on the old system of issuing Commissions of Array to the lord-lieutenants and sheriffs. In themselves Commissions of Array, especially when addressed to these particular officers, were nothing extraordinary; they had been in use to the reign of Queen Mary, and though more or less superseded by the appointment of lord-lieutenants, were by implication sanctioned by a statute of Henry the Fourth. 1640.

Now, however, these Commissions at once raised a storm. The deputy-lieutenants of Devon promptly approached the Council with an awkward dilemma. To which service, they asked, were the gentry to attach themselves, to the trained bands or to the feudal service implied in the Commissions of Array; since both were equally enjoined by proclamation? The Council answered that the service in the trained bands must be personal, and the feudal obligation satisfied by deputy or by pecuniary composition; in other words, if the gentry halted between two services, they could not go wrong in performing both. A second question from the deputy-lieutenants was still more searching: how were the bands levied under the Commissions to be paid? The reply of the Council pointed out that the laws and customs of the realm required every man, in the event of invasion, to serve for the common defence at his own charge. Here Charles was strictly within his rights; and the plea of invasion was sound, since the Scots had actually passed the Tweed. Parliament, however, seized hold of the Commissions of Array, and after innumerable arguments as to their illegality, took

[1]640. final refuge under the Petition of Right. Stripped of all redundant phrases, the position of the two parties was this: Charles asked how he could raise an army for defence of the kingdom, if the powers enjoyed by his predecessors were stripped from him; and Parliament answered that it had no intention of allowing him any power whatever to raise such an army.[1]

[A]ugust 28. The campaign in the north was speedily ended by the advance of the Scots and by the rout of the small English detachment that guarded the fords of the Tyne at Newburn. The Scots then occupied Newcastle, and England to all intent lay at their mercy. Nothing could have better suited the opponents of the king. A treaty was patched up at Ripon which amounted virtually to an agreement to subsidise the Scotch army in the interest of the Parliament. The Scots consented to stay where they were in consideration of eight hundred and fifty pounds a day, failing the payment of which it was open to them to continue their march southward and impose their own terms. Charles could not possibly raise such a sum without recourse to Parliament, and the assembly with which he had now to do was that which is known to history as the Long Parliament. [1]641, [M]ay. Within seven months it had passed an Act to prevent its dissolution without its own consent, and having thus secured itself, it allowed the English army to be disbanded, while the Scots, having played their part, retired once more across the Tweed.

It would be tedious to follow the widening of the breach during the year 1641. Both parties saw that war was inevitable, and both struggled hard to keep the militia each in its own hands. The scramble was supremely ridiculous, since it was all for a prize not worth the snatching. Charles has been censured for throwing the whole military organisation out of gear because he wished to employ it for other objects than the safety of the kingdom, but it would be difficult, I

[1] The whole of the controversy may be read at large in Rushworth.

think, for any one to explain what military organisation existed. By the showing of the Parliamentary lawyers themselves, there was no statute to regulate it except the Statute of Winchester; in strictness there was no legal requirement for men to equip themselves otherwise than as in the year 1285. It was to the party that first made an army, not to that which preferred the sounder claim to regulate the militia, that victory was to belong. Strafford had perceived this long before, but three years were yet to pass before Parliament should realise it. The few movements worth noting in the scramble may be very briefly summarised. The king reluctantly consented to transfer the power of impressment to the justices of the peace with approval of Parliament, and abandoned his right to compel men to service outside their counties. But he refused to concede to Parliament the nomination of lord-lieutenants or the custody of strong places, and Parliament therefore simply arrogated to itself these privileges without further question. In July the Commons resolved to levy an army of ten thousand men, in August the King unfurled the Royal Standard at Nottingham; and so the Civil War began. 1641-2. 1642.

The lists of the two opposing armies of 1642 are still extant: the King's, of fourteen regiments of foot and eighteen troops of horse, and the Parliament's, of eighteen regiments of foot, seventy-five troops of horse, and five troops of dragoons; but it would be unprofitable to linger over them, for except on paper they were not armies at all. Two names however must be noticed. The first is that of the commander of the royal horse, Prince Rupert, a son of the Winter-King. He had now been domiciled in England for seven years, in the course of which he had found time to serve the Dutch, as we have seen, at the siege of Breda in 1639, and the Swedes in the following year, commanding with the latter a regiment of horse in more than one dashing engagement. He was now three-and-twenty, not an unripe age for a General in those days, as Condé was

1642. presently to prove at Rocroi. The second name is that of the Captain of the Sixty-Seventh troop of the Parliamentary horse, Oliver Cromwell, a gentleman of Huntingdon, not inconspicuous as a member of Parliament but unknown to military fame. He was already forty-three years of age, and so far was little familiar with the profession of arms.[1]

On the 23rd of October these two men met at Edgehill, the first important action of the war, on which I shall not dwell further than to notice the part that they played therein. Rupert, knowing the deficiency of fire-arms in the royal cavalry, before the battle gave his horsemen orders to keep their ranks and to attack sword in hand, not attempting to use their pistols till they had actually broken into the enemy's squadrons. Here was an improvement on the Swedish system, a step nearer to shock-action, which was crowned by complete success. Oliver Cromwell having seen the havoc wrought by the Royalist cavalry, sought and found after the battle the cause of the inferiority of the Parliament's. "Your troops," he said to John Hampden, "are most of them old decayed serving-men and tapsters: their troops are gentlemen's sons and persons of quality. Do you think the spirits of such base and mean fellows will ever be able to encounter gentlemen who have courage, honour, and resolution in them? You must get men of a spirit that is likely to go as far as gentlemen will go, or you will be beaten still." Hampden heard and shook his head; he was a wise and worthy person, but he had probably an idea that no men except such as those which had been swept into the ranks by the King and the King's father could possibly be induced to become soldiers. So he said that it was a good notion but impracticable. Captain Cromwell set to work to show that it was not impracticable, and began to raise men who, in his own words,

[1] His name indeed appears as an ensign in the list of a company of foot raised for service in Ireland (printed in June 1642), but this does not count for much.

made some conscience of what they did, and to teach them discipline. 1642.

Meanwhile the helplessness of the Parliament in the early stages of the war was almost ludicrous; and though indeed few things are more remarkable than the rapid growth of administrative ability between the years 1642 and 1658, it must be admitted that at first the civil leaders of the people were little better than children. Nearly the whole nation, and with it the majority of legislators, had made up their minds that the first battle would decide the contest, and they were woefully disappointed when it did not do so. Failing at first to realise the elementary principle that money is the sinew of war the Houses trusted at first to irregular contributions for its support, nor was it until pressed to extremity that they determined to employ general taxation. Money was the first and eternal difficulty, which however pressed even harder on the King than on the Parliament. The next obstacle was the utter collapse of the existing military organisation. The county levies were ready enough to fight in defence of their own homes, but they were unwilling to move far from them; and when the enemy had left their own particular quarter they thanked God that they were rid of him and returned to their usual avocations. This again was a difficulty that beset both sides and was never overcome by the King. The Parliament tried to meet it by the establishment of associations of counties, which December. were virtually military districts, and did something, though not much, to widen the narrow sympathies of the militiamen. But these associations, though a step in the right direction, depended too much on the individual energy of the men at their head to attain uniform success; and one only, the Eastern, wherein Cromwell was the moving spirit, did for a time really efficient work.

A third and most formidable danger was the superiority of the Royalist cavalry. The long neglect of the mounted service left the supremacy to the ablest

1642. amateurs, and the majority of these, though there were hundreds of gentlemen on the Parliamentary side, were undoubtedly for the King. Nor was it only the courage, honour, and resolution of which Cromwell had spoken that favoured them; they had from the nature of the case better horses, a higher standard of horsemanship and equipment, a quicker natural intelligence and a higher natural training. The thousand lessons which the county gentlemen learned when riding with hawk and hound were of infinite advantage in the casual and irregular warfare of the first two or three years; and whatever may be said of Rupert's ability on the battlefield, there can be no question that the work of his innumerable patrols was admirably done. The dashing character of Rupert was also an advantage in a sense to the King's cause, for it attracted to him a group of fellow hot-heads similar to those that had followed Thomas Felton under the Black Prince. One fatal defect however marred what should have been a most efficient cavalry, the blot that had been hit by Cromwell, indiscipline.

1643. The campaign of 1643 found Parliament little wiser than before as to the true method of conducting a war. Though it had named Lord Essex as General it gave him no control over the operations of any army but his own, and there was consequently no unity either of design or of purpose. Charles, on the contrary, had a definite plan, which had been mapped out for him by some unknown hand and was within an ace of successful execution. He himself with one army fixed his headquarters at Oxford; a second army under Newcastle was to advance from the north, a third under Prince Maurice and Sir Ralph Hopton from the extreme west, both converging on Charles as a centre; and the united forces were then to advance on London. Hopton, an experienced soldier and as noble a man as fought in the war, executed his part brilliantly, advancing victoriously into Somerset from Cornwall, and finally defeating the force specially sent to meet him by the Parliament at

Roundway Down. This action is memorable for the appearance, and it must be added the defeat, of what was probably the last fully mailed troop of horse ever seen in England, Sir Arthur Hazelrigg's "Lobsters," so called from the hardness of their shells. Hopton's advance was only stayed by the unwillingness of his Western levies to move any further from their homes. In the north again the Parliament had suffered disaster; the Fairfaxes, who were the mainstay of the cause, sustained a crushing defeat, and but one man stood in the way to bar Newcastle's march upon London. 1643.

That man however was Oliver Cromwell. Already he had begun to put in practice the scheme which Hampden had pronounced impracticable. He had chosen his recruits from the Puritan yeomen and farmers of the Eastern Counties, men who had thrown themselves heart and soul into the religious struggles of the time, who made some conscience of what they did, "who knew what they were fighting for and loved what they knew," and who thought it honourable to submit to rigid discipline for so noble a cause. Cromwell was now a colonel, and he had already shown the mettle of his force, while it was still incomplete, by defeating a body of twice its numbers in a skirmish at Grantham. This too he had done not by any novelty in tactics, for he admits that he attacked only at a pretty round trot, but by superiority of handling and of discipline. With the same troops strengthened and improved he now advanced and met a strong force of Newcastle's advanced horse at Gainsborough; and by skilful manœuvring and full appreciation of the principle, as yet unwritten, that in the combat of cavalry victory rests with him that throws in the last reserves, he routed it completely. Following up his success he came, unexpectedly as he admits, upon the main body of Newcastle's army, both horse and foot. Horses and men were weary after a hard day's work and a long pursuit, but they showed a bold front; and Cromwell, drawing them off by alternate bodies, once again a movement which was not to be

1643. found in the text-books,[1] safely effected his retreat. In truth the man was a born soldier, and probably a great deal fonder of the profession of arms, late though he had entered upon it, than he would have cared to admit. "I have a lovely company," he wrote shortly after this action, with the genuine pride of a good regimental officer; and in spite of the rigour of his discipline his troops increased until they were sufficient to fill two complete regiments.

The danger from the north was averted for the moment, but the situation was so critical that the Parliament authorised the impressment of men and raised Essex's army to a respectable total. But meanwhile negotiations had been opened with the Scots for the advance of their army against the King's forces in the north, and by September the conditions, military, financial, and religious, were agreed upon. This treaty brought home to the Parliament the necessity for immediately opening up its communications with the north and making a way whereby the Scots might penetrate further southward. The difficult task was achieved by the united efforts of two men who here fought their first action together, Thomas Fairfax and Oliver Cromwell. The day of Winceby must for this reason remain memorable in the history of the Army, not the less so because it brought Cromwell nearer to his death than any action before or after it.

By the close of the year Parliament began to realise that if the war were to be carried to a successful issue, some more effective force than mere trained bands must be called into existence. It accordingly voted that Essex's army should be fixed at a permanent establishment of ten thousand foot and four thousand horse with a regular rate of monthly pay. This was progress in the right direction, but in the disorder of the financial administration it was extremely doubtful whether the scheme would not be wrecked by its cost. Meanwhile

[1] I have however found an early instance of it in the French religious wars, but have unfortunately mislaid the reference.

the Scots had crossed the Tweed and fairly entered as partners with the Parliament in the rebellion. This new factor led to the formation of a Committee of Both Kingdoms for the subsequent conduct of the war, an important step towards unity of design and administration but clogged by one fatal defect, namely, that the military members—Essex, Manchester, Waller, and Cromwell—were all absent in the field, and that the direction of operations therefore fell entirely into the hands of civilians. A Committee was better than a whole House, and that was all that could be said, for the new directorate soon came into collision with its officers in the field. On the invasion of the Scots, Charles of necessity altered his plan of campaign and detached Rupert to the north, who marked the line of his advance in deeper than ordinary lines of desolation and bloodshed. The Parliamentary generals in the north, Fairfax and Manchester, were at the time engaged upon the siege of York. The Committee, scared by the terror of Rupert's march, ordered them to raise the siege and move southward to meet him. They flatly refused; and their persistence in their own design led to the greatest military success hitherto achieved by the Parliament, the victory of Marston Moor. 1644.

Of no battle are contemporary accounts more difficult of reconciliation than those of Marston Moor, but the main features of the action are distinguishable and may be briefly set down. Both armies consisted of about twenty-three thousand men, and were drawn up in two lines, the infantry in the centre and the cavalry in the flanks. On the Royalist side Rupert, as was usual for the Commander-in-Chief, led the right wing,[1] five thousand horse in one hundred troops; his centre, fourteen thousand foot, was under Eythin, a veteran officer imported from Germany; his left, four thousand cavalry, was led by Goring. On the Parliamentary July 2.

[1] He is said to have posted himself opposite Cromwell, but he only took his usual place at the right of the line; he occupied the same position at Naseby and took no pains to meet Cromwell there.

1644, July 2. side Ferdinand, Lord Fairfax, commanded the right wing of horse, the first line consisting of English, the second of Scots; the centre was composed principally of Scottish infantry under old Alexander Leslie, Earl of Leven; the left wing of horse was commanded by Cromwell, his first line being composed of English, and the second of Scots under the leadership of David Leslie.

With extraordinary rashness and folly Rupert led his army down close to the enemy and posted it within striking distance, trusting that a ditch which covered his front would suffice to protect him from attack. The two forces having gazed at each other during the whole afternoon without moving, he at last dismounted between half-past six and seven and called for his supper, an example which was followed by several of his officers. The Parliamentary army seized the moment to advance with its whole line to the attack. Cromwell on the left led his cavalry across the ditch, and, though Rupert was quickly in the saddle to meet him, routed the leading squadrons of the Royalists. Rupert's supports however were well in hand, and falling on Cromwell threw his troops into disorder[1] till David Leslie, an excellent officer, brought up the Parliamentary supports in their turn and routed the Royalists. Then superior discipline told; Cromwell's men quickly rallied and the whole of Rupert's horse fled away in disorder. In the centre the Parliamentary infantry was for a time equally successful, but the horse on the right wing came to utter disaster. The ground on the right was unfavourable for cavalry, being broken up by patches of gorse; and although Thomas Fairfax with a small body of four hundred men, armed with lances, broke through the enemy and rode in disorder right round the rear of the Royalist army, the main body was hopelessly beaten. Goring, after the Swedish fashion, had dotted bodies of

[1] All kinds of reasons have been advanced to account for the (supposed) extraordinary fact that Cromwell's troopers at one moment were at a disadvantage. The explanation is quite simple, being no more than the usual swing of the pendulum in a combat of cavalry.

musketeers among his horse, who did their work 1644, July 2.
admirably. Part of Goring's troopers galloped off first to pursue, and then to plunder the baggage, while the remainder turned against the Scotch infantry and pressed them so hard that, in spite of Leven's efforts, almost every battalion was broken and dispersed. Three alone behaved magnificently and stood firm, till in the nick of time Cromwell returned from the left to rescue them. His appearance turned the scale, and the victory of the Parliament was made certain and complete.

Rupert after the action gave Cromwell the name of Ironside; he had never encountered so tough an adversary before. Marston Moor may indeed be termed the first great day of the English cavalry. We find, curiously enough, examples of three different schools in the field, the old school of the lance under Thomas Fairfax, the Swedish of mixed horse and musketeers under Goring, and the new English of Rupert and Cromwell; but the greatest of these is Cromwell's. He alone had his men under perfect control, and had trained them not only to charge, but what is far more difficult, to rally.

Little more than a week later came the first sign of an entirely new departure in the Parliament's conduct of the war. In spite of Marston Moor the general position of its affairs was anything but favourable. The inefficiency of local committees and the narrow self-seeking of local forces, combined with the jealousy of rival commanders and the absence of a commander-in-chief, threatened to bring swift and sudden dissolution to the cause. Time had aggravated rather than diminished the evil, and unless it were remedied forthwith, it would be useless to continue the war. Sir William Waller, an able commander, who had frequently suffered defeat less from his own incapacity than from the impossibility of keeping a force together, gave the authorities plainly to understand that unless they formed a distinct permanent army of their own, properly organised, properly disciplined, and regularly paid they could not hope for success.

Mutiny, desertion, and indiscipline had dogged every

1644. step of the local levies, as the Parliament very well knew; but experience still more bitter was needed before it could be induced to take Waller's advice. For the present it voted the formation of an army of ten thousand foot and three thousand horse and ordered it to be ready to march in eight days. Ignorance and infatuation could hardly go further than this. Shortly after came a great disaster in the west, nothing less than the capitulation of Essex's whole army. Then came the second battle of Newbury, which left the King in a decidedly improved position. Finally at the close of the campaign the Parliamentary forces sank into a condition which was nothing short of deplorable, the dissensions among the commanders rose to a dangerous height, and as a crowning symptom of the general collapse the Eastern Association, the strongest of all the local bodies, declared that its burden was heavier than it could bear and threw itself upon the Parliament. In the face of such a crisis the Houses could hesitate no longer, and on the 23rd of November they made over the whole state of the forces to the Committee of Both Kingdoms, with directions to consider a frame or model of the whole militia.

Thus the work that should have been done years before by Elizabeth was at length taken in hand; and the broken-down machinery of the Plantagenets was at last to be superseded. There was of course jealousy as to the hands in which so powerful an engine should be placed, and the difficulty was overcome only by the Self-denying Ordinance, which debarred members of both Houses of Parliament from command, and laid the ablest soldier in England aside as impartially as inefficient peers like Manchester and Essex. But such an evil as this could be easily remedied, for something more than an ordinance is required at such times to exclude the ablest man from the highest post. To bring the New Model into being was the first and greatest task; and this was done by the Ordinance of the 15th of February 1645. The time was come, and England had at last a regular, and as was soon to be seen, a standing army.

BOOK III

CHAPTER I

EVEN before the Ordinance for the establishment of the New Model Army had been passed, Parliament had voted, on the motion of Oliver Cromwell, that the chief command should be given to Sir Thomas Fairfax. There is little difficulty in discovering the reason for this choice. If by the Self-denying Ordinance all members of both Houses were to be excluded from command in order to rid the country of incompetent officers, there could be no doubt that Fairfax was the man best fitted to be captain-general. He had been the soul of the Parliamentary cause in the north, and, though by no means uniformly successful in the field, had shown vigour in victory, constancy in defeat, and energy at all times. Though not comparable to Cromwell in military ability, and perhaps hardly equal either to Rupert on the one side or to George Monk on the other, he was none the less a good soldier and a gallant man, though if anything rather too fond of fighting with his own hand when he should have been directing the hands of others. He knew the value of discipline and was strong enough to enforce it, but he understood also the art of leading men as well as driving them to obedience. Heir of a noble family and born to high station, he could fill a great position with naturalness and ease; being above all things a gentleman, honourable, straightforward, disinterested, and abounding in good sense, he could occupy it without provoking envy or jealousy. No higher praise can be given to Fairfax than that every one was not only contented but pleased to serve under him. 1645.

1645. Joined with him as sergeant-major-general, and therefore not only as commander of the foot but as chief of the staff, was the veteran Philip Skippon. His long experience of war in the Low Countries, and the respect which such experience commanded, doubtless prompted his selection to be Fairfax's chief adviser. The post of lieutenant-general, which carried with it the command of the cavalry, was left unfilled. Every one knew who was the right man for the place, and there could be little doubt but that, notwithstanding all self-denying ordinances, he must sooner or later be summoned to hold it. For the present he was employed, pending the expiration of the forty days of grace allowed him by the Ordinance, in watching the movements of the Royalist forces in the west. Though there had been trouble even with his famous regiments in the general collapse at the close of 1644, yet it was noticed that in January 1645 no troops had appeared so full in numbers, so well armed, and so civil in their carriage as Colonel Cromwell's horse. "Call them Independents or what you will," said one newspaper, "you will find that they will make Sir Thomas Fairfax a regiment of a thousand as brave and gallant horse as any in England."

This however was not to happen at once. Fairfax, having obtained the Parliament's approval of his list of officers, was busily engaged with Skippon in hewing rougher material than Cromwell's troopers into shape. Many of the disbanded regiments of Essex lay ready to his hand, but they had lately shown a mutinous spirit which it required all Skippon's tact and firmness to curb. The old man, however, as he was affectionately called, knew how to manage soldiers, and the promise of regular pay, notwithstanding that one quarter of the same was deferred as security against desertion, soon brought them cheerfully into the service. Nevertheless there were, even so, not voluntary recruits enough to supply the twenty-two thousand men required by the Ordinance; more than eight thousand were still wanting, and the Committee of Both Kingdoms could think of no better means for raising them than the press-gang. This was 1645. the system which, when enforced by Charles the First, had been denounced as an intolerable grievance, and it was not less violently resisted when sanctioned by Parliament. The Government, however, carried matters with a strong hand, and a couple of executions soon brought the recalcitrant recruits to submission.

The scene of the making of the New Army which was destined to subdue the King was, by the irony of fate, royal Windsor. It is on the broad expanse of Windsor Park and on the green meadows by the Thames, before the wondering eyes of the Eton boys, that we must picture the daily parade of the new regiments, the exercise of pike and musket and the assiduous doubling of ranks and files, old Skippon, gray and scarred with wounds, riding from company to company and instituting mental comparisons between them and the English soldiers of the Low Countries, and the younger sprightlier Fairfax, still but three-and-thirty, watching with all a Yorkshireman's love of horseflesh the arrival of troopers and baggage-animals. Every day the scene grew brighter as corps after corps received its new clothing, for the whole army, for the first time in English history, was clad in the familiar scarlet. Facings of the colonel's colours distinguished regiment from regiment; and the senior corps of foot, being the General's own, wore his facings of blue.[1] Thus the royal colours, as we now call them, were first seen at the head of a rebel army.

The senior regiment of horse was also in due time to be clothed in the same scarlet and blue. For Cromwell's two regiments of horse had been selected, as was their due, to be blent into one and to take precedence, as Sir Thomas Fairfax's, of the whole of the English cavalry. In this same month of April the regiment was in the field, turning out quicker than any other corps on the sounding of the alarm, while the "lovely company" of which the colonel had boasted,

[1] *Perfect Passages*, 30th April 1645.

1645. now called the General's troop, was distinguishing itself above all others. Modern regiments of cavalry that wear the royal colours need not be ashamed to remember that they perpetuate the dress of Oliver Cromwell's troopers. Excluded though Cromwell was from the making of the New Model Army, he was none the less its creator, for it was he who had shown the way to discipline and regimental pride.

It is now necessary briefly to sketch the organisation of the New Model. Beginning therefore with the infantry, the foot consisted of twelve regiments, each divided into ten companies of one hundred and twenty men apiece. As all the field-officers, even if they held the rank of general, had companies of their own, the full number of officers to a regiment was thirty : colonel, lieutenant-colonel, major, seven captains, ten lieutenants and ten ensigns. Each company included moreover two sergeants, three corporals, and one, if not two, drums.[1] The privates were divided as usual into an equal number of pikemen and musketeers: the weapons of officers being, for a captain, a pike ; for a lieutenant, the partisan ; and for an ensign, the sword. Since Skippon, a veteran of the Dutch school, was at the head of the infantry, it can hardly be doubted that the Dutch system of drill was preferred to the Swedish. Gustavus Adolphus, it must be remembered, was chiefly concerned with the Scots ; while the contemporary drill books of the English prefer the teaching of Maurice of Nassau. It is therefore reasonably safe to conclude that the normal formation of the infantry of the New Model was not less than eight ranks in depth.

The cavalry consisted of eleven regiments, each of which contained six troops of one hundred men. Here

[1] The drum-calls were six in all : 1, Call ; 2, March ; 3, Troop ; 4, Preparative ; 5, Battle ; 6, Retreat. The trumpet-calls were also six : 1, Butte sella, corrupted since into "Boot and Saddle" ; 2, Monte cavallo (mount) ; 3, Tucket (warning for march) ; 4, Carga (charge) ; 5, Alla Standarda (to the Standard) ; 6, Auquet (watch-setting).—Ward, *Animadversions of Warre.*

again every field-officer had a troop of his own, so that 1645. the full complement of officers to a regiment numbered eighteen, namely, colonel, major, four captains, six lieutenants, and six cornets. Three corporals and a trumpeter were included among the hundred men; and the admirable system which sorted each troop into three divisions, each under special charge of an officer and a corporal, was in full working order. In the matter of drill and tactics, the English cavalry was before rather than behind the times. The modified shock-action of Gustavus Adolphus had, under the influence of Rupert and Cromwell, been virtually superseded. The men indeed were still armed, according to the old fashion, with iron helmet and cuirass, and still carried each a brace of pistols as well as a sword; but they were instructed to trust to their swords in the charge, and to use their fire-arms only in the pursuit. Gustavus had formed his horse as a rule in four ranks; Rupert fixed the depth at three;[1] the Parliamentary officers went so far as to reduce the ranks to two, sacrificing depth to frontage, and trusting to speed, we cannot doubt, to overcome weight. Last and most daring innovation of all, they abolished the file as the tactical unit of the troop and substituted the rank in its place.[2] No better testimony to the improvement of English discipline could be found than this reduction in the depth of the ranks of cavalry. For once it may be said that the English horse stood in advance of all Europe.

As regards the duties of reconnaissance, not a treatise on cavalry omits to mention that it is the function of the horse to scour the ways in advance of an army; but there are no precise directions as to the manner of fulfilling it. Cromwell's constant references

[1] *The Young Horseman and Honest Plain-dealing Cavalier*, by John Vernon, 1644. A short drill-book in pamphlet form, prepared by a cavalier-officer in small compass for officers "to weare in their pocket." This is the first soldier's pocket-book for field service in our language. It is among the King's Pamphlets in the British Museum.

[2] Barriffe.

1645. to a "forlorn" of horse show that he employed advanced parties regularly, and attention has already been called to the efficiency of Rupert's patrols. There is no evidence, however, that the men received any instruction in the matter of reconnaissance, and it is only from the Royalist Vernon that we learn that vedettes were posted then, as now, in pairs.

The dragoons of the New Model seem, in spite of a resolution of the Commons that they should be regimented, to have been organised in ten companies, each one hundred strong. Their officers were a colonel, a major, eight captains, ten lieutenants, and ten ensigns. The dragoons were mounted infantry pure and simple, riding for the sake of swifter mobility only, and provided with inferior horses. They were armed with the musket and drilled like their brethren of the foot; their junior subalterns were called ensigns and not cornets, and they obeyed not the trumpet but the drum. Their normal formation was in ten ranks of ten men abreast. For action, nine out of the ten dismounted, and linking their horses by the simple method of throwing the bridle of each over the head of his neighbour in the ranks, left them in charge of the tenth man.[1]

Next we must glance at the Artillery which, together with the transport, was comprehended under the head of the Train. The only organised force of which we hear as attached to the train is two regiments of infantry and two companies of firelocks, which were used for purposes of escort only. The firelocks were distinguished from the rest of the army by wearing tawny instead of scarlet coats, and seem therefore to have been a peculiar people, but the immediate connection of flint-lock muskets with cannon is not apparent.

[1] Sometimes however the dragoons seem to have taken with them ten extra men per company simply to hold the horses. There are fugitive references to light dragoons even at this early period, but no clear account of them. After a few years it was as usual to speak of troops as of companies of dragoons.

The truth seems to be that the English were behind the times in respect of field artillery, and indeed we hear little of guns, except siege-cannon, during the whole period of the Civil War. English military writers of the period rarely make much of artillery in a pitched battle. They recommend indeed that the enemy's guns should be captured by a rush as early as possible, and they generally agree that cannon should be posted on an eminence, since a ball travels with greater force down-hill than uphill. On the other hand, it was objected even to this simple rule that if guns were pointed down-hill there was always the risk of the shot rolling out of the muzzle, so that in truth the gunner seems to have been sadly destitute of fixed principles for his guidance in action. 1645.

The neglect of field artillery in England is the more remarkable inasmuch as English gun-founders enjoyed a high reputation in Europe. The cannon of that day were necessarily heavy and cumbrous, since the bad quality and slow combustion of the powder made great length imperative; but there was no excuse for not imitating the light field-pieces of Gustavus Adolphus. The probable reason for the backwardness of the English was the peculiar organisation of the Dutch artillery, which gave no opening for the instruction of English gunners in the school of the Low Countries. Nevertheless there was a distinct drill for the working of guns, with thirteen words of command for the wielding of ladle and sponge and rammer. A gun's crew consisted of three men—the gunner, his mate, often called a matross, and an odd man who gave general assistance; and the number of little refinements that are enjoined upon them show that the artillerymen took abundant pride in themselves. Thus the with-drawal of the least quantity of powder with the ladle after loading was esteemed a "foul fault for a gunner to commit," while the spilling even of a few grains on the ground was severely reprobated, "it being a thing uncomely for a gunner to trample powder under his

1645. feet." Lastly, every gunner was exhorted to "set forth himself with as comely a posture and grace as he can possibly; for the agility and comely carriage of a man in handling his ladle and sponge is such an outward action as doth give great content to the standers-by." Nevertheless artillerymen seem nowhere, and least of all in England, to have been very popular. They had an evil reputation all over Europe for profane swearing, a failing which is attributed by one writer to their enforced commerce with infernal substances, but which was more probably due to the fact that, being less perfectly organised than other branches of the army, they were less amenable to rigid discipline.

But if the gunners were but a casual and ill-administered force, much more so were the drivers. Over a thousand draught-horses were collected for the general use of the New Model, but how many, if any, of these were set apart for the artillery, it is impossible to say. Ordinary waggoners with their teams were impressed or hired to haul the guns, and it is recorded that the hackney-coachmen of London performed the duty more than once. The chief use of the escort of infantry was therefore to prevent the drivers from running away. It is doubtful whether the guns themselves travelled on four wheels or on two, contemporary drawings showing instances of both; but in either case there was no approach to what is now called the limber, the horses being harnessed simply to the trail.[1] The ammunition again was transported in ordinary waggons, the powder being indeed occasionally made up into cartridges, but more often carried simply in barrels which were unloaded behind the gun when it was posted for action. It was the function of the odd man of the gun's crew to cover up the powder-barrel between each discharge of the gun, to avert the danger of a general explosion. In fact, one principal link alone connects the artillery of the New Model with the artillery of to-day, the gun-carriages were painted of a fair lead-colour.

[1] Which was then called the limber.

Lastly we come to the Engineers, a corps which is more obscure to us even than the Artillery. Even in the days of the Plantagenets the English kings had taken Cornish miners with them for their sieges; and in the war of Dutch Independence Yorkshire colliers were specially employed for the digging of mines. But, although by the middle of the sixteenth century the Germans had already organised a corps of sappers,[1] no such thing existed in England. In truth, the British were not fond of the spade. The English indeed handled it often enough under Vere and his successors, while the Scots, though sorely against the grain, were forced to do the like by Gustavus Adolphus. But considering the schools wherein the British were trained, nothing is more remarkable in the Civil War than the neglect of field-fortification and the extreme inefficiency with which at any rate the earlier sieges were conducted. It is significant that the pioneers,[2] who are the only men that we hear of in connection with the unorganised corps of engineers, were the very scum of the army, and that degradation to be "an abject pioneer" was a regular punishment for hardened offenders. It is still more significant that the principal engineers of the New Model Army bear not English but foreign names. 1645.

So much for the various branches of the military service: it remains to say a few words of the Army as a whole. Of the organisation of what would now be called the War Department, it is extremely difficult to speak. There was a parliamentary Committee of the Army, which seems to have enjoyed at first an intermittent and later a continuous existence, and which was entrusted with the general direction of its affairs and in particular with the business of recruiting. There were also Treasurers at War, who were charged with the financial administration, and there was the already venerable Office of Ordnance, which was responsible for arms and equipment. Speaking generally, though the functions of the

1 Schanzbauern. *Fronsperger*.

2 They stood on much the same level in France.

1645. Committee and of the treasurers seemed to have overlapped each other at various points, the military administration seems to have tended to the following allocation of responsibility: that the Committee of the Army took charge of the men, the Office of Ordnance of the weapons and stores, and the Treasurers at War of the finance, while the Commander-in-Chief was answerable for the discipline of the Army.

Passing next to purely military organisation, which of course fell within the province of the Lord-General, it is to be remarked that the makers and commanders of the New Model knew of no better distribution of command than under the three heads of Infantry, Cavalry, and Train. There was no such thing as a division comprehending a proportion of all three arms under the control of a divisional commander; and though we do hear frequently of brigades, the word signifies merely the temporary grouping of certain corps under a single officer, rarely an essential part of the general organisation. The subjoined list gives a tolerable idea of the allotment of functions among the members of the staff. It is only necessary to add that all orders of the commander-in-chief were issued through the sergeant-major-general, distributed by him to the sergeant-majors or, as they are now called, majors of the different regiments, and by the sergeant-majors in their turn to the sergeants of every company and the corporals of every troop.

COMMANDER-IN-CHIEF.

His Excellency Sir Thomas Fairfax, Knight, Captain-General.

HEADQUARTER STAFF.

(*Chief of the Staff*)—Major-General[1] Skippon.
Commissary-General of the Musters.—Comm.-Gen. Stone (with two deputies).
Commissary-General of Victuals.—Comm.-Gen. Orpin.

[1] So in Sprigge, more properly Sergeant-Major-General.

Commissary-General of Horse Provisions.—Comm.-Gen. Cooke.
(*Transport*) *Waggon-Master-General.*—Master Richardson.
(*Intelligence*) *Scout-Master-General.*—Major Watson.
(*Military Chest*) Eight Treasurers at War (civilians), (with one deputy).
Judge Advocate-General.—John Mills (civilian).
(*Medical*) *Physicians to the Army.*—Doctors Payne and Strawhill.
" *Apothecary to the Army.*—Master Web.
Chaplain to the Army.—Master Boles.
(*Military Secretary*) *Secretary to the Council of War.*—Mr. John Rushworth (civilian), with two clerks.
(*Aides-de-Camp*) *Messengers to the Army.*—Mr. Richard Chadwell, Mr. Constantine Heath.

FOOT.[1]

Major-General	Skippon.
Quartermaster-General	Spencer.
Assistant-Quartermaster-General	Master Robert Wolsey.
Adjutant-General	Lieutenant-Colonel Gray.
Marshal-General	Captain Wykes.

Ten regiments of foot; each regiment of ten companies; each company of one hundred and twenty men, exclusive of the officers.

REGIMENT.	COLONEL.	REGIMENT.	COLONEL.
1st.	Sir Thomas Fairfax. Lieut.-Colonel Jackson.	5th.	Harley.
		6th.	Montague.
2nd.	Major-General Skippon. Lieut.-Colonel Frances.	7th.	Lloyd.
		8th.	Pickering.
3rd.	Sir Hardress Waller.	9th.	Fortescue.
4th.	Hammond.	10th.	Farringdon.

HORSE.

Lieutenant-General	Oliver Cromwell.
Commissary-General	Henry Ireton.
Quartermaster-General	Fincher.
Adjutants-General	Captains Fleming and Evelyn.
Marshal-General	Captain Laurence.
Mark-Master General	Mr. Francis Child.

Eleven regiments of horse; each of six troops; each troop of one hundred men, besides officers.

[1] In Sprigge's list the foot take precedence of the horse; and this was the rule in the English, though not in the French, army.

Regiment.	Colonel.	Regiment.	Colonel.
1st.	Sir Thomas Fairfax. Major Disbrowe.	6th.	Lieut.-General Cromwell.
		7th.	Rich.
2nd.	Butler.	8th.	Sir Robert Pye.
3rd.	Sheffield.	9th.	Whalley.
4th.	Fleetwood.	10th.	Graves.
5th.	Rossiter.	11th.	Comm.-General Ireton.

The captain-general's bodyguard consisted of one troop, taken from his regiment of horse, under Colonel Doyley.

Dragoons.

Colonel Okey.

Ten companies each of one hundred men, besides officers.

Train.

Lieut.-General of the Ordnance	Lieut.-General Hammond.
Controller of the Ordnance .	Captain Deane.
Engineer General . . .	Peter Manteau van Dalem.
Engineer Extraordinary . .	Captain Hooper.
Chief Engineer . . .	Eval Tercene.
Engineers	Master Lyon, Master Tomlinson.
Master Gunner of the Field .	Francis Furin.
Captain of Pioneers . .	Captain Cheese.

A Commissary of Ammunition.
A Commissary of the Draught Horses.

Two Regiments of Infantry { Colonel Rainborough's. Colonel Weldon's.

Two companies of Firelocks.

1645, April 30. The regiments of the New Model were not yet complete when Fairfax received orders from the Committee of Both Kingdoms to march westward to the relief of Taunton. It is extraordinary that this presumptuous body of civilians, even after it had provided the General with an efficient army, still took upon itself to direct the plan of campaign. It is still more extraordinary that Fairfax, who had disregarded it before Marston Moor, should now have meekly obeyed. Charles, whose chief hopes rested in a junction with the gallant and victorious Montrose, was actually moving northward to meet him while Fairfax was tramping away to Taunton. Nay, even after Taunton had been relieved, the sage Committee could think of no better employment for the New Model than to set it down to the siege of Oxford. Fatuity could hardly go further than this. There were in the field on both sides four armies in all, ranged alternately, so to speak, in layers from north to south. Northernmost of all was Montrose, below him in Yorkshire lay Leven with the Scots, south of Leven was Charles, and south of Charles the New Model. And yet the Committee proposed to keep Fairfax inactive before Oxford while Charles and Montrose crushed Leven between hammer and anvil. 1645.

A brilliant victory of Montrose at Auldearn brought matters to a crisis. Leven was compelled to retreat into Westmoreland; and the Scots insisted that Fairfax must break up from before Oxford and move up towards the King. Charles, meanwhile, with his usual indecision had suspended his march northward for the sake of capturing Leicester, and was now lying at Daventry, uncertain whither to go next. Fairfax called a council of war, which decided to seek out the enemy and fight him wherever he could be found, and, more important still, requested the appointment of Cromwell to the vacant post of lieutenant-general. The Parliament meanwhile had come to its senses, and resolved that the General should henceforth conduct his own campaign without the advice of a committee of civilians. Having done so, it could hardly refuse to sanction the return of Cromwell. He was therefore summoned to headquarters; and Fairfax began to work in earnest. So energetic were his movements, when once the paralysing hand of the Committee was withdrawn, that the Royalists at once jumped to the conclusion that "Ironside" had rejoined the army. May 9.

He had not yet rejoined it, and yet the Royalists were right, for it was his spirit, the spirit of discipline, that was abroad in the army. The New Model was by no means perfect when it marched from Windsor at the end of April 1645. The old failings of insubordina-

1645. tion, desertion, and plunder, natural enough among a body of men largely recruited by impressment, showed themselves abundantly at the outset of the march to Oxford, but they were put down with a strong hand, not by preaching, but by hanging. Nor was it by severity only that Fairfax brought men to their duty. According to custom, every regiment was told off in succession to furnish the rearguard, but when the turn of Fairfax's regiment came, the men claimed that, being the General's own, they had a right to a permanent place in the van. Fairfax said nothing, but simply jumped off his horse and tramped along in the midst of them in the rearguard ; and after this there were no more quarrels over precedence. After a month in the field the newspapers could report that oaths, quarrelling and drunkenness were unknown in the New Model. "Yea, but let Cromwell be called back," they added ; and before long this too was done. At six o'clock on the morning of the 13th of June, while Fairfax was sitting at a council of war, Cromwell marched into the camp at Kislingbury at the head of his regiment. It was but a small reinforcement of six hundred troopers, but as they rode in a cheer rose from the cavalry which was taken up by the whole army, as the word ran round the camp that Noll was come.

1645, June 14. Next day was fought the battle of Naseby. It was not a well-managed fight. After considerable shifting of position, so much prolonged that Rupert came to the conclusion that Fairfax wished to decline an engagement, the New Model Army was finally drawn up on the plateau of a ridge about a mile to the north-east of Naseby village. It lay behind the brow of a hill which slopes down somewhat steeply to a valley below called the Broadmoor, and was formed according to the usual fashion of the time. Six regiments of three thousand six hundred horse formed the right wing, seven thousand infantry under Skippon made up the centre, two thousand four hundred more horse under Ireton made the left. Ireton's flank was covered by a hedge, which by Cromwell's direction was lined with dismounted dragoons. 1645, June 14.

The disposition of the Royalists was of the same kind, though their force was of little more than half the strength of the New Model. The right wing of cavalry was under Rupert, the centre of infantry under old Sir Jacob Astley, the left wing of cavalry under Sir Marmaduke Langdale. Each army held two or three regiments of infantry in reserve.

Rupert, conspicuous in a red cloak, opened the action by a rapid advance with his horse against Fairfax's left. Ireton thereupon drew over the brow of the hill to meet him, and Rupert, evidently rather astonished to find so large a force in front of him, incontinently halted. Ireton then made the fatal mistake of halting likewise. Whether he was hampered by the ground or unequal to the task of handling so large a body of horse, is uncertain; but, whatever the reason, his wing was in disorder, and instead of continuing the advance he began to correct his dispositions. Rupert at once seized the moment to attack. A few divisions under Ireton's immediate leadership charged gallantly enough and held their own until driven back by Rupert's supports, but the rest hung back, and Rupert pressing on, as was his wont, scattered them in confusion. Ireton, losing his head, instead of trying to rally them, plunged down with his few squadrons on the Royalist infantry, was beaten back, wounded and taken prisoner; and in fact the left wing of the New Model was for the time completely overthrown. Away went Rupert in hot pursuit with his troopers at his heels for a mile beyond the battlefield, and galloping up to the park of Parliamentary baggage, summoned it to surrender. He was answered by a volley of musketry, and then too late he recollected himself and rode back to the true scene of action.

In the centre also matters again had gone ill with the Parliament. Skippon was wounded early in the day, and though he refused to leave the field was unable

1645, June 14. actively to direct the engagement. Either his dispositions were incomplete, or his colonels were helpless without him; but the left centre, its flank exposed by Ireton's defeat, gave way and in spite of all the efforts of the officers could not be rallied. Fortunately Fairfax's regiment on the right centre stood firm; and the steadiness of three regiments in the reserve enabled the Parliamentary infantry to maintain the struggle.

But it was on the right that the best soldier in the field was stationed, and his presence counted for very much. He too was hampered by bad ground, patches of gorse and a rabbit-warren on his extreme right preventing all possibility of a general advance of his wing. But instead of halting like Ireton he took the initiative in attack. The leftmost troops under Whalley, having good ground before them, at once moved down, fired their pistols at close range,[1] and fell in with the sword. Langdale's horse met them gallantly enough, but were beaten back and retired in rear of the King's reserve, where they rallied. But Whalley's supports came up quickly to second him, and meanwhile the rest of Cromwell's wing came up as best it could over the broken ground, and falling on the opposing bodies of Royalist horse routed all in succession. The Royalists retreated for a quarter of a mile and rallied; and Cromwell, detaching part of his horse to watch them, rode down with three regiments against the King's reserve of horse. Charles, to do him justice, bore himself gallantly enough, but some one gave the unlucky word, "To the right turn—march!" whereupon the whole of his men turned tail and sweeping the King along with them joined their beaten comrades in rear. Thither also presently came Rupert with such a following of blown and beaten horses as he could collect. Ireton's wing had rallied, and was pressing so close on his rear that he dared not stop; and Rupert's foolish

[1] This incident shows that shock-action was not yet wholly the rule.

and premature pursuit had squandered his squadrons as effectually as a defeat. 1645, June 14.

The whole of Charles's army was now beaten or dispersed except his centre, and against this the whole force of the Parliamentary army was now directed. Okey, who commanded the dragoons, finding the ground clear before him, made his men mount and attacked it in flank; Fairfax's regiment of foot engaged it in front, and Ireton's rallied troopers in rear. All soon laid down their arms excepting a single battalion,[1] which stood alone with incredible courage and resolution till it was fairly overwhelmed. Even so, however, Fairfax dared not advance further till he had reformed his whole line of battle. But the Royalists could not face a second attack; they turned and fled; and the Parliament's cavalry pursued the fugitives for fourteen miles, capturing the whole of the King's artillery, his baggage, and practically his entire army. It was a decisive victory though not a very glorious one. But for Cromwell, who alone after Skippon's fall seems to have kept his wits about him and his men in hand, Naseby would probably have added one more to the indecisive battles of the Civil War.

Nevertheless the New Model had won its first action, and Fairfax now started on a campaign to the west, which did not end until he had penetrated through Somerset, Devon, and Cornwall, and crushed Royalism under foot even to the Land's End. It was a long march of incessant and at first of severe fighting, which taxed the mettle even of his best soldiers, but the army gathered strength, in spite of constant hardships, in its swift progress from victory to victory, and by the summer of 1646 it had finished the work begun at Naseby and was virtually master of England. Meanwhile the persistent folly of the King had raised it from a partisan to a national army. Charles, who had no

[1] Called by the name of a *tercio* in the contemporary plans, being formed probably in the old Spanish formation which Tilly had used at Leipsic.

1646. spark of patriotic feeling in him, had from the first striven not only to set nationality against nationality within the British Isles, but had appealed to foreigners from France, Lorraine, and Holland to uphold his rights. All these transactions had been revealed by the capture of his baggage at Naseby; and his defiance of all the insular prejudice of the English damaged him unspeakably even with those who were most sincerely attached to his cause. Margaret of Anjou was not yet forgotten; and if men coupled Charles's name with hers, it was no more than he deserved. Now, however, he was beaten, beaten on every side. In the first six months of 1645 Montrose, perhaps the most brilliant natural military genius disclosed by the Civil War, had scored success after success with a handful of Scots and Irish. A woman in emotion and instability, a man in courage, and a magician in leadership, he was an ideal leader for such untameable, combative spirits, the stuff of which Dundonalds are made. Yet Montrose's work had been undone at Philiphaugh, and Charles's last hope
Sept. 13. was gone. A few more ineffectual struggles to divide England against herself, and he was to be purged away as a public enemy by the ever victorious army.

CHAPTER II

On the subjugation of the west the English Parliament 1646. thought for the present only of securing its position within England itself. It has been seen how at the first outbreak of the war the Parliamentary leaders had taken the Scottish army into pay, and how even after the formation of the New Model they had tried to saddle it with the hardest of the work. In truth, the behaviour of the Parliament towards the Scots had been sufficiently shifty and ungracious; it had taken at any rate some care to pay its own troops, but it persistently neglected its allies, who had done excellent service in the north. Indeed, had Leven yielded to the English Parliament's wishes, had he not in fact been forced by the victory at Auldearn to retreat, the Scots instead of the English might have won the Naseby of the Civil War, an event which would have led to untold complications. Now however that the English army had done the work for itself, all parties in England became anxious to be rid of the Scots. Matters were somewhat confused by the fact that in 1646 Charles threw himself into the hands of Scotland; but by the close of the year it was agreed that the Scottish army should be paid off and withdrawn over the border, and that the King should be surrendered to the English, who had conquered him. The Parliament therefore gained its great object, a free hand for the management of its own affairs. It overlooked however in its calculations one important factor, the Army.

At the opening of 1647 there was a general cry 1647.

1647. throughout England for peace. The country was exhausted; the finance of the Parliament was in hopeless disorder; and the people groaned under the enormous expense of the war. Obviously the most natural item for retrenchment was the Army; its work was done, and there was no further reason for its existence; it should therefore be disbanded or at any rate very greatly reduced. Moreover economy was not the only motive that prompted such a policy. The Parliament, united for the moment in the general desire to get quit of the Scots, fell back, almost immediately after this was accomplished, into faction. Presbyterians and Independents were the original names of the two rival parties, but for our purpose it is simpler to narrow them forthwith to Parliament and Army; for among many of the Presbyterian members who had held commands in the first years of the war, there existed a professional as well as a political and religious jealousy of the successful officers who had supplanted them. Parliament having created the Army by a vote thought that it could extinguish it by the same simple process; having used it as a ladder whereon to rise to undisputed supremacy it now proposed to kick it down. But such an Army was not disposed to make itself a plaything of Parliament.

Petitions from various quarters for the disbandment of the New Model turned the heads rather than strengthened the hands of the two Houses. The only safe and honest course, if the Army must be disbanded, was to discharge the whole of the country's obligations to it in full. Now the pay of the foot was eighteen weeks and of the horse forty-two weeks in arrear, and the total debt due to the forces amounted to three hundred and thirty thousand pounds. The Parliament was in straits for money and by no means inclined to make the necessary effort to raise this sum. It proposed as an alternative to turn twelve thousand of the soldiers into a new army for the pacification of Ireland, and this without a word as to the terms on which the men had taken service, and without the least mention of a settlement of arrears. Further, as if it were not enough to irritate the men, the Parliament did its best to alienate the officers. It passed resolutions insulting to the army, insulting to Fairfax, insulting to Cromwell. So deeply injured indeed was Oliver by this ungrateful treatment, that he thought seriously of carrying his sword and such troops as he could raise to the wars in Germany. Such was the pitch of disgust to which the Parliament had driven the ablest of its servants. 1647.

The Army raised its first protest in the form of a respectful petition from the men: the Parliament met it with violent and ungracious censure. Certain officers who had supported this petition then tendered a vindication of their conduct: the Commons refused even to read it. Finally, as if to aggravate the Army to extremity, the Lords proposed to grant the troops six weeks' pay in temporary satisfaction of arrears. This was too much. Discontent grew apace in the ranks, the men refused to have anything to do with service in Ireland, and finally the Army, by the election of two representatives for each regiment, organised itself for the orderly maintenance of its just claims. These representatives were called agitators, a name which in those days signified simply agents. The degradation of the term in our own time into a synonym for political busy-bodies must not mislead us, nor blind us to the dignified patience, under extreme provocation, of this irresistible body of disciplined men.

For the moment the Parliament was awed into concessions and promises, but its leaders did not lightly submit to humiliation, and rather than yield to the Army looked about for a force to countervail it. First they turned to the City of London, which was strongly Presbyterian, and sought an armed force in the City train-bands. Next they resorted to Scotland, which was intensely jealous of the New Model, and formed a coalition with it in favour of the King, thereby

1647. sowing the seeds of a quarrel between North and South Britain. Finally, after stultifying itself by a promise May 25. of attention to the Army's complaints, it passed an Ordinance for its disbandment without further ado. This was past endurance. The soldiers broke into open mutiny; and Fairfax and Cromwell, having striven in vain to gain justice for their men, and at the same time to keep them in subordination to the Parliament, placed themselves at the head of a movement which they could no longer repress. It was indeed high time, for the Presbyterian leaders had already invited the Prince of Wales to place himself at the head of the Scots for an invasion of England.

On the 4th of June the Army assembled about four miles from Newmarket at Kentford Heath. There in the course of the next few days it erected a general council, composed of the general officers who had taken the side of the men and of two officers and two privates from each regiment, and made a written declaration of its policy. Still the Parliament remained obstinate, and now endeavoured to enlist the discharged soldiers of the earlier armies in order to meet force with force. The Army advanced to Triplow Heath, whither Parliament sent a last message to propose terms for an agreement. The overtures were rejected, and the Army continued its advance. In panic fear the Parliament now offered bribes to any officers or men who would desert the Army. This contemptible device was a total failure. It then tried to raise troops, to reopen negotiations with the Army, to call out the London trained bands, to forbid the Army's further advance, to gain certain troops, which were not of the New Model, from the north; all was in vain. Irresistible as fate, the Army marched on. At St. Albans it halted and issued a manifesto demanding the expulsion of eleven of its enemies from the Commons, and receiving no encouragement advanced to Uxbridge. There again it halted and spent three weeks in the hopeless effort to arrange a peaceful settlement with the King; and finally it marched straight into London and occupied the capital. 1647, August 6.

Still the Commons persevered in opposition to the Army; and at last Cromwell, without the orders and in spite of the unwillingness of Fairfax, gave the Presbyterian majority a strong hint to convert itself into a minority. His arguments consisted of one regiment of horse, stationed in Hyde Park, and a small party of foot at the door of the House; and they were sufficient and conclusive. The House thus purged, Cromwell turned to the task which was to occupy the remainder of his life and drive him worn-out to his grave, a final settlement of the original quarrel. Wisely enough he thought that this could be effected only by agreement with the King; and it was to negotiation with Charles Stuart for this object that he now devoted the whole of his energy. But negotiation with a man who was constitutionally incapable of straightforward and honourable dealing could have but one end. The lower ranks of the Army, not more far-seeing but less sanguine than their leader, again interposed. A section of extremists, known at that time by the name of Levellers, began, as is usual at such times, to raise its head, and condemning all further traffic with the King boldly put forward a revolutionary scheme of its own.

Herein, however, the Levellers mistook their man. However Cromwell might be distracted by the difficult questions of a settlement, he was perfectly clear on one point, that the discipline of the Army must be maintained. Symptoms all too significant appeared that that discipline was impaired, and he lost no time in restoring it. One regiment refusing to obey his orders, Cromwell promptly drew his sword and rode single-handed straight into the middle of the malcontents. His resolution speedily convinced the men that he would not be trifled with; the mutineers yielded, and a single execution sufficed to re-establish order.

Then as usual the portentous folly of the King united all parties not only in the Army but in England 1648, January.

1648. against himself. He might have made honourable terms with Cromwell; he preferred to throw himself into the arms of the Scots. Both Houses of Parliament thereupon broke with their North British allies, and the dispute assumed the new phase of a quarrel between English and Scots. English refugees inflamed national feeling at Edinburgh, and on the 11th of April the Scottish Parliament pronounced the treaty between the two nations to be broken. By the first week in May the army which was to invade England began slowly to assemble, and on the 8th of July it crossed the border, ten thousand five hundred strong, and occupied Carlisle.

July. Meanwhile the energies of the English had been distracted by Royalist risings in Kent and in Wales which kept Fairfax and Cromwell both busily employed; and it was not till the 11th of July that Cromwell was able to leave Pembroke and march to the north. Even then his force, after a trying campaign in very inclement weather, was in no very good state. He was entirely destitute of artillery, and his men were most of them both shoeless and stockingless. In one principal respect, however, the force was strong, for it was perfect in spirit and in discipline. I shall not dwell on the details of Cromwell's dash from Wales into Yorkshire. The Scots, embarrassed by a multitude of commanders, suffered him to attack their far more numerous army in detail, when it was divided on opposite banks of the Ribble; and after one sharp engagement at Preston the campaign resolved itself into a mere pursuit of the beaten Scots. How hotly Cromwell pressed the chase, and with what hardships to his own little army, may be read in his own despatches. Unfavourable weather, torrents of rain, and the miserable state of the roads brought men and horses to the last stage of exhaustion. "The Scots," wrote Cromwell, "are so tired and in such confusion that if my horse could but trot after them we could take them all, but we are so weary we can scarce be able to do more than walk after them . . . my horse are miserably beaten out, and we have ten thousand prisoners." The memory of this swift raid into Yorkshire, and of the unrelenting chase that followed it should be treasured by the British cavalry that fought through the Pindarri war and the Central Indian campaign of 1857-58. 1648.

With the close of the pursuit after Preston, the second Civil War came to an end. The operations of Fairfax in the south had shown him at his very best, swift, active, and resolute, and had been brilliantly successful. Those of Cromwell in the north, though they were directed against Royalist Scotland only, not yet the sterner Scotland of the Covenant, had been crushing. England was now completely under the sway of the Parliament; but it became a question whether Parliament was its own master. A movement arose in the Army for the punishment of the men who had brought all this bloodshed upon the country, and in particular of the chief delinquent, Charles Stuart, who was guiltiest of all. By a final overture for a settlement the Army gave the King a last chance, and on its failure appealed to Parliament to bring him to justice.

Ireton seems to have been the moving spirit in the actions that followed, though there can be no doubt that Cromwell was in full sympathy with them. Oliver was intensely English in spirit, and had been greatly exasperated by the English Royalists who had called the Scots over the border. He was vehement for justice upon them, and upon the King as the chief of them. Parliament, on the other hand, was engaged in nominal negotiations with Charles; and it was therefore not to be expected that it would comply with the Army's request that he should be brought to trial. But the Army was not to be stopped. The King's person was seized; the Parliament was purged of recalcitrant members; and from these actions to the High Court of Justice the march was short. One leading soldier, Fairfax, did indeed recoil from the final step, but the majority of the officers pressed on; and on the 30th of

1649. January 1649, the King was brought out into the ring of red coats to meet his death. He had done his worst against the British Isles. He had invited foreign armies against England, and when he failed had roused Welsh, Scots, and Irish to a hopeless effort to subdue her. But he succeeded only in establishing her strength; and the fall of his head was but the first instalment of the great work done by Cromwell and the Army towards the unity of the islands under the supremacy of England.

We have a pleasant glimpse of Oliver in his lighter moods before he next unsheathed his sword. On the evening of the 23rd of February, as he and Ireton were returning from dinner with Bulstrode Whitelocke, their coach was stopped by the soldiers who were in charge of the streets. They explained who they were, but the captain of the guard would not believe them and threatened to put them into the guard-room. Ireton began to lose his temper, but Cromwell laughed, and pulling out twenty shillings gave them to the men as a reward for doing their duty. Less than three weeks later he was summoned to take command of the army that was collecting for the reconquest of Ireland; for that unlucky island had been chosen by the Royalists as the base of operations for the invasion of England. Rupert, now turned admiral, had already sailed to Kinsale to enlist Irish sailors, and the faithful Ormonde had invited Charles the Second to place himself at the head of the loyal party in Ireland. Cromwell was not unwilling to undertake the duty. He had no idea of yielding England either to Scots or Irish, least of all to the Irish, whose land was regarded rather as a colony than as an integral part of the realm, and was also a stronghold of papistry. Still he declined to accept the command until he had assured himself that all the wants of his troops should be satisfied; he loved his men and would not suffer them to be enticed by the magic of his name to thankless or unprofitable service.

Four regiments of foot and one of horse were then

chosen by lot, and the men were informed that they 1649.
need not go to Ireland unless they wished, but that if they refused they would be discharged from the Army. Several hundred men thereupon at once threw down their arms and were dismissed; but by some blunder, which was none of Cromwell's, not a word was said about the payment of the arrears that were due to them. The idea spread through the ranks that they must either go to Ireland or forfeit those arrears; discontent was naturally aroused and presently burst out into formidable mutiny. Fairfax and Cromwell, however, could depend on their own regiments, and faced the danger with extraordinary swiftness and energy. The mutineers were suppressed with a strong hand. One ringleader was executed in St. Paul's Churchyard, a cornet and a corporal were shot before the eyes of their comrades against the walls of Burford Church, and discipline was again restored. Shortly after, Parliament passed an Ordinance to relieve the financial difficulties of the soldiers, and the preparations for the Irish campaign were resumed. It is curious to note the extreme slowness with which the civilians learned that soldiers were after all men of flesh and blood, not puppets to be hugged or broken according to the caprice of the hour.

The details of the preparations for the war in Ireland may still be read in the State Papers of the time. There are still to be seen the orders for fifteen thousand cassocks, "Venice-red colour, shrunk in water," the like number of pairs of breeches "of grey or other good colour," ten thousand shirts, ten thousand hats and bands,[1] one thousand iron griddles, fifteen hundred kettles, giving a curious picture of the equipment of the first English regular army for what was then esteemed to be foreign service. But I shall not follow the red coats through the terrible Irish campaign of 1649. It was not, like the later war with the Scots, an honourable

[1] This item furnishes indirect evidence that either few pikemen were employed, or that if employed they were stripped of defensive armour. The pike was already falling obsolete.

1649. contest for supremacy: it was rather the stern suppression of a rebellion, wherein the spirit of the masters was inflamed by the insolence of long superiority, by the bitterness of religious hatred, and by the recollection of past outrages which, even if truly reported, would have kindled men to vengeance, and when exaggerated by rage and fear fairly blinded them to mercy. If any Englishman doubted whether the Irish could fight with desperate gallantry he was undeceived at the storm of Drogheda and at Clonmel: but they could not stand, untrained and unorganised as they were, against the veterans of the New Model. Much has been said about Cromwell's cruelty, and that he was ruthlessly severe there can be no question; but when we speak of cruelty we should take at any rate some account of the standard of humanity in the warfare of the seventeenth century. The Irish War was a war of races, a war of creeds, and a war of vengeance. That there should therefore have been such slaughter as at Drogheda and at Wexford is nothing surprising,[1] however deplorable. What is really remarkable in such a war is that Cromwell, from the moment of landing, should have paid his way, visited plunder with the sharpest penalties, and upheld the sternest and most inflexible discipline. Forty years later, when the conquest of Ireland was undertaken by a former marshal of France and a king long schooled in war against the first generals of the time, they were glad to search out Cromwell's plans for his Irish campaign and follow them at such a distance as they might.

1650, January 8. Cromwell was still in full career of victory when the alarming news of a treaty between Charles the Second and the Scots moved the Parliament to recall him to watch over its own safety. He arrived in London on the 1st of June, and was joyfully welcomed not only by Fairfax and the officers of the Army but by all ranks and all classes. It was now almost certain that the Scots would invade England in the King's

[1] See the very pertinent extract from Wellington's despatches, quoted by Mr. Gardiner—*Commonwealth*, vol. I, pp. 132, 147.

name, and no time was lost by the Council of State in 1650.
appointing Fairfax and Cromwell to command the June 12.
English army in the north. That they would work loyally together in the field no one could doubt; but when the Council consulted the two generals as to plan of campaign, their opinions were found to be diametrically opposed to each other. Cromwell was for taking the bull by the horns and carrying the war into Scotland before the Scots could cross the border; Fairfax, never quite at his ease since the establishment of the Commonwealth, thought such aggressive action unjustifiable. It is impossible to believe that this was his true military opinion, but not all the arguments of the Council nor the pressing entreaty of Cromwell could prevail with him to alter it. Despite all protests he resigned his commission on the plea of physical infirmity, and from this moment passes out of the history of the Army. June 26.
Never perhaps has that Army possessed a more popular and deservedly popular commander-in-chief.

Only one man could be his successor. On the selfsame 26th of June Cromwell received his commission as captain-general and commander-in-chief; and two days later he started on his journey to the north. Charles Fleetwood was his lieutenant-general, John Lambert, an excellent soldier, his major-general; and joined to his staff was another officer whom we saw fighting in the Low Countries many years ago, Colonel George Monk. He had served in the Civil War first with the Royalists, and had been taken prisoner by Fairfax at Nantwich in January 1645; he had then passed some time in confinement in the Tower, and finally had taken service with the Parliament in Ireland, where his merit had attracted the attention of Cromwell. Oliver was now anxious to provide him with a regiment; but the corps which he had designed for him was unwilling to receive a Royalist for colonel. Five companies were therefore taken from Sir Arthur Hazelrigg's regiment at Newcastle and as many more from Colonel Fenwick's at Berwick; and the ten com-

1650. panies were united into Monk's regiment of foot. Thus was formed the oldest of our existing national regiments, the one complete relic of the famous New Model,[1] the one surviving corps which fought under Oliver Cromwell, itself more famous under its later name of the Coldstream Guards.

On the 19th of July Cromwell halted near Berwick, where he mustered sixteen thousand men, a third of them cavalry; and on the 22nd he crossed the Tweed and marched up the coast upon Edinburgh. A fleet on the east coast provided him with supplies as he advanced, which furnishes an interesting precedent for the system that was to be seen later under Wellington in the Peninsula. On the 28th of July he was at Musselburgh, and on the following morning he came in sight of the Scottish army, which was entrenched along the line from Leith to the Canongate.

The Scottish force comprehended a nominal total of twenty-six thousand men, of which eighteen thousand were foot and eight thousand horse. It was under the command, in deed if not in name, of David Leslie, the same excellent officer who had routed the brilliant Montrose at Philiphaugh and had handled his cavalry so efficiently at Marston Moor. His troops however were inferior in quality to the English. It is true that in 1647 the Scotch had followed the example of England in remodelling their army, but the total strength of this force was but five thousand foot and fifteen hundred horse; and this, even supposing the whole of it to have been efficient, was but a small leaven among twenty-six thousand men. Leslie therefore stood carefully on the defensive and resisted all Cromwell's temptations to a pitched battle. After a couple of days Cromwell was compelled to fall back to Musselburgh for supplies.

[1] The pedigree of Monk's regiment is as follows: Weldon's Regiment of the New Model became first Robert Lilburn's, and in 1649-50 Sir A. Hazelrigg's. Lloyd's of the New Model passed in succession to Herbert, Overton, and in 1649 to Fenwick. I am indebted for this information to the kindness of Mr. C. H. Firth.

He then determined to march round Edinburgh and push on to Queensferry, where he could regain touch with his fleet on the northern side of the town. Political reasons however induced him to linger in the execution of this project, and the delay enabled Leslie to take up a position which rendered it impossible. Unable to force Leslie to an engagement, and not daring to attack him with inferior numbers, Cromwell found himself completely outmanœuvred. Dysentery broke out in the English troops; supplies began to fail; and he was compelled to fall back by Haddington and Musselburgh to his ships at Dunbar. There he arrived on the 1st of September with "a poor, shattered, hungry, discouraged army." The Scots had pressed the pursuit very closely, the rearguard had been constantly engaged, and, most significant of all, the English discipline even under Oliver himself had begun to fail.[1] Having driven his enemy into the peninsula of Dunbar, Leslie sent forward a force to bar a defile on the road to Berwick at Cockburnspath, and cut off his retreat. The situation of the English was desperate, and Cromwell was at his wits' end. His army was reduced by sickness to eleven thousand men, while the Scots still numbered twenty-three thousand; he could expect no relief from Berwick; and Leslie lay in a strong position, from which it was hopeless to attempt to dislodge him, between him and the Tweed. 1650.

Leslie on his side might well feel confident that he held his enemy in the hollow of his hand. He had but to remain on his hill-side and watch the English army melt away, or wait for the most favourable moment to attack it either in the effort to embark or while struggling through the defile in retreat. He was however not his own master, but was controlled by an Aulic Council called the Committee of Estates, which urged him to descend from his weather-beaten position on the hill and move to the ground below, where he would not only find greater convenience of supplies but stand

[1] Hodgson.

1650, Sept. 2. within closer striking distance of his enemy. Down therefore he came, not altogether unwillingly, and took up a new position on a triangle of ground enclosed between the sea, the hill which he had just left, and a small stream called the Broxburn. This stream, which runs at the bottom of a course from forty to fifty feet deep, covered the whole of his front. On his extreme left it runs close under the steep declivity of the hill and forms with it, so to speak, the apex of the triangle; but further down it quits the slope and takes its own course to the sea, leaving plenty of space between it and the hill for a camping-ground. Half-way between the open space and the sea, by the grounds of Broxmouth House, the deep banks of the stream give place, as is usual with such waters, to gentle inclines, not unfavourable to the action of cavalry. This point by Broxmouth House formed Leslie's extreme right. The whole position, as he judged, was not ill suited to a force with great superiority in cavalry. He could post his foot on his centre and on his left behind the deep trench dug by the Broxburn, and mass his horse on the right where it could dash down the gradual incline and across the shallow water without risk or difficulty. By four o'clock in the afternoon of the 2nd of September his new dispositions were complete.

Cromwell from the other side of the stream followed every movement with intense attention. At last turning to Lambert he said that he thought the enemy gave him an opportunity. Lambert replied that the very same idea had occurred to him. Monk, who had probably received higher military training than any officer in the army, was next appealed to, and cordially agreed. If Leslie's right, at the base of the triangle, could be turned, the whole of his force must be pent up between the hills and the burn, his horse hurled on to the backs of his foot, and the entire army forced up to the gorge at the apex of the triangle in ever increasing confusion, and, in a word, lost. The time of attack was fixed for the morrow before dawn, and the details of the English dispositions were entrusted to Lambert. 1650, Sept. 3.

Rain fell in torrents all through the night, and the Scotch picquets laid themselves down to sleep with what comfort they could among the corn-shocks. The English, as ever even during the worst and most disorderly of retreats, had recovered themselves at the prospect of battle. At four the moon rose and found Lambert already hard at work. The bulk of the force, six out of eight regiments of horse and three and a half regiments of foot, was moved down to the extreme English left. Five regiments of horse under Lambert were to cross the burn by Broxmouth House and attack the Scottish cavalry in front; three regiments of foot and one of horse, all picked corps, were to cross the water farther down and sweep round upon its right flank. Cromwell himself took command of this turning movement, and the regiment of horse which he took with him was that which he had made six years before on the model of his own "lovely company." The remainder of the force with the artillery was stationed along the edge of the trench of the Broxburn to check any movement of the enemy's centre and left.

The light was beginning to creep over the sea before Lambert had posted the artillery to his liking. There was some stir in the Scotch camp; a trumpet sounded *boute-selle;* and Cromwell, fearful lest the enemy should gain time to change position, grew impatient for Lambert's coming. At last he came, and both columns moved off. Lambert's regiments of horse advanced to the burn; and then the trumpets rang out, and the troopers dashed across the water and poured up the opposite slope to the attack. The Scots, though unprepared, met them gallantly enough. Foreigners would have called them ill-equipped, for they carried lances, an obsolete weapon, in their front rank; but the lance was in place in the shock-combat which Cromwell had taught to the English cavalry, and the first onset of the English horse was borne back

1650, Sept. 3. across the burn. The supports came quickly up and the fight was renewed, though against heavy odds, for the Scots could bring infantry and guns to the aid of their horse, which the English could not yet. But while the combat of cavalry was still swaying to and fro, the infantry of Cromwell's turning column came up steady and inexorable upon the flank of the Scots. Still Leslie's gallant men fought on for a short time undismayed. They had been faultily disposed, as Cromwell had noted, and could not easily change front,[1] but they met the new attack as best they might and even checked the leading regiment of English infantry. But Cromwell's own regiment of foot came up in support, strode grimly forward straight to push of pike, and swept the stoutest corps of Scottish infantry into rout.

Then the Scots lost heart and wavered; the English, horse and foot, gathered themselves up for a final terrible charge; and the Scottish cavalry, reeling back upon the foot, carried it away in choking disorder towards the gorge. Meanwhile Cromwell was urging his third regiment of foot to the left, always farther to the left; and as, panting and breathless, they climbed the lower slopes of the hill they saw the whole length of the battle spread out before them and the Scotch all in confusion. "They run, I profess, they run!" cried Oliver as he looked down. And while he spoke the sun leaped up over the sea, and flashed beneath the canopy of smoke on darting pikes and flickering blades and glancing casques and swaying cuirasses, as the redcoats rolled the broken waves of the Scottish army before them. "Now let God arise and let His enemies be scattered," cried Cromwell in exultation, for the victory was won. The Scots, wedged tighter and tighter between hills and stream, were caught like rats in a pit, and like rats they ran desperately and aimlessly up the steep slope, only to be caught or turned back by the English skirmishers above them. Their horse fled as best they could with the English cavalry spurring

[1] Hodgson.

after them, till Cromwell ordered a rally. While the broken ranks were reforming he sang the hundred and seventeenth Psalm, the chorus swelling louder and louder behind him as trooper after trooper fell into his place. Then the psalm gave way to the sharp word of command, and the horse trotted away once more to the pursuit past Dunbar and Belhaven even to Haddington. Three thousand of the Scots fell in the field; ten thousand prisoners, with the whole of the artillery and baggage and two hundred colours, were taken. It was the greatest action fought by an English army since Agincourt.

1650, Sept. 3.

Cromwell lost no time in following up his success. On the day after the battle he sent Lambert forward with six regiments of horse to Edinburgh, and occupied the port of Leith and the whole of the town, except the Castle, without resistance. Leaving sufficient men to blockade the Castle and hold the works at Leith he pushed on against Leslie, who had entrenched himself with five thousand men at Stirling; but finding his position unassailable he returned to Edinburgh and busied himself with the reduction of the Castle, while Lambert completed the subjugation of the West. In the middle of September the Castle surrendered, and therewith all Scotland south of the Forth and Clyde was subject to the English.

At Westminster the joy over the victory of Dunbar was enthusiastic, and found vent in the grant of a medal[1] and of a gratuity to every man who had fought in the campaign. This, the first medal ever issued to an English army, bore, in spite of his protests, the effigy of Cromwell upon the obverse, no unfitting memorial of the first founder of our Army of to-day. But the struggle even now was not yet over. Royalist Scotland had been beaten at Preston, the Scotland of the Covenant at Dunbar; but Charles Stuart was able, by

1 This again seems to be borrowed from the French. Vieilleville issued medals bearing the King's effigy to his troops in 1558, with a ribbon of his own colours (see *Memoires de Vieilleville*).

1651. unscrupulous lying and shameless hypocrisy, to unite both for a last effort in his cause, and to gather a new army around that of David Leslie at Stirling. Accordingly on the 4th of February 1651 Cromwell left his winter-quarters for Stirling, but was compelled by the severity of the weather to retreat, with no further result to himself than a dangerous attack of fever and ague, which kept him on the sick-list until June.

On the 25th of June the English army was concentrated on the Pentland Hills, and from thence marched once more to Stirling. Leslie, true to the tactics which had proved so successful in the previous year, had occupied an impregnable position which no temptation could induce him to quit. After a fortnight's manœuvring, therefore, Cromwell decided, like Surrey before Flodden, to move round Leslie's left flank and to cut off his supplies from the north. It is plain, from the fact that Monk had been engaged in operations for the reduction of Inchgarvie and Burntisland on the northern shore of the Firth of Forth, that Cromwell's plans for this movement were fully matured.

July 19-20. The first step was to send Lambert across the Firth with four thousand men to entrench himself at Queensferry. Leslie met this move by detaching a slightly inferior force against Lambert, which was utterly and disastrously routed, with a loss of five-sixths of its numbers. Ten days later Inchgarvie and Burntisland fell into Cromwell's hands, and, his new base being thus secured, he advanced quickly into Fife. Meanwhile he sent orders to General Harrison, whom he had left at Edinburgh with a reserve of three thousand horse, that he was to move at once to the English border in the event of Leslie's marching southward. By the 2nd of August he had received the surrender of Perth, but, even before he could sign the capitulation, intelligence reached him that the Scots had quitted Stirling two days before and were pouring down to the border. Leaving five or six thousand men with Monk to reduce Stirling, he at once hurried off in pursuit.

Two days sufficed to bring his army to Edinburgh, 1651,
where he halted for forty-eight hours. Harrison had August 4.
already marched for the Border, and with ready intelligence had mounted some of his infantry to strengthen his little force. Lambert was now despatched with three thousand horse to hang upon the enemy's rear; a letter was despatched to the Speaker exhorting the Parliament to be of good heart; and on the 6th of August Cromwell resumed his advance. Both armies, English and Scots, were now fairly started on their race to the south. Charles, in the hope of picking up recruits, stuck to the western coast and the Welsh border, moving by Carlisle, Lancaster, and the ill-omened town of Preston. Cromwell's course lay farther east; he passed by Newburn, a scene of English defeat, and by the more famous field of Towton, where the south had first taught a lesson of respect to the north. Lambert and Harrison united, and on the 16th of August obtained contact with the enemy at Warrington, but not venturing to attack retired eastward to cover the London road and to draw closer to the line of Cromwell's march.

The Ribble and the Aire once passed, the two armies began to converge. On the 22nd of August Charles halted with the Scots at Worcester and proceeded to fortify the town, and four days later Cromwell occupied Evesham. Charles had but sixteen thousand men; while Cromwell by a masterly concentration had collected no fewer than twenty-eight thousand. The militia, which had been reorganised by the Parliament in the previous year, had been called out and had answered admirably to the call. There could be little doubt of the issue of an action where the advantages both of numbers and of quality were all on one side, and there is no need to dwell on the battle fought on
the anniversary of Dunbar at Worcester. It was a Sept 3.
victory in its way as complete as Sedan: hardly a man of the Scottish army escaped. But it was also the crown of the great work of the Army, the establishment of England's supremacy in the British Isles.

CHAPTER III

THE victory had not long been reported to Parliament when the House began to consider the question of reducing the forces. Silently and almost imperceptibly the strength of the Standing Army had grown since 1645 until it now amounted to thirty regiments of foot, eighteen of horse and one of dragoons, or close on fifty thousand men. Besides these there were independent companies in garrison to the number of seven thousand more, and several more regiments which were borne permanently on the Irish establishment. Five whole regiments, thirty independent companies, and two independent troops were ordered to be disbanded forthwith; other regiments were reserved for service in Ireland or to replace the disbanded companies in garrison; and the establishment for England and Scotland was fixed at eighteen regiments of foot and sixteeen of horse. It appears too that the actual strength of companies was reduced from one hundred and twenty to eighty, and of troops from one hundred to sixty, thus diminishing the number of men while retaining the name of the corps intact. The system is no novelty in these days, but this is the first instance of its acceptance in the history of the Army.

1652. A revolutionary Government, however, does not easily find peace. By June 1652 the recruiting officers were abroad again, and regiments were increasing their establishment owing to the outbreak of the Dutch War. The quarrel with the United Provinces was curious, inasmuch as the English commonwealth had expected sympathy from the sister-republic which had been made 1652-53. by English soldiers, and had even sought to unite the two republics into one. But there is no such thing as national gratitude; and the discourtesy of the Dutch soon led the English to exchange friendly negotiations first for the Act of Navigation and very shortly after for war. The story of that war belongs to the naval history of England, wherein it forms one of its most glorious pages. Never perhaps has more desperate fighting been seen than in the six furious engagements which brought the Dutch to their knees. Yet in these too the red-coats to the number of some two thousand[1] took part, under the command of men who had made their mark as military officers—Robert Blake, Richard Deane and, not least, George Monk. The last named was so utterly ignorant of all naval matters that he gave his orders in military language—"Wheel to the right," "Charge"—but he made up for all shortcomings by his coolness and determination. When Deane, his better-skilled colleague, was cut in two by a round shot at his side he simply whipped his cloak over the mangled body and went on fighting his ship as though nothing had happened. Finally, in the last action of the war he boldly met the greatest admiral of the day, and one of the finest sailors of all time, with but ninety ships against one hundred and forty, fought him not only with superb gallantry but with skilful manœuvre, and wrenched from him the supremacy of the sea.

And meanwhile the Army ashore had done the deed whereof the Nemesis has never ceased to pursue it. So far, except for a few intervals too brief to be worth noting, the Commonwealth had been occupied with the business of war, and the principal function of the Parliament had been to provide ways and means for the conduct of war. Incapable of dissolution save by its own act, the House of Commons had resolved just before the execution of the King that it would put an end to itself

[1] The men were drawn from three Dunbar regiments: Cromwell's own, Goff's and Ingoldsby's, not, alas! from Monk's.

1652-53. in three months; but this had been rendered impossible by the Irish and Scotch campaigns. After the victory of Worcester Cromwell as a private member again brought forward the question of dissolution, but the Rump, as the small remnant that remained after several purgings was called, now showed no disposition to part with the authority which it had so long enjoyed. Frequent conferences were held between the officers of the Army and the members of the House, with the only result that the latter introduced a Bill which, while providing in some fashion or another for the settlement of the nation, reserved to themselves a perpetuity of power. The Army did not conceal its objections to this Bill; and the climax came when certain members tried to smuggle it through the House before the officers could interfere. 1653, April 20. Then Cromwell went down to Westminster, and with twenty or thirty musketeers quickly settled the whole matter.

It is difficult to see how things could have ended otherwise. The House had been sufficiently warned at the close of the first civil war that the Army would not submit to do all the hard work in order that a handful of civilians might reap the profits. The prestige of that Parliament rested and still rests on the achievements of its armed forces, and it depended for its life on the exertions of men who had subjected themselves for its sake to the restraint of military discipline and to the hardships and dangers of war. The Parliament itself had shown no such devotion and self-sacrifice. While soldiers were in distress for want of the wages due to them, corrupt members were making money; while soldiers were flogged and horsed for drunkenness or fornication, drunkards and lewd livers passed unpunished in the House. Even in matters of administration, if we judge by financial management, the Parliament had not shown extraordinary capacity. Its difficulties were certainly enormous, but not a few of them had been evaded rather than honestly met. The Army, on the other hand, for once contained more than its share of the brains of the nation, and comprehended not less administrative talent and far more patriotic feeling than was to be found in the Parliament. It was therefore too much to expect that it would resign all share in the settlement of the nation to such a body as the Rump. If the question of legality be raised, a House of Commons indissoluble without its own consent, and working without the checks of lords and sovereign, was as unknown to the Constitution as a standing army, and at least as dangerous a menace to liberty. If the Long Parliament taught a salutary lesson to kings, the Army taught a lesson no less salutary to parliaments. It would have been better perhaps for the future of the British Army had Cromwell suffered the Rump to remain in power until it should be dissolved in anarchy and confusion, instead of taking the initiative and keeping stern order during the next five dangerous years. But it would have been incomparably worse for England. 1653.

Nine months later, after the Little Parliament had been summoned and had in despair resigned its powers, the soldier who had ousted the Rump and taken over its authority to himself was installed as Lord Protector of Dec. 16. the Commonwealth of England, Scotland, and Ireland. Since 1652 he had been Commander-in-Chief, the first in our history, of the forces in all three Islands; in virtue of that command he now took over the general government. As was to be expected, he chose his deputies and chief advisers from the officers of the Army; and if thereby he placed the realm under military rule we must not allow ourselves to be scared by the phrase from recognition of the worthiness of the administration. There is nothing to make a soldier blush, unless with pride, in the military government of the Protectorate.

Let us begin first with Scotland, which at the close of the Dutch War had been placed under the charge of George Monk. The country was as yet by no means quiet. Agents of Charles Stuart were busy making mischief in the Highlands: and the English found

1654. themselves confronted for the first time with the difficulties of a mountain campaign. Monk's predecessor, Robert Lilburn, had essayed the task with but sorry results; Monk himself accomplished it with a success that suffices of itself to stamp him as a great soldier.

Without going into elaborate detail it is worth while to notice his plan for reducing the Highlands. The Royalist forces and their Highland allies were gathered together principally in two districts, in Lochaber under Glencairn, and in Sutherland under Middleton. Monk's design was to cut the Highlands in twain along the line of the present Caledonian Canal, that he might pen his enemy at his will into either half of the country thus divided, and deal with his forces in detail. North of this line the country was sufficiently circumscribed by nature; south of it he was compelled to fix his own boundaries. The east and south was already guarded by a strong chain of posts running from Inverness through Stirling to Ayr, while one corner to the south-west was secured by the neutrality of the Campbells, which had been gained by diplomacy. Monk now established three independent bases of operations, one at Kilsyth to southward, two more at Perth and Inverness. He then left one column at Dingwall, under Colonel Thomas Morgan, an officer of whom we shall hear more, to hinder the junction of Middleton and Glencairn; and arranged that another column, under Colonel Richard Brayne, of whom also we shall hear more, should sail with all secrecy from Ireland and seize Inverlochy, which was to be his fourth independent base to westward. This done he advanced himself with a third column into the hills from Kilsyth, attacked and defeated Glencairn, and closed the one gap in the net which he had drawn round the Highlands between Loch Lomond and the Clyde.

Then hearing that Middleton had eluded Morgan and passed into Lochaber, he suddenly shifted his base to Perth and advanced into the heart of the mountains.

In two days he had established an advanced magazine at Loch Tay, where the news reached him that the Northern clans had been summoned to assemble at Loch Ness. He at once gave orders that the enemy should be allowed to pass to the southward, and concerted a combined advance of himself, Brayne, and Morgan from the south-west and east to crush him. Unfortunately Morgan, in his eagerness to close in behind the Highlanders, arrived before them and headed them back again to northward. Monk, however, pursued them even thither, hunting them for a week from glen to glen by extraordinary marches, such as the Highlanders had not looked for from mere Englishmen. 1654.

Retiring after this raid to Inverness Monk sent Morgan away by sea to threaten the Royalist headquarters at Caithness. The feint was successful. Middleton, who was again in command in the north, at once came down towards the south. His march was seen and reported from the English station at Blair Athol, and Monk was presently on his track over the Grampians. The chase lay through the Drumouchter Pass, Badenoch, Athol, and Breadalbane, thence westward to the head of Loch Awe and back again into Perthshire and over the mountains to Glen Rannoch; and there, as Monk had arranged, Middleton ran straight into the jaws of Morgan's column and was utterly routed. He fled to Caithness with Morgan hard at his heels; while Monk dispersed the few remaining forces of Glencairn in the hills and destroyed every Highland fastness about Loch Lomond. By August 1654 the work was done; and the Highlands, if ever they may be said to have been conquered, were conquered by George Monk. The English who now wander in thousands over that rugged and enchanting land should remember that the first of their kind that were ever seen therein were Monk's red-coats.[1]

[1] I am indebted for the elucidation of this campaign to Mr. Julian Corbett's *Monk* (Men of Action Series), an admirable sketch of a remarkable man. Monk's letters may be read in Thurloe.

1654. Such very briefly was the first English mountain campaign, admirably designed and admirably executed. The difficulties of military operations in so wild and mountainous a tract were extraordinarily great, and were increased by constant rain and tempest; yet Monk's movements were amazingly rapid. His column on one occasion covered sixty miles in the twenty-four hours. Still more remarkable is his recognition of the fact that in such a campaign success depends mainly on the efficiency of advanced parties and outposts. He never moved without a cloud of scouts on front and flanks; he made it a rule never to march after mid-day; and when he halted he marked out the camp, and posted every picquet and every sentry himself. He showed himself to be the first English exponent of the principle of savage warfare. He invaded the enemy's country, carrying his supplies with him, and sat down. If he was attacked he was ready in a strong position; if not, he made good the step that he had taken, left a magazine in a strong post behind him, and marched on, systematically ravaging the country and destroying the newly-sown crops. The enemy was obliged to move or starve, and wherever they went he swiftly followed. If they turned and fought, he asked for nothing better than the chance of dispersing them at a blow; if they dodged, he brought forward another column from another base to cut them off, while he destroyed the fastnesses which they had deserted. Finally, when his work was done he settled down quietly to govern the country in a conciliatory spirit. He was able gradually to reduce his military establishment, and, ruling at once with mildness, firmness, watchfulness, and unflagging industry, showed himself to be not less able as an administrator than as a general. Scotland has known many worse rulers and few better than her first English military governor.

In Ireland, after Cromwell's departure, the reduction of the country to order was carried on also by a number of flying columns. Of their leaders but two of the most successful need be named, namely Robert Venables and John Reynolds, the latter Cromwell's kinsman by marriage and sometime captain in his regiment of horse. Ireton had been appointed Lord Deputy on Cromwell's departure, but dying in November 1651 was succeeded by another soldier, Charles Fleetwood. Though a valuable man when under the command of a strong officer Fleetwood was soon found to be useless when invested with supreme control, and he was soon practically superseded by Henry Cromwell, the Protector's 1655. second surviving son. Henry had entered the army at sixteen, had fought with his father in Ireland, and had become a colonel at two-and-twenty. He was appointed 1657. Lord Deputy of Ireland at the age of twenty-eight. The country was quiet enough at his accession so far as concerned open rebellion; the Tories had been mercilessly hunted down from bog to bog, and the Irish fighting men had been transported in thousands by recruiting officers to the armies of Spain and of France. What gallant service they did under Lewis the Fourteenth, for they did not greatly love the service of Spain, has been told with just pride by Irish writers; and we too shall encounter some of their regiments before long. Henry Cromwell's difficulties lay not with the native Irish but with his own officers, the veterans of the Civil War, who were alike jealous of his appointment and insubordinately minded towards the Protector. Immediately on Henry's arrival some of these malcontents held a meeting, wherein they put it to the question whether the present government were or were not according to the Word of God, and carried it in the negative. The very members of the Irish Council, old field-officers who should have known better, were disloyal to him, but being old comrades of Oliver's could not be dismissed. Young as he was, however, Henry gave them clearly to understand that he intended to be master, and therewith proceeded to the difficult, nay impossible, task of executing what is known as the Cromwellian settlement of Ireland. He

1654-1658. showed conspicuous ability in extremely trying circumstances, abundant firmness and foresight, and a tolerance of spirit towards the men of other creeds, even Catholics, which was as rare as it was politic. The military governor of Ireland under the Commonwealth was assuredly not a man of whom the British Army need feel ashamed.[1]

Lastly we come to England, where Oliver Cromwell himself sat at the head of the Provisional Government which he was honestly and unceasingly striving to settle on a permanent basis. He defined his own position accurately enough : he was a good constable set to preserve the peace of the parish. But that parish was in a terribly disturbed condition. All that the most visionary could have dreamed of in the subversion of the old order had been accomplished, had even been crowned by the execution of the King ; yet still the expected millenium was not yet come. All factions of political and religious dissent, all descriptions of dreamers, of fanatics, of quacks, and of self-seekers had been welded together for the moment by the pressure of the struggle against Royalism and against the rule of alien races. That pressure removed, the whole mass fell asunder into incoherent atoms of sedition and discontent, for which Royalism, as the one element which strove for definite and attainable ends, formed a general rallying-point. Good and gallant soldiers who had followed Cromwell on many a field—Harrison, Okey, Overton—fell away into disloyalty. Sexby, who had brought the news of Preston to Westminster, became the most dangerous of conspirators. There is nothing more pathetic in history than the desertions from Cromwell after the establishment of the Protectorate. Nevertheless the misfortune was inevitable, for an army which meddles with politics cannot hope to escape the diseases of politics. Yet, through all this,

[1] The best contemporary account of Henry Cromwell's administration will be found in his own letters in Thurloe's *State Papers*.

Cromwell on one point was resolute; he would not allow successful rebellion to be followed by a riot in anarchy. Come what might, he would not suffer indiscipline. 1654-1658.

To preserve the peace, however, in such a hot-bed of plots and conspiracies was no easy matter; and before he had been eighteen months Protector, Cromwell brought military government closer home to the people by parcelling England into at first ten and then twelve military districts, each under the command of a major-general. The force at the disposal of these officers for the suppression of disorder varied in the different districts from one hundred to fifteen hundred men, and was composed almost exclusively of cavalry. It amounted on the whole to some six thousand men, all drawn from the militia, who received pay to the amount of eighty thousand pounds annually. Strictly speaking, therefore, it was rather a force of mounted constabulary than of regular cavalry, and there can be no doubt that, if order was to be preserved, such a body of police was absolutely necessary. Yet it is probable that no measure brought such hatred on the Army as this. The magnates of the counties were of course furious at this usurpation of their powers, and the poorer classes resented the intrusion of a soldier and a stranger between themselves and their old masters. After little more than a year the major-generals were abolished, to the general relief and satisfaction. Their brief reign has been forgotten by the Army, which can hardly believe that it once took complete charge of the three kingdoms and administered the government on the whole with remarkable efficiency. But the major-generals have not been forgotten by the country. The memory of their dictatorship burned itself deep into the heart of the nation, and even now after two centuries and a half the vengeance of the nation upon the soldier remains insatiate and insatiable.

CHAPTER IV

It is now time to pass to the foreign wars of the Protectorate; for though they be little remembered they fairly launched the Army on its long career of tropical conquest, and of victory on the continent of Europe.

It is not easy to explain the motives that prompted Cromwell to make an enemy of Spain. He was eagerly courted by both French and Spaniards, and it was open to him to choose whichever he pleased for his allies. The probability is that he was still swayed by the old religious hatred of the days of Elizabeth, and, like her, looked to fill his empty treasury with the spoils of the Indies. He did not perceive that the religious wars of Europe were virtually ended, and that nations were tending already to their old friendships and antagonisms as they existed before the Reformation. Be that as it may, he was hardly firm in the saddle as Protector when he began to frame a great design against the Spanish possessions in the New World. His chief advisers were one Colonel Thomas Modyford of Barbados, who had his own reasons for wishing to ingratiate himself with the Protector, and Thomas Gage, a renegade priest, who had lived long in the Antilles and on the Spanish Main and had written a book on the subject. The most fitting base of operations was obviously Barbados, which, from its position to windward of the whole Caribbean Archipelago, possessed a strategic importance which it has only lost since the introduction of steam-vessels. It lay ready to Crom-

well's hand, having been an English possession since 1654.
1628, and was, if Modyford were to be believed, ready to give active assistance in the enterprise. There remained the question whether the expedition should be directed against an Island or against the Main. Gage was for the latter course, and named the Orinoco as the objective: Modyford recommended Cuba or Hispaniola,[1] and Modyford's opinion prevailed.

Gradually the design matured itself, and presently assumed gigantic proportions. A footing once established on one of the Spanish Islands to leeward, there was to be a general contest with the Spaniards for the whole of the South Atlantic. Two fleets were to be employed, one in seconding the army's operations on the Islands and making raids upon the Main, the other in cruising off the Spanish coast so as to interrupt both plate-fleets from the west and reinforcements from the east. Lastly, not England only, but New England was to play a part in the great campaign. Supplies would be one principal difficulty, but these could be furnished from English America, and not only supplies but settlers, who, trained to self-defence by Indian warfare, should be capable of holding the territory wrested from Spain. Thus the English from both sides of the Atlantic were to close in upon the Spanish dominions in the New World, and turn Nova Hispania into Nova Britannia. There was no lack of breadth and boldness in the design.

All through the latter half of 1654 mysterious preparations went forward with great activity in the English dockyards, and France, Spain, and Holland each trembled lest they might be turned against herself. But the existing organisation in England was unequal to the effort. To equip two fleets of forty and of twenty-five ships for a long and distant cruise was a heavy task in itself; but to add to this the transport of six thousand men over three thousand miles of ocean for an expedition to the tropics was to tax the resources of the naval and

[1] St. Domingo.

1654. military departments to excess. The burden of the duty fell upon John Desborough, major-general and commissioner of the Admiralty, who was not equal to thinking out the details of such an enterprise nor disposed to give himself much trouble about them. His difficulties were increased by the rascality of contractors, and by the composition of the expeditionary force. By a gigantic error, which has not yet been unlearned, Cromwell, instead of sending complete regiments under their own officers, made up new corps, partly of drafts selected by various colonels and probably containing the men of whom they were most anxious to be rid, and partly of recruits drawn from the most restless and worthless of the nation. He returned in fact to the old system that had so often been found wanting in the days of Elizabeth, of James, and of Charles.

The distribution of command was also faulty. The military commander-in-chief was Robert Venables, who had made a reputation as a hunter of Tories in Ireland; the Admiral joined with him was William Penn, who is unjustly remembered rather as the father of a not wholly admirable Quaker than as one of the ablest and bravest naval officers of his day. But as if two commanders were not already sufficient, there were joined with them three civil commissioners, one Gregory Butler, an officer who had served in the Civil War, Edward Winslow, a civilian and an official, and the Governor of Barbados, Daniel Searle. There was of course nothing new in the presence of civil commissioners on the staff, and a general's instructions since the days of Henry the Eighth had usually bound him to act by the advice of his Council of War only; but it is abundantly evident that Winslow was employed not only as a commissioner, but as a spy on his colleagues, or on some one of them whose loyalty was suspected. It is strange that so sensible a man as Cromwell should have made such a mistake as this. Monk was the man whom he had wished to send, could he have spared him from Scotland; but failing Monk, Penn and Venables were both of them men who had shown ability in their previous service. 1654.

With immense difficulty the expedition was got to sea at the end of December 1654, just two months too late. Even so it sailed without a portion of its stores, which Desborough promised faithfully to send after it without delay. The fleet reached Barbados after a good passage on the 29th of January 1655; and then the troubles began. From too blind faith in the promises of Thomas Modyford, the Protector had trusted to Barbados in great part to equip his army, and to help it on its way. Barbados, from its Governor downwards, refused to move a finger. It had no desire to denude itself of arms or of men, and so far from assisting the English threw every possible obstruction in their way. The planter upon whom Venables had been instructed chiefly to depend was found to be entirely under the thumb of his wife. She was averse to the expedition; and the commissioners, observing her, as they said, to be very powerful and young, abandoned all hope of co-operation from that quarter. Every day too brought fresh evidence of the rotten composition of the force at large, which was without order, without coherency, and without discipline. Unfortunately Venables was not the man to set such failings right. He showed indeed some spasmodic energy, called the Barbadian planters a company of geese, improvised rude pikes of branches of the cabbage-palm, organised a regiment of negroes and a naval brigade, and after several weeks' stay sailed at last for St. Domingo. On the way he picked up a regiment of colonial volunteers which had been collected by Gregory Butler at St. Kitts, and on the 13th of April the expedition was in sight of St. Domingo.

The naval officers were for running in at once and taking the town by a sudden attack. Winslow, the civilian, objected: the soldiers, he said, would plunder the town, and he wanted all spoil for the English treasury. This order against plunder raised something like a mutiny among the troops; but eventually a new plan was chosen,

1655. which was probably based on the precedent of Drake in 1586. Venables with three thousand five hundred men sailed to a landing-place thirty miles west of the town, and there disembarked ; leaving fifteen hundred more men under a Colonel Buller to land to the eastward of it and march on it from that side. Buller, however, finding it impracticable to obey his instructions, after two days' delay also landed to the westward of the town, though but ten miles from it, at a point called Drake's landing. Elated by a trifling success against a handful of Spaniards who had opposed his disembarkation, he laid aside all thought of co-operation with Venables and pushed on hastily into the jungle to take St. Domingo by himself. No sooner was he gone, past call or view, when up came Venables to the identical spot where Buller had landed. He had for two days pursued a terrible march of thirty miles through jungle-paths, in the sultry steam of the tropical forest. The men's water-bottles had been left behind in England, and they were choked with thirst; they had torn the fruit from the trees as they passed and had dropped down by scores with dysentery. Hundreds had fallen out, sick and dead, and the column was not only weakened but demoralised.

Next day Venables effected a junction with Buller, and the force, though heartless and spiritless, made shift to creep up to a detached fort which covered the approach to the town. On the way it fell into an ambuscade, and though it beat off the enemy, it lost in the action the only guide who knew where water was to be found, and was compelled to retire ten miles to Drake's landing. There it remained for a week, eating bad food from some scoundrelly contractor's stores, drinking water that was poisoned by a copper mine, and soaked night after night by pouring tropical rain. Dysentery raged with fearful violence, and Venables himself did not escape the plague. Unfortunately, instead of sharing the hardship with his men in camp, he went on board ship to be nursed by Mrs. Venables, who had accompanied him on the voyage. Thus arose open 1655. murmurs and scandalous tales, which cost him the confidence of the army.

Neverthless after six days' rest he again advanced by the same line to the fort from which he had been forced to retreat. To prevent repetition of mishaps from ambuscades he gave strict orders that the advanced guard should throw out flanking parties on each side of the jungle-path. The injunction was disobeyed. The advanced guard walked straight into an ambuscade, two officers fell dead, the third, Adjutant-General Jackson, who was in command, turned and ran; the advanced guard fled headlong back on to the support; the support tumbled back on to the main body, and there, wedged tight in the narrow pass, the English were mown down like grass by the guns of the fort and the lances of the Spanish cavalry. At last an old colonel contrived to rally a few men in the rear, and advancing with them through the jungle fell upon the flank of the Spaniards and beat them back. He paid for his bravery with his life, but he assured the retreat of the rest of the force, which crept back beaten and crest-fallen to the ships, leaving several colours and three hundred dead men behind it.

Venables and his men were now thoroughly cowed by failure and disease. Penn in vain offered to take the town with his sailors, but Venables and Winslow would not hear of it. All ranks in the fleet now abused the army for rogues, and the worst feeling grew up between the two services. Finally, on the 7th of May, the expedition sailed away in shame to Jamaica. Arrived there, Penn, openly saying that he would not trust the army, led the way himself at the head of the boats of the fleet; and after a trifling resistance the Island was surrendered by capitulation. Then fleet and army began to fight in earnest, officers as well as men; and at last, after the commissioners in command had spent six weeks in incessant quarrelling, Venables and Penn sailed home, leaving the troops and a part of the squadron behind them.

1655. Cromwell's disappointment and chagrin over the failure of his great enterprise were extreme. Both the returned commanders were forthwith sent to the Tower, and though presently released, remained throughout the whole of the Protectorate in disgrace. Still Jamaica had been won and must be held. The command after Venables' departure had devolved on Richard Fortescue, a colonel of the New Model, who, without concealing his infinite contempt for those who had gone home, set himself cheerfully to turn the new possession to account. To him Cromwell wrote letters of encouragement and thanks, with promise of speedy reinforcement. But now a new enemy appeared in Jamaica, one that has laid low many tens of thousands of red-coats, the yellow fever. In October 1655 the first reinforcements arrived, under command of Major Sedgwicke. He had hardly set foot on the island before Fortescue succumbed, and he could only report that the army was sadly thinned and that hardly a man of the survivors was fit for duty. Then the recruits began to fall down fast, and in a few days the men were dying at the rate of twenty a day. Sedg-
1656. wicke was completely unnerved; he gave himself up for lost, and in nine months followed Fortescue to the grave. Fresh reinforcements, including all the vagabondage of Scotland, were hurried across the Atlantic to meet the same fate. Colonel Brayne, who had served with Monk in Scotland, arrived to succeed Sedgwicke in December 1656. He lasted ten months, surviving even so two thirds of the men that he brought with him, and then went the way of Sedgwicke and Fortescue. Finally a Colonel D'Oyley, who had sailed with the original expedition, took over the command, and being a healthy, energetic man, soon reduced things to such order that when in May 1658 the Spaniards attempted to recapture the island, he met and repulsed them with brilliant success. Thus at length was firmly established the English possession of Jamaica.

So ended the first great military expedition of the English to the tropics, the first of many attempts, nearly all of them disastrous, to wrest from Spain her Empire in the West. I have dwelt upon it at some length, for it is the opening chapter of a long and melancholy story, whereof one recitation will almost serve for the whole. We have still to go with Wentworth to Carthagena and with Albemarle to Havanna; we shall accompany Abercromby and Moore to St. Vincent and St. Lucia, and other less noted officers to Demarara and Surinam; we shall even see Wellington himself drawing up a plan for operations on the Orinoco: but in spite of a hundred experiences and a thousand warnings we shall find the mistakes of Oliver Cromwell eternally repeated, and though we may never again have to tell so disgraceful a story as that of the repulse from St. Domingo, yet we shall seldom fail to encounter such mournful complaints as were made by Fortescue, Sedgwicke, and Brayne, of regiments decimated as soon as disembarked, and annihilated before the firing of a shot. We have now well-nigh learned how to conduct a tropical expedition, and life in the tropics is a thing familiar to tens of thousands of Englishmen; but it is worth while to give a thought to these poor soldiers of the Commonwealth. They were the first Englishmen who went to the tropics, not like Drake's crews as fellow-adventurers, but simply as hired fighting men. Yet the traditions of Drake's golden voyages were strong upon them, and they landed, big with expectations of endless gold told up in bags.[1] We can picture their joy at coming ashore, bronzed healthy Englishmen, and their open-mouthed wonder at all that they saw; and then after a few hours the first cases of sickness, the puzzled surgeons with busy lancets, the first death and the first grave; the instant spread of fever on the turning of the virgin soil, and then a hideous iteration of ghastly symptoms, and, sundown after sundown, the row of silent forms and shrouded faces. Englishmen had faced such terrors in the flooded leaguers of Flanders, but it was hard to find them in a fruitful and pleasant land, where the sun shone brighter

[1] Fortescue's own expression. See his letters in Thurloe.

and the forest grew greener than in England, the loved England that lay so far away over the glorious mocking blue of the tropic sea.[1]

The aggressive attack on St. Domingo at once decided the hostility of Spain towards the Commonwealth, and drove her to take Cromwell's most formidable enemy, Charles Stuart, to her heart. The Protector, 1655, Sept. 9. on his side, hastened to make treaty of peace and friendship with France, which he presently expanded into an 1657, March. offensive and defensive alliance. Mazarin, who had to encounter not only Spain but Condé, was only too glad to welcome the English to his side. By the terms of the treaty it was agreed that the French should provide twenty thousand men, and the English six thousand men, as well as a fleet, for the coming campaign against the Spaniards in Flanders. Of the English six thousand half were to be paid by France, but the whole were to be commanded by English officers, and reckoned to be the Lord Protector's forces. The plan of campaign was the reduction of the three coast-towns of Mardyck, Dunkirk, and Gravelines, of which the two first were to be made over to England and the third retained by France. Cromwell's great object was to secure a naval station from which he could check any attempted invasion of England by Charles Stuart from Spanish Flanders, and he was therefore urgent that Dunkirk should be first attacked. Turenne disliked this design, and even threatened to throw up his command if it should be insisted on. To beleaguer Dunkirk without first securing Nieuport, Furnes, and Bergues would, he said, be to be besieged while conducting a siege. But Cromwell had made up his mind that the thing should be done, and, as shall soon be seen, it was done.

1657. Throughout the spring of 1657 therefore preparations for the expedition kept both military and naval

[1] The story of the West Indian expedition is very fully told in Thurloe's *State Papers*. There are a few supplementary papers in *Cal. S. P., Col.*, and two accounts in Ogilvy's *History of America* and in the *Harleian Miscellany*.

departments busily employed, for the fleet was not only 1657.
to supply the army but to second its operations. The six thousand men, though for the most part old soldiers, were made up of drafts and of new recruits, and were distributed into six regiments. Turenne would gladly have preferred complete corps from the standing Army, but in the existing menace of invasion Cromwell was indisposed to spare them. Nevertheless the new regiments were in perfect order and discipline when they embarked on the 1st of May from Dover for Boulogne. The general in command was Sir John Reynolds, whom we saw lately in Ireland; the major-general was Thomas Morgan, Monk's right-hand man in the Highland war, an impetuous little dragoon known by the name of the "little colonel,"[1] and justly reputed to be one of the best officers in the British Isles.

The arrival of the six thousand English foot, all dressed in new red-coats, created a great sensation in France. They were cried up for the best men that ever were seen in the French service; they took precedence of the whole French army, even of the famous Picardie, excepting the Swiss and Scottish body-guards; and they were welcomed by emissaries from the King and Mazarin and inspected by the royal family. It is significant of the difference between the French and English even in their civil wars that the six thousand were amazed to see all the villagers fly from their houses at their approach. They were told that the French soldiery were dreaded as much by their countrymen as by their enemies; and yet Reynolds admitted that the discipline of the French troops was good, for France. "But we," he added proudly, "can lie in a town four days without a single complaint." One thing alone went amiss with the English: they quarrelled with the French ammunition-bread, and clamoured loudly for beef and beer.

By the ill-faith of Mazarin, Reynolds' force instead of marching to Dunkirk was moved inland, and found

[1] See the pamphlet, *The Bloudie Field*, in King's Pamphlets, British Museum.

1657. itself engaged at the siege of St. Venant. Here it gave the Spaniards a taste of its quality. It seems that the English, who were never very happy in handling the spade, were working in some confusion at the advanced trenches when Count Schomberg, a man whom readers should bear in mind, and a few more foreign officers came up and began to pass criticisms. Morgan, wincing under their remarks, impatiently called for a party of fifty men to come to him ; whereupon every English soldier in the trenches, incontinently jumped up and without further ado assaulted the town, captured three redoubts, and forced the Spaniards to capitulate. Such blundering gallantry had distinguished the nation since Cocherel, and was to be repeated on a grander scale at Minden. But Cromwell was not the man to allow his regiments to be wasted in such operations as these. Dismissing all of Mazarin's excuses as "parcels of words for children," he insisted that the true business of the campaign should be taken in hand at once. In September, therefore, Turenne moved slowly up to the coast ; and Cromwell to give him encouragement sent him a reinforcement of two thousand men. Mardyck was easily taken on the 29th of September ; but there Turenne stopped. Lockhart, the English ambassador, in vain offered him five of the old regiments of the standing Army if he would proceed at once to the siege of Dunkirk ;[1] the great General would not move ; and with the capture of Mardyck the campaign of 1657 came to an end.

The English undertook to garrison Mardyck and the town of Bourbourg close to it, and while engaged in this duty incurred the strong censure of Turenne. They kept, he complained, very bad guards, and seemed unable to stand the work of watching ; and the failing, it seems, was no new one, for Monk expressed no surprise at hearing of it. Nevertheless, when on one night in October the Spaniards attempted to surprise Mardyck with five thousand men, they found this unwatchful garrison formidable enough and were repulsed

[1] Thurloe, vol. vi. p. 18.

with heavy loss. The truth was that the condition of things in the town was what would now be thought appalling. The winter was unusually severe and the troops very imperfectly protected against it. Pestilence had broken out among them and men were dying at the rate of ten or twelve a day: once indeed the death-roll within twenty-four hours ran as high as fifty. Reynolds protested in vain, and at last in December he sailed for England to represent matters in person to the Protector. He was cast away on the Goodwin Sands and never seen again. By the time when the season opened for active operations the English had lost since their disembarkation their General and not far from five thousand men. 1657-1658.

Lockhart, who took over the command after Reynolds' death, found the remnant of the army in a very bad state. Discipline was decidedly lax; and the French complained bitterly of the insolence of their allies. This of course was no new thing. So far back as 1603, in the wars of Dutch Independence, a dispute about some firewood had set an English and a French regiment fighting; and the quarrel had ended in the flight of the French to their ships, leaving their Colonel and sixteen of their comrades dead behind them.[1] The English now, probably on some equally trivial occasion, fell at variance with the French guards and killed several of them; nor could all the frenzy of French indignation avail to obtain the least redress. Lockhart attributed this insubordinate spirit to the dearth of chaplains; but the true explanation was that over eighty of the officers, disliking the tedium of winter-quarters, had absented themselves, as was customary, from their regiments. When they returned, and four thousand fresh troops with them, Morgan seems to have had little difficulty in restoring discipline. 1658.

Morgan opened the campaign before the arrival of Lockhart by the capture of two small redoubts that lay on the road to Dunkirk; but it was not till the 4th of March.

[1] Collins, *State Papers* (July 1603), p. 277.

1658. May that Turenne broke up his quarters at Amiens, and after a very difficult march to Dunkirk, on the 27th (May 16/27.) invested the town. A brilliant repulse of a Spanish sortie by the English put him in good humour with his allies, and he was fain to confess that they had done right well.[1] He was to appreciate them still higher within a week; for on the 2nd of June (May 23/June 2.) the Spanish army, fifteen thousand strong, under Don John of Austria, Condé, the Marquis Caracena, and James, Duke of York, drew down to within a mile of his head-quarters, with the evident design of forcing the besiegers' lines.

We must pause for a moment over the composition of the motley Spanish host, for there is a part of it under James, Duke of York, with which we are nearly concerned. Five regiments in all, amounting to some two thousand men, were entrusted to the Duke's command. Three of these, James's own, Lord Ormonde's, and Lord Bristol's, were Irish, the relics of the loyal party that had been scattered by Cromwell; one, Middleton's, was Scotch, and represented fragments of the force that had been broken up by Monk; and one, which readers must not omit to mark, was English, made up of refugees mostly of gentle birth. It comprehended the last shreds of old English royalism, and was called the King's Regiment of Guards.

Nor must we omit to throw a passing glance at the army of Turenne. First and foremost there were the six regiments sent out by Cromwell. Then there was a regiment with which we parted last after the battle of Verneuil, the Scottish body-guard of the kings of France. Next, there was a regiment which we saw pass from the Swedish to the French service in 1635, Regiment Douglas, some time the Scots Brigade of King Gustavus Adolphus. It had passed through many campaigns and absorbed other corps of British within the past twenty years, and could now add the names of Rocroi, Lens and Fribourg to its records; but here it was, newly

[1] "Les Anglais y firent fort bien." See his letter in Thurloe.

recruited from Scotland by the Protector's permission, 1658.
marching side by side with the red-coats, though quite May 23
unconscious how soon it was destined to take its place June 2.
among them, to fight the battle of Dunkirk Dunes. Lastly, an Irish regiment, known by the name of Dillon, and made up of men who had fled from the wrath of Cromwell, completed the strange representation of the united Commonwealth.[1]

It was evening of the 2nd of June before Turenne could satisfy himself that the whole of the Spanish army was present before him, but no sooner was he assured of it than he resolved to fight on the morrow. The English were still at Mardyck, and the orders reached Lockhart so late and came as such a surprise that the marshal politely intimated his wish to give reasons for his determination. "I take the reasons for granted," answered Lockhart, "it will be time to hear them when the battle is over." At ten o'clock the English marched off, Lockhart, who was suffering agonies from stone, driving in his carriage at their head, and at daybreak reached Turenne's headquarters. The next three hours were spent in drawing up the line of battle, which was of the mathematical precise type that prevailed in those days. In the first line there were thirteen troops of cavalry on the right wing, as many on the left, and eleven battalions of infantry in the centre; in the second line there were ten troops on the right, nine on the left, and seven battalions in the centre. Five troops of horse were posted midway between the two lines of infantry, and four more were held in reserve. The whole force was reckoned at six thousand horse and nine thousand foot, of which latter the English contingent made more than half. The place assigned to the red-coats was the left centre, which, if not the post of honour, was assuredly the post of danger.

Don John's line of battle was widely different. He

[1] It must be remembered that this was no figure of speech. Cromwell was the first who gathered in representatives of Scotland and Ireland to Westminster.

1658. May 24/June 3. had taken up a strong position among the sand hills, facing west, his right resting on the beach, his left on the Bruges Canal; and the whole of his infantry was drawn up in his first line. A sand hill higher than the rest on his right was regarded as the key of the position, and was strongly held, as the place of honour, by four Spanish regiments. Next to them on their left stood the five regiments under the Duke of York, with one battalion in reserve, and the line was continued by battalions of Germans and Walloons. The Spanish horse was massed behind the foot in columns according as the sand hills permitted; and the whole force numbered between fourteen and fifteen thousand men.

Notwithstanding that they had marched all night, and in spite of Turenne's orders that the line should dress by the right, the English outstrode the French in the advance and began the action alone. The position occupied by the Spaniards in their front was so strong, that Lockhart by his own confession despaired of carrying it. Lieutenant-Colonel Fenwick however, who commanded Lockhart's regiment, undertook the task without the General's instructions. Covered by a cloud of skirmishers he advanced steadily with his pikes to the foot of the sand hill, and while the musketeers wheeling right and left maintained a steady fire, he calmly halted the pikes to let the men take breath. Then with a joyful shout they swarmed up the treacherous sand and went straight at the Spaniards. Fenwick fell at once, mortally wounded by a musket shot; his major, Hinton, took his place, and was also shot down. Officer after officer fell, but the men were not to be checked, and though the Spaniards, backed by a company of the English guards, fought hard and well, they were fairly swept off the sand hill, and retired in confusion, leaving nine out of thirteen captains dead on the ground. James, Duke of York, tried to save the rout by charging Lockhart's victorious regiment with his single troop of horse, but he was beaten back, and though at a second attempt he succeeded in breaking into its flank he met with so sturdy a resistance from every isolated man as convinced him that his effort was hopeless. Meanwhile the rest of the English regiments advanced quickly in support; the French horse on the left wing came up likewise, and the rout of the Spanish right was complete. 1658. May 24/June 3.

With the uncovering of its right flank the whole of Don John's line wavered, and few regiments, except those under the immediate direction of Condé, far away on the left, showed more than a feeble resistance to the advancing French. Very soon the whole force—Spaniards, Walloons and Germans, Scots and Irish—were in full retreat, and a single small corps of perhaps three hundred men stood isolated and alone in the position among the sand hills. A French officer rode forward and summoned the little party to surrender. "We were posted here by the Duke of York," was the answer, "and mean to hold our ground as long as we can." The Frenchman explained that resistance was hopeless. "We are not accustomed to believe our enemies," was the reply. "Then look for yourself," rejoined the Frenchman; and leading the commander to the top of a sand hill he showed him the retreating army of Spain. Thereupon the solitary regiment laid down its arms: it was the English King's Royal Regiment of Guards.[1]

The losses of the victorious English were very severe. In Lockhart's regiment but six out of the whole number of officers and sergeants had escaped unhurt; and the honours of the day were admitted by all to lie with the red-coats. The action led to the speedy fall of Dunkirk; and Lockhart, being reinforced by two regiments from England, was able to detach four to continue the campaign under the command of Morgan. Bergues, Dixmuyde, and Oudenarde fell in quick succession, and little opposition was encountered until the siege of Ypres, where the English delivered so daring and brilliant an assault that Turenne, over-

[1] Clarke's *James II.*

1658. come with admiration, embraced their leader, Morgan, and called him one of the bravest captains of the time. The capture of Ypres was the last exploit of the six thousand—the immortal six thousand, as they were styled in the admiring pamphlets of the day. After an advance almost to the walls of Brussels, the campaign came to an end; Morgan returned to England to receive knighthood, and the English retired to Dunkirk to spend another winter in cold and misery and want, and worst of all in deep uncertainty for the future.[1]

For even while Morgan was watching the Spanish garrison march out of Ypres, the soldier who had made the English Army was lying speechless and unconscious at St. James's, worn out with many campaigns and with the work of keeping the peace in England. Before tattoo sounded on the 3rd of September 1658, Oliver Cromwell was dead, and no man could say who should come after him. Richard Cromwell, his son, held two trump-cards in his hand—Henry Cromwell and the army in Ireland, George Monk and his army in Scotland.
1659, April 21. He was afraid to play either, and yielded up his power to a clique of his father's old officers—Fleetwood, Desborough, and others—who brought back the Rump of the Long Parliament to reign in his stead. Henry Cromwell resigned his command, and the power of the Cromwells was gone. The Rump now took over Cromwell's body-guard for its own protection, and to make the Army thoroughly subservient decided that all officers should be approved by itself, and all commissions signed by the Speaker. So large was the military establishment that this work of revising the list of officers was never completed. George Monk, however, accepted the Speaker's commission without a word.

It was not in the nature of things that the English

[1] The best English source for the account of the campaign in Flanders is Thurloe's *State Papers*; there are also some curious details in a tract in the *Harleian Miscellany*, which, however, I have accepted only when confirmed by newspapers. Bussy Rabutin's *Memoires*, and Clarke's *James II.* are among other authorities.

generals should long submit to the junto of politicians 1659. which it had set over England. In a very short time the leaders of the Army for the second time cleared away the Rump, and took the supreme power into their own hands; but herein they overlooked the existence of the ablest soldier left in Great Britain. Monk was ready enough to take his orders from Oliver Cromwell, but not from such small men as Lambert and Desborough. No sooner did the news of the new October 17. departure reach him at Dalkeith than with amazing rapidity he secured every garrison in Scotland, seized the bridge over the Tweed at Berwick, purged his troops of all officers disloyal to the Parliament, and gave orders for his whole force to concentrate at Edinburgh. Morgan, with the glories of Flanders still fresh on him, presently came to help him in the reorganisation of his army, and by the middle of November he began to move slowly south. Negotiations with the English leaders had been in progress ever since Monk first took decided action, and, though fully aware that they must come to nothing, he was not sorry to gain a little time in order to establish discipline thoroughly in the force under his command. By the end of November he had fixed his headquarters at Berwick.

There, at one o'clock on the morning of the 7th of December, he was surprised by the news that, in spite of much peaceful profession, the English general Lambert had besieged Chillingham Castle and had marched within twenty miles of the Border. One hour sufficed for Monk to write the necessary orders for the movement of the troops, and at two o'clock he was in the saddle and away to inspect the fords of the Tweed. The night was stormy and pitch dark, and the roads were sheets of ice, but on he galloped, despite the entreaties of his staff, through wind and sleet, up hill and down, at dangerous speed. "It was God's infinite mercy that we had not our necks broke," wrote one who was an unwilling partaker of that ride.[1] By eleven

[1] Gumble, the chaplain, from whose *Life of Monk* this account is taken.

1659. o'clock the inspection was over and headquarters were fixed at Coldstream. A regiment of foot had already arrived there to guard the ford before the General came, and had cleared away every scrap of provisions. His staff-officers dispersed to find food where they could, but George Monk put a quid of tobacco into his cheek and sat down contented with a good morning's work. He had occupied every pass from Berwick to Kelso, and had so thought out every detail that he could concentrate his whole force at any given point in four hours. The bulk of his troops under Morgan were stationed on the exposed flank at Kelso; he himself was in the centre at Coldstream. Lambert might attack his front or turn his flank if he dared.

For three weeks Monk's army lay in this position, four regiments of horse and six of foot,[1] waiting for the moment to advance. The cold was intense, and the quarters in the little village of Coldstream were very strait. The General occupied a hovel wherein he had hardly space to turn round, and the men suffered greatly from privation and hard weather, but Monk's spirit kept them all in cheerfulness, and those who had shared his hardships never ceased to boast themselves to be Coldstreamers. At last, on the 31st of December, came the news that the army which had deposed the Rump was up in mutiny; and at daybreak
1660. of the 1st of January 1660 Monk's army crossed the Tweed in two brigades and began its memorable march to the south. All day they tramped knee-deep through the snow, full fifteen miles to Wooler, while the advanced-guard of horse by a marvellous march actually covered the fifty miles to Morpeth. At York they were met by Fairfax, who had roused himself at such a crisis for a last turn of military duty, and picking up deserters on every side from Lambert's regiments they increased their strength at every march. On the 31st of January Monk received at St. Albans the Parliament's

[1] According to the usual establishment, 9600 men besides officers.

confirmation of his commission as General, and three 1660. days later he occupied London. His own regiment of foot was quartered for the first time in and about St. James's.

It is unnecessary to dwell on the intricate movements in the political world during the three following months; it must suffice to say that Monk was finally obliged to coerce the Rump as all other soldiers had coerced it. In spite of all engagements to dissolve itself without delay, this pretentious little assembly still clung, notwithstanding its unpopularity, to power; but a letter from the General was sufficient to bring it to reason without a file of musketeers. Such a letter arrived on the 6th of April; and though the House resolved not to read it until it had gratified its vanity by a little further debating, yet it decided after opening it to make the question of dissolution its very next business. Before evening it had ceased to exist. One last desperate attempt of Desborough and Lambert to divide the Army was suppressed with Monk's habitual promptitude, and on the 1st of May the General, sitting as member for his native county in a new House of Commons, moved that the King should be invited to England. Three weeks later Monk's life-guard and five regiments of horse escorted the restored monarch into London; and the work of the New Model Army was done.

CHAPTER V

It is strange that our historians have for the most part taken leave of the New Model without a tinge of regret, without estimation of its merits or enumeration of its services. Mountains of eulogy have been heaped on the Long Parliament, but little has been spared for this famous Army; nay, even military historians by a strange perversity begin the history of the Army not from its foundation but from its dissolution. Much doubtless besides the creation of a standing Army dates from the great rebellion, though few things more important in our history, unless indeed it be the cant that denies its importance. The bare thought of militarism or the military spirit is supposed to be unendurable to Englishmen. As if a nation had ever risen to great empire that did not possess the military spirit, and as if England herself had not won her vast dominions by the sword. We are accustomed to speak of our rule as an earnest for the eternal furtherance of civilisation; but we try to conceal the fact that the first step to empire is conquest. It is because we are a fighting people that we have risen to greatness, and it is as a fighting people that we stand or fall. Arms rule the world; and war, the supreme test of moral and physical greatness, remains eternally the touchstone of nations.

Surely therefore the revival of the military spirit, and on the whole the grandest manifestation of the same in English history, are not matters to be lightly overlooked. The campaigns of the Plantagenets had shown how deep was the instinct of pugnacity that underlay the stolid

English calm, but since the accession of the Tudors no sovereign had given it an outlet ashore in any great national enterprise. Elizabeth never truly threw in her lot with the revolted Netherlands; James hated a soldier, and shrank back in terror from the idea of throwing the English sword into the scale of the Thirty Years' War; Charles's miserable trifling with warfare contributed not a little to the unpopularity which caused his downfall. The English were compelled to sate their military appetite in the service of foreign countries, and as fractions of foreign armies.

Then at last the door of the rebellion was opened and the nation crowded in. It is hardly too much to say that for at any rate the four years from 1642 to 1646 the English went mad about military matters. Military figures and metaphors abounded in the language and literature of the day, and were used by none more effectively than by John Milton.[1] Divines took words of command and the phrases of the parade ground as titles for their discourses, and were not ashamed to publish sermons under such a head as "As you were." If anything like a review or a sham fight were going forward, the people thronged in crowds to witness it; and one astute colonel took advantage of this feeling to reconcile the people to the prohibition of the sports of May-day. He drew out two regiments on Blackheath, and held a sham fight of Cavaliers and Roundheads, wherein both sides played their parts with great spirit and the Cavaliers were duly defeated; and the spectacle, we are assured, satisfied the people as well as if they had gone maying any other way. It is true that the sentiment did not endure, that the eulogy of the general and his brave soldiers was turned in time to abuse of the tyrant and his red-coats; but when a nation after beheading a king, abolishing a House of Lords, and welcoming freedom by the blessing of God restored, still finds that

[1] It is not I think irrelevant in this connection to remind the reader of the military manœuvres of the rebel angels in *Paradise Lost*.

the golden age is not yet returned, it must needs visit its disappointment upon some one. The later unpopularity of the strong military hand does not affect the undoubted fact of a great preliminary outburst of military enthusiasm. Nor indeed even at the end was there any feeling but of pride in the prowess of Morgan's regiments in Flanders.

The rapid advance of military reform in its deepest significance is not less remarkable. For two years it may be said that opposing factions of the Civil War fought at haphazard, after the obsolete fashion of the days of the Tudors. The most brilliant soldier on either side was a military adventurer of the type that Shakespeare had depicted, a man who

> dreams of cutting Spanish throats,
> Of trenches, ambuscadoes, Spanish blades
> And healths five fathoms deep.

Against the wild, impetuous Rupert the primitive armies of the Parliament were powerless. From the first engagement Cromwell perceived that such high-mettled dare-devils could be beaten only by men who took their profession seriously, who made some conscience of what they did, who drew no distinction between moral and military virtues, who believed that a bad man could not be a good soldier, nor a bad soldier a good man, who saw in cowardice a moral failing and in vice a military crime. Cromwell's system is generally summed up in the word fanaticism ; but this is less than half of the truth. The employment of the phrase, moral force, in relation to the operations of war, is familiar enough in our language ; but the French term *morale* is now pressed into the service to signify that indefinable consciousness of superiority which is the chief element of strength in an army. Such narrowing of old broad terms is in a high degree misleading. It should never be forgotten that military discipline rests at bottom on the broadest and deepest of moral foundations ; its ideal is the organised abnegation of self. Simple fanaticism is in its nature undisciplined ;

it is strong because it assumes its superiority, it is weak because it is content with the assumption; only when bound under a yoke such as that of a Zizka or of a Cromwell is it irresistible. Cromwell's great work was the same as Zizka's, to subject the fanaticism that he saw around him to discipline. He did not go out of his way to find fanatics. "Sir," he once wrote, "the State in choosing men for its service takes no notice of their opinions; if they be willing faithfully to serve it, that satisfies." In forming his original regiment of horse he undoubtedly selected men of good character, just as any colonel would endeavour to do to-day. But Fairfax's was by no means an army of saints. One regiment of the New Model mutinied when its colonel opened his command with a sermon, and the Parliament with great good sense prohibited by Ordinance the preaching of laymen in the Army. It is time to have done with all misconceptions as to the work that Cromwell did for the military service of England, for it is summed up in the one word discipline. It was the work not of a preacher but of a soldier.

That the discipline was immensely strict and the punishments correspondingly severe followed necessarily from the nature of his system. The military code took cognisance not only of purely military offences, but of many moral delinquencies, even in time of peace, which if now visited with the like severity would make the list of defaulters as long as the muster-roll. Swearing was checked principally by fine, drunkenness by the wooden horse. This barbarous engine, imitated from abroad, consisted simply of a triangular block of wood, like a saddle-stand, raised on four legs and finished with a rude representation of a horse's head. On this the culprit was set astride for one hour a day for so many days, with from one to six muskets tied to his heels; and that degradation might be added to the penalty, drunkards rode the horse in some public place, such as Charing Cross, with cans about their necks. A soldier who brought discredit on his cloth by public misconduct paid the penalty with public disgrace. Fornication was commonly punished with the lash, the culprit being flogged so many times up and down the ranks of his company or regiment according to the flagrancy of the offence. It is small wonder that men forced by such discipline to perpetual self-control should have scorned civilians who allowed themselves greater latitude, and despised a Parliament which, in spite of many purgings, was never wholly purged of loose livers.

Towards the unfortunate Royalists the feelings of the Parliamentary Army after 1645 were of unutterable contempt. It was not only that it felt its moral superiority over the unhappy cavaliers; it mingled with this the keenest professional pride. No sergeant-major of the smartest modern cavalry regiment could speak with more withering disdain of the rudest troop of rustic yeomanry than did the Parliamentary newspapers of the prisoners captured at Bristol.[1] It is instructive, too, to note the patronising tone adopted by Reynolds towards the army of Turenne, his criticism of the discipline that was "good, for France," and his observations as to the proverbial inefficiency of a French regiment at the end of a campaign. Beyond all doubt the English standing Army from 1646 to 1658 was the finest force in Europe. It is the more amazing that Cromwell should have suffered its fair fame to be tarnished by the rabble that he sent to the West Indies.

Such an army will never again be seen in England; but though its peculiar distinctions are for ever lost, the legacies bequeathed by it must not be overlooked.

[1] "First came half-a-dozen of carbines in their leathern coats and starved weather-beaten jades, just like so many brewers in their jerkins made of old boots, riding to fetch in old casks; and after them as many light horsemen with great saddles and old broken pistols, and scarce a sword among them, just like so many fiddlers with their fiddles in cases by their horses' sides. . . . In the works at Bristol was a company of footmen with knapsacks and half pikes, like so many tinkers with budgets at their backs, and some musketeers with bandoliers about their necks like a company of sow-gelders."—*Newspaper*. (Reference unfortunately lost.)

Enough has been said of the institution of the new discipline, and of the virtual extinction of the old stamp of military adventurer; it remains now briefly to summarise the minor changes wrought by the creation of a standing Army. First comes the incipient organisation of a War-Department as seen in the Committee of the Army working with the Treasurers at War on one side and the ancient Office of Ordnance on the other, and in the appointment of a single commander-in-chief for all the forces in England, Scotland, and Ireland. And here it must be noted in passing that the division of the Army into an English, Scotch, and Irish establishment, which lasted until the three kingdoms were one by one united, becomes fully defined in the years of the Protectorate. Next must be mentioned the organisation of regiments with frames of a fixed strength, regiments of horse with six troops, and of foot and dragoons with ten companies, and the maintenance of a fixed establishment for services of artillery and transport.[1] Further, to combine the unity of the Army with the distinction of the various corps that composed it, there was the adoption of the historic scarlet uniform differenced by the facings of the several regiments.

Clothing however, leads us to the more complicated question of the pay of the Army. The regular payment of wages was, as has been seen, the first essential step towards the establishment of a standing force; and with it came concurrently the system of clothing, mounting and equipping soldiers at the expense of the State. It should seem, however, that the rules for regulating the system were sufficiently elastic, for we find quite late in the second Civil War that troopers generally still provided their own horses, and received a higher rate of pay, and that colonels were permitted to make independent contracts for the clothing and equipment of their regiments.

[1] This is evident from the mention of the "train" in the list in the *Commons Journals*, September 1651. The field-train was then transferred to Scotland bodily, where we find it still in December 1652 and again in 1659 (April). See *Commons Journals*.

The stoppages from the soldiers' pay at this period are also instructive. The deduction of a fixed sum for clothing dates, as has been already told, from the days of Elizabeth if not from still earlier times. But to this was now added the principle of withholding a proportion of the wages, under the name of arrears, as security against misconduct and desertion; while it was a recognised rule that both men and officers should forfeit an additional proportion so long as they lived at free quarter. An allowance for billet-money, and a fixed tariff of prices to be paid by soldiers while on the march within the kingdom, contributed somewhat to lighten the burden of all these stoppages, and made a precedent for the Mutiny Act of a later day. It is worthy of remark that the garrison of Dunkirk found in the town special buildings, constructed by the Spaniards for their troops and called barracks,[1] and that it was duly installed therein in the autumn of 1659. The reader, if he have patience to follow me further, will be able to note for himself how long was the time before English soldiers exchanged life in alehouses for the Spanish system of life in barracks.

But there is another and more interesting aspect of the question of pay, when we pass from that of the men to that of the officers. The extinction of the old military adventurer brought with it the total abolition, for the time, of the system of purchase. In the Royalist regiments that gathered around Charles Stuart in Flanders, we find that companies and regiments still changed hands for money, but in the English standing Army the practice seems utterly to have disappeared. Promotion was regulated not necessarily by seniority but by the recommendation of superior officers, and, as external evidence seems to indicate, ran not in individual regiments but in the Army at large. The arrears of

[1] Thurloe, vol. vii. p. 714. This is the first passage in which I have encountered the word thus spelt: "certain buildings . . . called the barracks or Spanish quarters." But there is mention of a *baraque* in the besiegers' lines before Ostend in 1604. *Grimeston.*

officers, especially of those who possessed means of their own, often remained, through their patriotic forbearance, not only many months but many years overdue ; and it is interesting to mark that their inability to watch over their own interests while they were engaged on active service led to the appointment of regimental agents, who drew their pay and transacted their financial business with the country on their behalf. The Army Agent may, therefore, justly boast himself to be a survival of the Civil War.

Nor can I leave this subject without reference to yet another remarkable feature in the New Model Army, which unfortunately has not passed into a tradition. I allude to the great and sudden check on the ancient evil of military corruption. To say that corruption came absolutely to an end would be an excessive statement, for the minutes of courts-martial on fraudulent auditors are still extant, but it is probable that during the Civil War it was reduced to the lowest level that it has touched in the whole of our Army's history. The abolition of purchase and the higher moral tone that pervaded the whole force doubtless contributed greatly to so desirable an end. It is, however, melancholy to record that the evil was evidently but scotched, not killed. Before the Protector had been dead a year, there was seen, at the withdrawal of part of the garrison of Dunkirk, a deliberate and disgraceful falsification of the muster-rolls, aggravated by every circumstance that could encourage fraud and injure good discipline. Contact with foreign troops was probably the immediate cause of this lamentable backsliding, but it furnishes a sad commentary on the fickleness of Puritan morality.

Finally, let us close with the greatest and noblest work of the New Model Army, the establishment of England's supremacy in the British Isles as a first step to their constitutional union. No achievement could have stood in more direct antagonism to the policy of Charles Stuart, who strove with might and main to set nation against nation and kingdom against kingdom, and paid for his folly with his life. It may be that the greatness of this service will in these days be denied. There were not wanting in the Long Parliament men who intrigued with Scotland against England rather than suffer power to slip from their hands, and it is not perhaps strange that the type of such men should be imperishable. Those, however, who call England the predominant partner in the British Isles should not forget who were the men that made her predominant.[1] The Civil War was no mere rebellion against despotic authority. It accomplished more than the destruction of the old monarchy; it was the battle for the union of the British Isles, and it was fought and won by the New Model Army.

Authorities.—In so slight a sketch of the Civil War and the Protectorate as is given in these pages any lengthy enumeration of the authorities would be absurd. Readers will find them for themselves in the exhaustive history of Mr. Gardiner, to whose labours, as well as to those of Mr. C. H. Firth, I am very greatly indebted. Such collections of documents as the *Calendars of State Papers*, Rushworth, Thurloe, and Carlyle's *Cromwell's Letters and Speeches* are almost too obvious to call for mention. The Clarke Papers are of exceptional value for purposes of military history, and Sprigge's *Anglia Rediviva* is of course an indispensable authority as to the New Model. But even in such fields as the newspapers and the King's Pamphlets Mr. Gardiner and Mr. Firth have left little harvest ungleaned. Of the military writers of the time Barriffe is the most instructive, particularly in respect of certain comments added in the later editions. A French folio volume, *Le Mareschal le Bataille* (1647), gives excellent plates of the drill of pike-men and musketeers, and beautiful diagrams of the evolutions.

---

[1] It is curious to note that a vote for a statue of Oliver Cromwell was in 1895 moved by the party that proposes to undo his work, and was defeated by the party that wishes to continue it. The supporters of the Union deliberately refused this tardy honour to the man who did more than any other to accomplish the Union, and who actually was the first to summon representatives from Scotland and Ireland to Westminster. Whether either party was sincere may well be considered doubtful.

BOOK IV

CHAPTER I

THE restoration of the Stuarts had been to all outward semblance effected, Charles had been escorted through the streets of London by the horse of the New Model, and yet the power which had practically ruled England since 1647 was still unbroken. The problem which the Long Parliament had treated with such disastrous contempt in that year was still unsolved; and there could be no assurance of stability for the monarchy until the Army should be disbanded. As to the manner in which this most difficult task must be accomplished the events of 1647 had given sufficient warning, for an army of sixty-five thousand men was even less to be trifled with than the comparatively small force of the second year of the New Model. Disbandment must not be hurried, and all arrears of pay must be faithfully discharged. Still the work could not but be both delicate and dangerous, requiring good faith and a tact that could only be found in a soldier who understood soldiers and a man who understood men. Fortunately such a man and such a soldier was to hand in the person of George Monk. 1660.

His scheme was soon prepared and adopted by Parliament. The regiments were to be broken up gradually, the order of disbandment being determined by lot, with the reservation that Monk's own regiments of horse and foot, together with two others that had been taken over by the Dukes of York and Gloucester,[1] should be kept until the last. An Act copied from an

[1] The Duke of Gloucester died in the same year.

Ordinance of the Commonwealth was passed, to enable discharged soldiers to engage in trades without preliminary apprenticeship, and thus to facilitate their return to civil life. By extraordinary exertions the needful money was raised, and the work proceeded apace. It seemed as if the close of the year 1660, according to the old reckoning which began the new year on the 25th of March, would have seen it completed, for by the first
1661. week in January the hand of disbandment had reached Monk's regiment of horse.

There however it was stayed. On the 6th of January an insurrection of fifth-monarchy men, a fanatical sect which had felt the might of Cromwell's repressing arm, not only saved the last relic of the New Model, but laid the foundation stone of a new Army. The rising was not suppressed without difficulty, not indeed until the veterans of Monk's regiment of foot, to whom such work was child's play, came up and swept it contemptuously away. The outbreak showed the need of keeping a small permanent force for the security of the King's person. The disbandment of this regiment and of the troop of horse-guards which had been assigned to Monk on his first arrival in London was thereupon countermanded, and the King gave orders for the raising of a new regiment of Guards in twelve companies, to be commanded by Colonel John Russell; of a regiment of horse in eight troops to be commanded by the Earl of Oxford; and of a troop of horse-guards, to be commanded by Lord Gerard. The Duke of York's troop of horse-guards, the same which he had led to an unsuccessful charge at Dunkirk Dunes, was also summoned home from Dunkirk.

The first stones of the new army being thus laid, there remained nothing but formally to abolish, in accordance with the letter of the Act of Parliament, the last remnant of the New Model. On the 14th of February, 1661 Monk's regiment of foot was mustered on Tower Hill, where it solemnly laid down its arms, and as solemnly took them up again, with great rejoicing, as the Lord General's regiment of Foot-Guards. But to

England at large this corps had but one name, that 1661.
which still survives in its present title of the Coldstream Guards. Though ranking second on the list of our infantry, this is the senior regiment of the British Army. Other corps may boast of earlier traditions, but this is the oldest national regiment and the sole survivor of the famous New Model. Well may it claim, in its proud Latin motto, that it is second to none.

Colonel Russell's regiment, being the King's own regiment of Guards, and raised specially for the protection of his person, obtained precedence not unnaturally of its earlier rival, and presently, by absorbing the handful of gallant men who had refused to surrender at Dunkirk Dunes, established its claim to represent the defeated cavaliers, as the Coldstream represent the victorious Roundheads, in the long contest of the Civil War. It is the regiment once called the First Guards, and now the Grenadier Guards, and it has known little of defeat since it ceased to fight against its countrymen.

The two troops of Life-Guards—the first the King's, commanded by Lord Gerard, the second the Duke of York's own—took precedence in like manner of Monk's Life-Guard; and after long existence as independent troops, blossomed at last into the First and Second regiments of Life-Guards that now stand at the head of our Army list. They were composed of men of birth and education, and for more than a century were rightly called gentlemen of the Life-Guards. Cromwell too had possessed such a guard, for he knew the value of gentlemen who had courage, honour, and resolution in them. Thus they stood apart from Lord Oxford's regiment of horse, which is still known to us from the colour of its uniform by its original name of the Blues. This corps was almost certainly made up of disbanded troopers of the New Model, of which there was no lack at that time in England;[1] while its colonel brought to it

[1] I find no sufficient ground for assuming that the regiment was Unton Crook's of the New Model, which had been disbanded two months before.

1661-1662. traditions of still earlier days in the honoured name of Vere.

But there was yet another regiment to be gathered in from the battlefield of Dunkirk Dunes, this time not from the defeated but from the victorious army. In view of the peril of the King from Venner's insurrection, Lewis the Fourteenth was requested to restore to him the regiment of Douglas, the representative of the Scots Brigade of Gustavus Adolphus; and this famous corps, having duly arrived in the year 1662, became the Royal or Scots regiment, and took the place which it still occupies at the head of the infantry of the Line under the old title of the Royal Scots. It returned to France in 1662 and did not return permanently to the English service until 1670, but it retained its precedence and it retains it still.

So far for the King's provision for his own safety. But it was also necessary for him to provide himself with money, and this he did in the simplest fashion by marrying an heiress, Catherine, Princess of Portugal, who brought him half a million of money, Bombay and Tangier, to say nothing of promises of pecuniary aid from Lewis the Fourteenth, who encouraged the match for his own ends. Tangier being in constant peril of recapture by the Moors was a troublesome possession, and required a garrison, for which duty a regiment of foot and a strong troop of horse were raised by the Earl

1661, October. of Peterborough, the recruits being furnished mainly by the garrison of Dunkirk. These corps also survive among us as the Second or Queen's regiment of Foot, and the First or Royal Dragoons.

Concurrently in this same year 1661 an Act was passed for the re-organisation of the militia. The obligations to provide horse-men and foot-men were distributed, following the venerable precedent of the statute of Winchester, according to a graduated scale of property, and the complete control of each county's force was committed to the lord-lieutenant. To him also were entrusted powers to organise the force into regiments and companies, to appoint officers, and to levy rates for the supply of ammunition. Finally, the supreme command of the militia, over which the Long Parliament had fought so bitterly with Charles the First, was restored to the King, together with that of all forces by sea and land. 1661-1665.

So much was accomplished in the first two years of Charles the Second. It sufficed for two years longer, when English commercial enterprise involved the restored monarchy in its first war. In truth it is hardly recognised how powerfully the spirit of adventure and colonisation had manifested itself under the Stuarts. The Empire indeed was growing fast. In 1661 England already possessed the New England States, Maryland and Virginia, as well as, for the time, Acadia, Nova Scotia, and Newfoundland. Off the American coast the Bermudas were hers; in the Caribbean Archipelago Barbados, Antigua, Montserrat, Nevis, St. Kitts, and Jamaica were settled; while Dominica, St. Lucia, St. Vincent, and Tobago, though not yet wrested from the Caribs, were reckoned subject to the British Crown. In 1663 one Company received a charter for the settlement of Carolina, and another, the Royal African, which enjoyed the monopoly of the trade in negro slaves, had fixed its headquarters at Cape Coast Castle. Nor must it be omitted that the East India Company, originally incorporated in 1599, received in 1660 a second charter conferring ampler powers, most notably in respect of military matters. 1665, February.

England, however, had abundance of rivals in distant adventure, whereof none was more jealous and more powerful than the Dutch federation which her own good arm had created. Cromwell had read the Dutch a lesson in 1653, and had imposed upon them restrictions which, if observed, would have checked their encroachments on English trade; but the Dutch not only evaded these obligations, but added to this delinquency wanton aggression both on the Guinea Coast and in the East Indies. The African Company at once commenced reprisals on the Gold Coast, and an

1665. expedition against the New Netherlands of America captured New Amsterdam and gave it its now famous name of New York. Meanwhile the complaints of English merchants were willingly heard by both King and Parliament. Charles had received no great kindness in his exile from the oligarchical faction which dominated the Dutch Republic; and now that the same faction had stripped the House of Nassau of its high dignities, to the prejudice of his nephew William, he was not sorry for the opportunity of revenge. Parliament voted liberal supplies for the war. A new regiment, called the Admiral's regiment, was raised by the Duke of York for service on board ship; large drafts were taken from the two regiments of Guards for the same purpose, and on the 3rd of June, James, Duke of York, won with them a great naval action off Lowestoft.

But there were English soldiers outside England who were troubled by this war. The descendants of the volunteers, who had followed Morgan in 1572 and had won an imperishable name under Francis Vere, were still in the Dutch service and were now comprised in seven regiments, three of them English and four Scotch, numbering in all three-and-fifty companies. As soon as war was declared the Pensionary De Witt forced upon the United Provinces a resolution that the British regiments must either take the oath of allegiance to the States-General or be instantly cashiered. This was the reward offered by the Dutch Republic to the brave foreigners who, with their predecessors, had done her better service than she could ever repay. Dismissal from the service meant ruin to the unfortunate officers, and want and misery to the men. Many Dutchmen were ashamed of the resolution, but they passed it; and it remained only to be seen whether British loyalty would stand the test. The English officers hesitated not a moment. They refused point blank to swear fealty to Holland, and were ruthlessly turned adrift. By the help of the English Ambassador, however, they made their way to England and were presently formed

into the Holland regiment, which now ranks as the 1665.
Third of the Line and is known from the facings which it has worn for more than two centuries, by the honoured name of the Buffs.[1]

The Scottish regiments behaved very differently. Though Charles was a Stuart and a Scot, only two officers had the spirit to follow the English example. The rest, who at first had made great protestation of loyalty, remained with their Dutch masters and, like all shamefaced converts, professed exaggerated love for the Dutch service and extravagant willingness to invade Great Britain if required. A century hence these regiments will be seen begging in vain to be received into the British service, and only accepted at last, after enduring sad insult from the Dutch, in time to become not the Fourth but the Ninety-Fourth of the Line. The corps finally ceased to exist in 1815, while the Buffs are with us to this day. It was a hard fate, but there is a nemesis even for unfaithful regiments.

In the following year Lewis the Fourteenth, seeing 1666.
therein an opportunity for furthering his darling project of extending his frontier to the Rhine, threw in his lot with the Dutch and declared war against England. The time is worthy of remark. For a century England in common with all Europe had abandoned traditional friendships and enmities, and sought out new allies by the guidance of religious sentiment. All this was now at an end, and the old jealousy of France was strong throughout the nation. But though the people were in earnest, the King was not; the policy of keeping France in check was after two years abandoned, and Charles, like a true Stuart, sold himself to Lewis the Fourteenth. False, wrong-headed, and unpatriotic, the dynasty was already preparing for itself a second downfall.

The next step was a declaration of war by France 1672.
and England against Holland. One hundred and fifty thousand men, under the three great captains, Turenne,

[1] For the return of the Buffs to England see the *Holland Papers* (Record Office), Bundles 233-235.

1672. Condé and Luxemburg, with Lewis in person at the head of all, swept down upon the United Provinces, mastered three of them almost without resistance, and actually crossed the Rhine. Six thousand English, grouped around a nucleus from the Guards, served with them under the command of James, Duke of Monmouth, and among the officers was a young captain named John Churchill. He had been born in 1650, less than three months before Dunbar, had been page to the Duke of York, and had received through him an ensigncy in the King's Guards. He had seen his first service, as became an English officer, in savage warfare at Tangier; he now enjoyed his first experience of a scientific campaign under the first General of the day. Soon he became known to Turenne himself not only as the handsomest man in the camp, but as an officer of extraordinary gallantry, coolness, and capacity. As Morgan had won the great captain's eulogy at Ypres, so did young Churchill at Maestricht; and it is worthy of note that on both of the two occasions when an English contingent served under Turenne the most brilliant little action of the war was the work of the red-coats.

But on the Dutch side also there was a young man, born in the same year as Churchill, who was to show lesser qualities indeed as an officer, though, as his opportunity permitted him, perhaps hardly inferior qualities as a man. William of Orange, long excluded by the jealousy of faction from the station and the duties of his rank, with firm resolution and unshaken nerve assumed the command of the United Provinces, and began the great work of his life, the work which was to be finally accomplished by the handsome English soldier in the enemy's camp, of taming the insolence of the French.

It is unnecessary to dwell further on the story of this campaign. The courage of William sufficed to tide Holland over the moment of supreme danger; and, the crisis once passed, Austria and Spain, alarmed at the designs of Lewis, hastened to her assistance.

Charles made peace with the Dutch in 1674, and, while 1674.
declining to withdraw the English troops in the French service, promised to recruit them no further. Churchill came home to be colonel of the Second Foot; and from the troops disbanded at the close of the war, were formed three English regiments for the service of the Prince of Orange. Among their officers was James Graham of Claverhouse. We shall meet with him again, and we shall see two of the regiments also return in due time, like their prototype, the Buffs, to take their place in the English infantry of the Line.

With the treaty of 1674 the wars of Charles the Second came to an end. It was not that the people of England were unwilling to fight. They were heart and soul against the French; and the Commons cheerfully voted large sums for army and fleet while the war lasted, asking only that the money might be expended on its legitimate object. But the crookedness and untrustworthiness of the King were fatal to all military enterprise, and indeed to all honest administration. Though the military force of England was far too small for the safety of her possessions abroad, Parliament never ceased to denounce the evils of standing armies, and to clamour for the disbanding of all regiments. In the days of Cromwell the burden of the red-coats had been grievous to be borne, but Oliver had at all events made England respected in Europe. Charles sought to impose a like burden, but without sympathy for England's quarrels, and without care for England's glory. He made shift, nevertheless, to keep his existing regiments throughout
his reign, and in 1680 even to add another to them for 1680.
the service of Tangier. In 1684 that ill-fated possession, 1684.
having cost many thousands of lives and witnessed as gallant feats of arms as ever were wrought by English soldiers, was finally abandoned; though not before the English had learned one secret of Oriental warfare. In March 1663, after long endurance of incessant harassing attacks from the Moors, the Governor, who had hitherto stood on the defensive, took the initiative and

launched the Royal Dragoons straight at them. So signal was the success of this first venture that it was repeated a fortnight later by the same regiment, and renewed on a grander scale after two months by a sally of the whole garrison, which after desperate fighting ended once more in victory. So much at least must be recorded of this first long lost settlement in Africa.[1] The new regiment, which had arrived too late for fighting, came home to take rank as the Fourth of the Line and to remain with us to this day.

In truth the little Army, which Parliament so bitterly hated, was busy enough from the day of the King's accession to the day of his death. In regiments or detachments it fought in Tangier, in Flanders, and in the West Indies ; it did marines' duty in four great naval actions, one of them the fiercest ever fought by the English, and it suppressed an insurrection in Scotland and a rebellion in Virginia. The reign gave it a foretaste of the work that lay before it in the next two centuries, and showed good promise for the manner in which that work would be done.

1685. Charles died on the 6th of February 1685. His brother James, who succeeded him, was a man of stronger military instincts than any English king since Henry the Eighth. He had served through four campaigns under Turenne and through two more with the Spaniards, and his narrative of his wars shows that he had studied the military profession with singular industry and intelligence of observation. Nor was he less interested in naval affairs. He had commanded an English fleet in two great actions without discredit as an Admiral, and with signal honour as a brave man. Moreover, he felt genuine pride in the prowess alike of the English sailor and the English soldier. Finally he had shown uncommon ability and diligence as an administrator. The Duke of Wellington a century and

[1] The historian of the Second regiment of Foot has printed a great deal of matter respecting Tangier. Details will also be found in Clifford Walton's *History of the British Standing Army*, p. 22.

a half later spoke with the highest admiration of the system which James had established at the Office of Ordnance, and actually restored it, as Marlborough had restored it before him, when he himself became Master-General. The Admiralty again acknowledges that his hand is still felt for good in the direction of the Navy. In fact, whatever his failings, James was an able, painstaking, and conscientious public servant, and as such has no little claim to the gratitude of the nation. 1685.

So far then the succession of a diligent and competent administrator to the shrewd but incorrigibly idle Charles promised advantages that were obvious enough. But there was another side to the question. Parliament had requited James's services to the public by excluding him as an avowed Catholic from all public employment, whether civil or military; and James was a narrow-minded, a vindictive, and, like all the Stuarts, essentially a wrong-headed man. Though valuable as the head of a department, he was totally unfit to administer a kingdom; though not devoid of constancy and patience in adversity, he was swift and unsatiable in revenge; though ambitious of military fame, proud of English valour, and not without jealousy for English honour, he saw no way to the greatness which he coveted in Europe except by the overthrow of English liberty. He longed to interfere effectively abroad, but with England crushed under his heel, not free and united at his back.

So he too sold himself to France, hoping to consolidate his power by her help and to turn it in due time to her own hurt; and meanwhile he sought to strengthen himself by the maintenance of a standing Army. For this design Monmouth's insurrection of 1685 afforded sufficient excuse.[1] The opportune return of the garrison of Tangier had already added two regiments of Foot and one of Horse to the English

[1] No reader, I am confident, will blame me for leaving him alone with his Macaulay for the account of this insurrection.

1685. establishment; and James seized the occasion of the outbreak to summon the six British regiments, three of them Scottish and three English, from Holland. These, though they presently returned to William's service, secured for two of their number on the invasion of England in 1688 the precedence of Fifth and Sixth of the Line. Simultaneously twelve new regiments of infantry and eight of cavalry were raised under the same pretext. Of the foot the first was an Ordnance-regiment, designed like the firelocks of the New Model to act as escort to the artillery, and was called from its armament the Regiment of Fusiliers. It is still with us as the Seventh of the Line. The remainder of the foot, some of them formed round the nucleus of independent garrison-companies, also abide with us, numbered the Eighth to the Fifteenth.[1] Of the cavalry six were regiments of horse, and are now known as the First to the Sixth Regiments of Dragoon Guards; the remaining two, which are now numbered the Third and Fourth, after having been successively dragoons and light dragoons, have finally become the two senior regiments of hussars. Add to these thirty independent companies of foot, borne for duties in garrison, and it will be seen that King James's army was increasing with formidable speed.

The King himself found genuine delight, not in the sinister spirit of an oppressor but in the laudable pride of a soldier, in reviewing his troops. In August 1685 he inspected ten battalions and twenty squadrons which were in camp at Hounslow, and wrote to his son-in-law, William of Orange, with significant satisfaction of their efficiency. In November he met Parliament, and required of it the continuance of the standing Army in lieu of the militia. The courtiers had received their cue, and pointed to the flight of the western militia before Monmouth's raw levies as proof sufficient of its

[1] It is worthy of note that but two of these regiments were raised in the districts indicated by their present titles, viz., the 11th (North Devon) and 12th (East Suffolk).

untrustworthiness. The fact indeed was self evident. 1685.
But Parliament was not disposed to welcome a royal speech which submitted no further measures than the maintenance of a standing army and the admission of popish officers to command therein. The memories of Oliver and of his major-generals was still vivid, and the revocation of the edict of Nantes was but a month old. Red-coats as saints had been bad; red-coats as papists would doubtless be worse. Edward Seymour, the head of that historic house, put the matter as Englishmen love to put it. The militia, he confessed, was in an unsatisfactory state, but it might be improved, and with this and the navy the country would be secure; but a standing army there must not be. Then as now, it will be observed, the House of Commons never stinted the navy, nor doubted its ability to repel invasion; and then as now it refused to remember that the British possessions are not bounded by the British Isles, and that a successful war is something more than a war of defence. But unfortunately it had but too good ground for opposing the King in this case. The debate lasted long. James had asked for £1,400,000 for the Army; the Chancellor of the Exchequer expressed his willingness to accept £1,200,000; the House voted £700,000, and even then declined to appropriate the sum to any specific purpose.

James was greatly annoyed. He answered the note of the Commons with a reprimand, and prorogued December.
Parliament; nor did he summon it again during the remainder of his reign. He then concentrated from thirteen to sixteen thousand men at Hounslow Heath, 1686,
and kept them encamped there for three years in the June.
hope of overawing London. Never did man make a more complete mistake. The Londoners, after their first alarm had passed away, soon discovered that the camp was a charming place of amusement. A new generation had sprung up since a Parliamentary colonel had held a sham fight to compensate the people for the loss of the sports of May-day, and there was a certain

1686-1688. novelty in military display. Hounslow camp became the fashion, and the lines were thronged with a motley crowd of all classes of the people; for then as now the women loved a red-coat, and where the women led the men followed them. The troops were doubtless well worth seeing, for James flattered himself that they were the best paid, the best equipped, and the most sightly in Europe.

Still, merry as the camp might be, there were not wanting signs of a graver spirit beneath the new red-coats. There were early rumours of quarrels between protestant and catholic soldiers, ominous to the catholic officers whom James had set in command against the law. Agitators scattered tracts appealing to the Army to stand up in defence of the liberties of England and the protestant religion; and the Londoners perceived, what James did not, that consciences cannot be bought for eightpence a day, nor flesh and blood extinguished by a red coat and facings. The Buffs had been the earliest English volunteers in the cause of liberty and protestantism; the Royal Scots had rolled back papistry under the Lion of the North, and, as if one presbyterian regiment were not sufficient, there was another, just brought into England for the first time from Scotland, and known by its present name of the Scotch or Scots Guards. Again, monks in the habit of their Order were among the visitors to the camp; and it was easy to ask how long it was since such men had been seen in England, and what was the cause of their disappearance. Cromwell's soldiers had made short and cruel work of monks in Ireland; yet soldiers, only one generation younger, were to be called upon to fight against their kith and kin for a king who openly favoured them, a king, too, who in the face of all law openly thrust papists into all places of authority.

It was not long before the seed sown by the agitators began to bear fruit. When the seven bishops who
1688, June. had refused to read the declaration which suspended the penal laws against catholics were committed to the

Tower, the guards drank their health; and when the 1688. news of their acquittal reached Hounslow Heath, it was received by the Army with boisterous delight. In alarm James broke up the camp and scattered the regiments broadcast over the country. Having thus isolated them he attempted to work upon them separately, and selected as the first subject for this experiment Lord Lichfield's Regiment, known to us as the Twelfth Foot. The men were drawn up on Blackheath in the King's presence, and were informed that they must either sign a pledge to carry out the royal policy of indulgence towards catholics, or leave his service forthwith. Whole ranks without hesitation took him at his word, and grounded their arms, while two officers and a few privates, all of them catholics, alone consented to sign. James stood aghast with astonishment and disgust. Dismissal meant something more than mere exclusion from the Army; it carried with it the forfeiture of all arrears of pay and of the price of the officers' commissions, but neither men nor officers took account of that. James eyed them in silence for a time, and then bade them take up their arms. "Another time," he said, "I shall not do you the honour to consult you."

Foiled in England, James turned, as his father had turned before him, to Ireland. The Irish speak of the curse of Cromwell; they might more justly speak of the curse of the Stuarts, for no two men have brought on them such woe as Charles and James. Already, in 1686, the King had sent a degenerate Irishman, the Earl of Tyrconnel, to ensure popish ascendency at any rate in Ireland; and no better man could have been found for such mischievous work than lying Dick Talbot. The army in Ireland consisted at the time of his arrival of about seven thousand men: within a few months Tyrconnel, by wholesale dismissal of all protestants, had turned it upside down. Five hundred men were discharged from a single regiment on the ground that they were of inferior stature, and their places shame-

1688. lessly filled by ragged, half-trained Irish, beneath them both in size and quality. In all four thousand soldiers were broken, stripped of the uniforms which they had bought by the stoppage of their pay, and dismissed half-naked to go whither they would. Three hundred protestant officers shared a like fate in circumstances of not less hardship. Many of them had fought bravely for the Stuarts in past days, the majority had purchased their commissions, yet all alike were turned adrift in ruin and disgrace. The disbanded took refuge in Holland, whence they presently returned under the colours of William of Orange, with such feelings against the Irish as may be guessed.

But James did not stop here. He now conceived the notion of surrounding himself with Irish battalions, and of moulding the English regiments to his will by kneading into them a leaven of Irish recruits. When we reflect that it was just such an importation of Irish that had turned all England against his father, we can only stand amazed at such folly. The English held the Irish for aliens and enemies; they knew them as a people who for centuries had risen in massacre and rebellion whenever the English garrison had been weakened, and that had sunk again into abject submission as soon as England's hands were free to suppress them. They did not know them, in spite of their occasional gallant resistance to Cromwell, as a great fighting race. They had not read, or, reading, had not believed, the testimony of Robert Munro to their merits as soldiers.[1] Lastly and chiefly the Irish were catholics and the English protestants.

The resentment against the new policy soon made itself manifest. The Duke of Berwick, the King's natural son, who had been appointed colonel of the Eighth Foot, gave orders that thirty Irish recruits should be enlisted in the regiment. The men said flatly that they would not serve with them, and the lieutenant-colonel with five of his captains openly

[1] *Expedition*, vol. ii. pp. 37, 73.

remonstrated with the Duke against the insult. They 1688.
had raised the regiment, they said, at their own expense for the King's service, and could procure as many English recruits as they wanted; rather than endure to have strangers forced upon them they would beg leave to resign their commissions. James was furious. He tried the six officers by a court-martial, which sentenced them to be cashiered; but the culprits none the less received the sympathy and applause of the whole nation. The prevalent feeling against the Irish found vent in a doggrel ballad, known, from the gibberish of its burden, by the name of Lillibulero. Partly from the nature of its contents, still more probably from the rollicking gaiety of its tune,[1] it became a great favourite with the Army, and if we may judge from Captain Shandy's partiality for it, was the most popular marching song of the red-coats in Flanders.

But meanwhile William of Orange had received his invitation to come with an armed force for the delivery of England from the Stuarts, and for some months had been making preparations for an invasion. It was long before James awoke to his danger, but when at last he perceived it he hastened to strengthen the Army. Commissions were issued for the raising of new regiments, of which two are still with us as the Sixteenth and Seventeenth of the Line, and of new companies for existing regiments. Four thousand men in all were added to the English establishment; three thousand were summoned from Ireland, and as many more from Scotland; and James reckoned that he could meet the invader with forty thousand men. On the 2nd of November William, after one failure, got his expedition safely to sea, and by a feint movement induced James to send several regiments north-

[1] The tune, which is in the key of G major and in $\frac{6}{4}$ time, may be found in modern editions of *Tristram Shandy*, at the end of chap. iii. of the second book. It is admirably suited for fifes and drums.

1688. ward to meet a disembarkation in Yorkshire. These regiments were hastily recalled on the intelligence that the armament had passed the Straits of Dover steering westward, and fresh orders were given for concentration at Salisbury.

In a short time twenty-four thousand men were assembled at the new rendezvous, but before James could join them, he received news that Lord Cornbury, the heir of his kinsmen the Hydes, had deserted to the enemy. Cornbury had attempted to take his own regiment, the Royal Dragoons, and two regiments of horse with him; but officers and men became suspicious, and with the exception of a few who fell into the hands of William's horse and took service in his army, all returned to Salisbury. Before setting out for the camp James summoned his principal officers to him—Churchill, since 1683 Lord Churchill, and recently promoted lieutenant-general; Henry, Duke of Grafton, colonel of the First Guards; Kirke and Trelawny, colonels of the Tangier Regiments. One and all swore to be faithful to him; and the King left London for Salisbury.

Arrived there, he learned from Lord Feversham, his general-in-chief, that though the men were loyal the officers were not to be trusted. It is said that Feversham proposed to dismiss all that he suspected and promote sergeants in their stead. His suspicions proved to be just. Within a week Churchhill, Grafton, Kirke, and Trelawny had all deserted to the Prince of Orange. Other officers were less open in their treachery; and it is said that one battalion of the Foot Guards was led into William's camp by its sergeants and corporals. The desertion of his own children finally broke the spirit of James. On the 11th of December he signed an order for the disbandment of the Army, and took to flight; and on the 16th he returned to London to find on the following night that the battalions of the Prince of Orange were marching down St. James's Park upon Whitehall. The old colonel of the Coldstream Guards,

Lord Craven, though now in his eightieth year, was for 1688.
resistance, but James forbade him. The Coldstream Guards filed off, and a Dutch regiment mounted guard at Whitehall. Five days later James left England for ever.

CHAPTER II

1660-1688. BEFORE entering on the reign of William we must pause for a time to study the interior administration of the Army. The reign of the two last Stuarts is rightly considered as marking the end of a period of English general history—the final fall of the old monarchy first overthrown with King Charles the First. But in regard to military history the case is different. It is a critical time of uncertainty during which the Army, a relic barely saved from the ruins of a military government, struggled through twenty-eight years of unconstitutional existence, hardly finding permission at their close to stand on the foundation which Charles and James, using materials left by Cromwell, had made shift to establish for it. Precarious as that foundation was, it received little support for nearly a century, and little more even in the century that followed, thanks to the blind jealousy of the House of Commons. It will therefore be convenient at this point to examine it once for all.

Beginning, therefore, at the top, it must be noted that the first commander-in-chief under the restored Monarchy was a subject, George Monk, Duke of Albemarle. His appointment was inevitable, for he had already held that command as the servant of the Parliament over the undisbanded New Model, and he was the only man who could control that Army. Charles, in fact, lay at his mercy when he landed in 1660, and could not do less than confirm him in his old office. The powers entrusted to Monk by his commission were very great. He had authority to raise forces, to fix the establishment, to issue commissions to all officers executive and administrative, and to frame Articles of War for the preservation of discipline; he signed all warrants for expenditure of money or stores, and, in a word, he exerted the sovereign's powers as the sovereign's deputy in charge of the Army. On his death in January 1670, Charles, by the advice of his brother James, did not immediately appoint his successor, and though in 1674 he issued a circular to all officers of horse and foot to obey the Duke of Monmouth, yet he expressly reserved to himself many of the powers formerly made over to Monk. Finally, when in 1678 he appointed Monmouth to be captain-general, he withheld from him the title of commander-in-chief. On Monmouth's disgrace in 1679 Charles appointed no successor, but became his own commander-in-chief, an example which was duly followed by James the Second and William the Third. Thus the supreme control of the Army, with powers far greater than have been entrusted to any English commander-in-chief of modern times, continued at first practically the same as it had been made by Oliver Cromwell. It was exclusively in military hands. 1660-1688.

The special branch of military administration in the hands of the commander-in-chief was that relating to the men. The care of material of war was committed to the ancient and efficient Office of Ordnance. At the Restoration the old post of Master of the Ordnance was revived with the title of master-general; and in 1683 the Department was admirably reorganised, as has been seen, by the Duke of York. At the head stood, of course, the master-general; next under him were two officers of two distinct branches, the lieutenant-general and the surveyor-general. The lieutenant-general was charged with the duty of estimating the amount of stores required for the Navy and the Army, and of making contracts for the supply of the same; he was also responsible for the maintenance of marching trains

1660-1688. for service in the field, and for the general efficiency of the artillery both as regards guns and men. His first assistant was named the master-gunner. The surveyor-general was responsible for the custody and care of all stores, and for all services relative to engineering; his first assistant was called the principal engineer. Transport of ordnance by land was the care of a waggon-master, transport by water of a purveyor. The laboratory was committed to a fire-master, whose duties included the preparation of fireworks for festive occasions. The only weak point of the office was the exclusiveness of its jurisdiction over artillery and engineers, which was carried to such a pitch that all commissions in the two corps were signed by the master-general, though that functionary and his staff received their own commissions from the commander-in-chief.

I turn next to the department of finance. Here in place of the old treasurers at war there was created a new officer called the paymaster-general. Parliament, I must remind the reader, never recognised the existence of the Army under the Stuarts, nor voted a sixpence expressly for its service. The force was paid out of the King's privy purse, or, in the case of James, out of sums intended for the payment of the militia. Thus the House of Commons through sheer perversity lost its hold upon the paymaster-general, and when it came to examine his office a whole century later, found, as shall be told in place, a system of corruption and waste which is almost incredible. The first paymaster-general, Sir Stephen Fox, received a salary of four hundred pounds a year, but this he soon supplemented by becoming practically a farmer of a part of the revenue. Knowing that Charles was chronically deficient in cash, he undertook to advance funds on his own private credit for the weekly pay of the Army, in consideration of a commission of one shilling in the pound. At the end of every four months he applied to the Treasury for reimbursement, and if his claims were not immediately satisfied, he received eight per cent on the debt owing to him, thus making a very handsome profit. This system was discontinued in 1684, but the deduction, or poundage as it was called, was still levied on the Army, for no reason whatever, for a full century and a half. For the care of all other military expenses there was an office called by the old title of Treasurer of the Armies. 1660-1688.

So much for the broad divisions of the administration, under the three heads of men, military stores, and finance. It is now necessary to trace the rise of a new department, which was destined to give to civilians the excessive share that they still enjoy in the direction of military affairs. While Charles the Second was yet an exile in Flanders in 1657, he had appointed a civilian, Sir Edward Nicholas, who had been Secretary of Council to Charles the First, to be his Secretary at War. It was not uncommon for such civilian secretaries[1] to be attached to a general's staff, and we have already seen John Rushworth taking the field with the New Model as secretary to the Council of War. After the Restoration, and within six months of the date of Monk's commission, one Sir William Clarke was appointed to be secretary to the forces. Though a civilian, he received a commission couched in military terms, which were preserved for fully a century unchanged, bidding him obey such orders as he should from time to time receive from the King, or the general of the forces for the time being, according to the discipline of war. In effect he was a civilian wholly subordinated to the military authorities and subject to military discipline so far as that discipline existed; little more, indeed, than a secretary to the commander-in-chief. His services were not estimated at a very high rate, for he received at first but ten shillings, and after 1669 one pound a day, as salary for himself and clerks. The appointment was of so personal a nature that Clarke accompanied Monk to

[1] It is possible that there was difficulty in finding ready writers among the military, and still more difficulty in persuading them to unite sword and pen.

1660-1688. sea in 1666, and was killed in the naval battle of the 1st of June, the first and last secretary at war who has fallen in action.

Monk then applied for the services of one Matthew Lock, whom he knew to be a good clerk, and Lock was appointed to be Clarke's successor with the title of sergeant or secretary at war. There is not a letter from him to be found in the State Papers until after Monk's death, which is sufficient proof that he was a person of no great importance; but in 1676, when there was no longer a single commander-in-chief, he was entrusted with the removal of quarters, the relief of the established corps, the despatch of convoys, and even with authority to quarter troops in inns, all of which duties had been previously fulfilled by military men. Thus early and insidiously arose once more that civil interference with military affairs which had with such difficulty been thrown off at the establishment of the New Model. The system was wholly unconnected with any question of Parliamentary control, for Parliament would have nothing to do with the standing Army. Most probably it was due simply to the indolence of the King, who would neither do the work of commander-in-chief himself nor appoint any other man to do it for him. Thus the Army was placed once and for all under the heel of a civilian clerk.

The staff at headquarters was based on the model of that which had prevailed under Cromwell, though of course on a scale reduced to the minute proportions of the Army. The duties must, at first, have been within the scope of a very few officials, and it is probable that Monk required little assistance. There was, however, a commissary of the musters, to whom in 1664 a scoutmaster-general, or head of the intelligence department, was added. The business of foreign intelligence in all its branches, diplomatic, naval, and military, had been conducted with admirable efficiency during the Protectorate by the Secretary of State, John Thurloe, but Pepys remarked a sad falling away in this department after the Restoration, due, as he admits, to the scanty allowance of funds allotted to the service. Charles was not the man to face the difficulties of establishing a great administrative office on a sound basis. James, on the other hand, began to grapple with them very early after his accession. He strengthened the staff by the addition of adjutants and quarter-masters-general of horse and foot, and strove hard to improve the efficiency of the office; but his time was too short and his distractions too manifold to permit him to do the work thoroughly. Had he reigned for ten years, his familiarity with the system of Louvois and his own administrative ability might have reduced our military system once for all to order. It is not too much to say that his expulsion was in this respect the greatest misfortune that ever befell the Army. 1660-1688.

Even he, however, would have found it a hard task to overcome the obstacles raised by Parliament, namely, the difficulties of regular payment of wages and of maintaining discipline. It was impossible to enforce military law on the troops, since Parliament steadily withheld its sanction to the same.[1] Nothing therefore remained but the civil law. A soldier who struck his superior officer or got drunk on guard could legally only be haled before the civil magistrate for common assault or for drunkenness, while if he slept on his post or disobeyed orders or deserted he was subject to no legal penalty whatever. Parliament never seems to have been the least alive to the danger of such a state of things, nor to have weighed it against its fixed resolution not to recognise the standing Army. As a matter of fact, however, military offences seem to have been punished as such throughout the reign of Charles, though without ostentation; and discipline appears to have been maintained without serious difficulty. The

[1] But indeed I have failed to discover by what legal authority martial law was enforced on the Parliamentary troops in the Civil War. There seems to have been no effort to give so much as a semblance of legality to the power of the generals.

1660-1688. number of the troops was, after all, but small; many of the men were already inured to obedience; the traditions of Oliver and of George Monk were still alive; and the men probably accepted service with a tacit understanding that they were subject to different conditions from the civilian. But when the three regiments returned from foreign service and savage warfare at Tangier, and Monmouth's rebellion had brought about a multiplication of regiments, the situation was altogether changed. James, who knew the value of discipline, determined to arrogate the powers that Parliament denied to him, but, like all weak men, endeavoured to effect his purpose by half measures. To secure the punishment of certain deserters he packed the Court of King's Bench with unscrupulous men; and though the culprits were hanged, discipline was only preserved at the cost of the integrity of the courts of law, a proceeding which damaged him greatly both in the Army and the country at large. It will presently be seen how this question of discipline was forced upon Parliament in a fashion that allowed of no further trifling.

The subject of pay opens a melancholy chapter in the history of English administration. It has already been related that Charles the Second let out the payment of the Army to a contractor for a commission of a shilling in the pound. This commission of course came out of the pockets of officers and men; they paid, in fact, a tax of five per cent for the privilege of receiving their wages, and this not to the State, to which the officers still pay sometimes an equal amount under the name of income-tax, but for the benefit of a private individual. If the mulcting of the Army had ended there, the evil would not have been so serious, but as a matter of fact it was but one drop in a vast ocean of corruption. I have already alluded to the immense service wrought by the Puritans towards integrity of administration, and towards raising the moral standard of the military profession. The destruction of the old traditions and the substitution of new principles was a

magnificent stroke, but it was unfortunately premature. 1660-1688.
The new principles might indeed have endured had they but been cherished and encouraged for another generation, but unfortunately no man better fitted to starve them could have been found than the merry monarch. His difficulties were doubtless very great, but he brought but one principle to meet them, that come what might he must not be bored. His indolent selfishness was masked by an exquisite charm of manner, and being a kind-hearted man, he always heard complaints with a sympathetic word; but to redress them cost more trouble than he could afford. Any man who would save him trouble was welcome; any shift that would stave off an unpleasant duty was the right one. There was abundance of deserving suitors to be provided for, still greater abundance of importunate favourites to be satisfied; administration was a bore and money was sadly deficient. All difficulties could be solved by the simple process of providing alike the impecunious and the greedy with administrative offices, or, in other words, with licences to plunder the public. If they chose to purchase these offices for money, so much the better for the royal purse. Thus the whole fabric built up during the Commonwealth was shattered almost at a blow.

The effect on the Army was immediate. A great many of the returned exiles, including Charles and James themselves, had served in the French army, where the system of purchasing commissions had never been abandoned, and where the abuses which had been shaken off by the New Model were still in full vigour. The old corrupt traditions had not been killed in thirteen years, and, reviving under the general reaction against Puritan restraint, they sprang quickly into new life. The old military centralisation of Oliver, upheld for a time by Monk, rapidly perished, and what might have still been an army sank into a mere aggregate of regiments, the property of individual colonels, and of troops and companies, the property of individual

1660-1688. captains. Every civilian of the military departments hastened to make money at the expense of the officers, and every officer to enrich himself at the cost of the men. The flood-gates so carefully closed by the Puritans were opened, and the abuses of three centuries streamed back into their old channel to flow therein unchecked for two centuries more.

At its first renewal the system of purchase was carried to such lengths that the very privates paid premiums to the enlisting officers; but the practice was speedily checked by Monk in 1663. In March 1684 the system received a kind of royal sanction through the purchase by the King himself of a commission from one officer for presentation to another. Then nine months later Charles suddenly declared that he would permit no further purchase and sale of military appointments. Whether he would have abolished it if he had lived may be doubted, but it is certain that the system continued in full operation under James the Second, gathering strength of course with each new year of existence.

Let me now attempt briefly to sketch the organised system of robbery that prevailed in the military service under the two last of the Stuarts. The study may be unpleasant, but it is less pathological than historic. First, then, let us treat of the officer. On purchasing his commission he paid forthwith one fee to the Secretary at War, and a second, apparently, to one of the Secretaries of State. After the institution of Chelsea Hospital, as to which a word shall presently be said, he paid further five per cent on his purchase money towards its funds, the seller of the commission contributing a like proportion from the same sum to the same object. He then became entitled to the pay of his rank, but this by no means implied that it was regularly paid to him. In the first place, his pay was divided into two parts, termed respectively his subsistence and his arrears, or clearings. The former sum was a proportion of the full pay, which varied according to the grade of the officer, it being obvious that an ensign, for instance, could not subsist if any large fraction was deducted from his daily pittance, whereas a major could be more heavily mulcted and yet not starve. This subsistence was therefore paid, or supposed to be issued, in advance from the pay-office and to be subject to no stoppage. The balance of the full pay, or arrears, was paid yearly after it became due, and after considerable deductions had been made from it. First of these deductions came the poundage, or payment of one shilling in the pound, to the paymaster-general, and the discharge of one day's full pay to Chelsea Hospital. These stoppages were more or less legitimate. Then the commissary-general of the musters stepped in to claim from the officer, as from every one else in the Army, one day's pay, a tax which caused much discontent, and was in 1680 reduced to one-third of a day's pay. Then came a vast number of irregular exactions. Every commissary of the musters claimed a fee, amounting sometimes to as much as two guineas for every troop or company passed at each muster, which, as musters were taken six times a year, was sufficiently exorbitant. Next the auditors demanded thirty shillings, or eight times their legal fee, for each troop and company on passing the accounts of the paymaster-general. Finally, fees to the exchequer, fees to the treasury, fees for the issue of pay-warrants, fees, in a word, to every greedy clerk who could make himself disagreeable, brought the tale of extortion to an end. Let the reader remember that this system of subsistence and arrears, with the same legitimate deductions and almost equal opportunities for irregular pilfering, was still in force when we began the war of the French Revolution, and let him not wonder that officers of the Army will still cherish unfriendly feelings towards the clerks at the War Office.[1]

1660-1688.

[1] It should not be forgotten meanwhile, in justice to the clerks, that their salaries were very irregularly paid and that they depended chiefly on their perquisites. We do not realise, in fact, how recently salaries have supplanted fees in the payment of officials.

1660-1688. Now comes the more distressing examinations of the officers' methods of indemnifying themselves. For this purpose let us study the pay of a private centinel, as he was called, of the infantry of the Line. This consisted, as it had been in Queen Mary's time, and was still to be in King George the Third's, of eightpence a day, or £12 : 13 : 4 a year. Of this, sixpence a day, or £9 : 2 : 6 a year, was set apart for his subsistence, and was nominally inviolable. The balance, £3 : 0 : 10 a year, was called the "gross off-reckonings," which were subject of course to a deduction of five per cent, or 12s. 2d., for the paymaster-general, and of one day's pay to Chelsea Hospital, whereby the gross off-reckonings were reduced to £2 : 8s. This last amount, dignified by the title of "net off-reckonings," was made over to the colonel for the clothing of the regiment, an item which included not only the actual garments, but also the sword and belt, and as time went on the bayonet and cartridge box. The system, as will be remembered, dated from the days of Queen Elizabeth, when half a crown a week was allowed to the men for subsistence and a total of £4 : 2 : 6 deducted for two suits a year. It is sufficiently plain that the sum now allowed for clothing was insufficient, and that a colonel who did his duty by his men must inevitably be a loser. Moreover, this was not his only expense. The clerical work entailed by his duties demanded assistance, for which he was indeed authorised to keep a clerk, but supplied with no allowance wherewith to pay him. This clerk presently became known as the colonel's agent, and though a civilian and the colonel's private servant, virtually performed the duties of a regimental paymaster.

The results of such an arrangement may easily be guessed. It was not in consonance with military tradition, certainly not in accordance with human nature, that colonels should lose money by their commands, and it is only too certain that they did not. The contractor was called in, and the door was opened wide to robbery at the expense of the soldier. Colonels took commissions or even open bribes from the contractors; the agent took his fee likewise; and in at least one recorded case a colonel actually accepted a bribe from his own agent to give him the contract. It may easily be imagined how the soldiers fared for clothing. But the mischief did not end here. The subsistence-money, though in theory subject to no deduction, was practically at the mercy of the colonel and his agent, who, under various pretexts, appropriated a greater or smaller share of the poor soldier's sixpence. As an additional source of profit, it was not uncommon for colonels to abstain from reporting the vacancy caused by an officer's death, to continue to draw the dead man's pay and to put it into his own pocket. 1660-1688.

Captains of companies, with such an example before them, were not slow to imitate it; and from them too the unfortunate soldiers suffered not a little. But their easiest road to plunder was the old beaten track of false musters, which was rendered all the easier by the corruption of the commissaries. Any vacancy in the ranks after one muster was left unfilled until the day before the next muster, and the captain drew pay for an imaginary man during the interval. Or again, the *passe-volant*, old as the days of Hawkwood, made his reappearance at musters and was passed, with or without the collusion of the commissaries, as a genuine soldier. Finally, Charles himself gave countenance after a manner to this fraud by reviving the practice of allowing officers so many imaginary men or permanent vacancies in each troop or company in order to increase their emoluments. And so the *passe-volant* became naturalised first as a "faggot," and later as a "warrant man" in the infantry and a "hautbois" in the cavalry, and survived to a period well within the memory of living men.[1] The remoter a regiment's quarters from

[1] The warrant men and hautbois can generally be found in old muster-rolls under the names of John Doe, Richard Roe, and Peter Squib.

1660-1688. home the grosser were the abuses that prevailed in it, and in Ireland they seem to have passed all bounds. Captains calmly appropriated the entire pay of their companies, and turned the men loose to live by the plunder of the inhabitants. It was a reversion to the evils rampant in Queen Elizabeth's army in the Netherlands, and, in justice to the officers, it must be added that those evils were brought about in both cases by the same cause. Officers were simply forced into dishonesty by the withholding of their own pay by civilian officials in London.

It must not be thought that these scandals passed unnoticed at headquarters. As early as 1663 orders were issued to put a stop to fraudulent musters, and two years later the salaries of the officers of the Ordnance were increased almost threefold to check the sale of places and to diminish the temptation to accept bribes. Similar orders were respectively promulgated from time to time, but with little or no effect; possibly they were issued mainly as a matter of form, to stop the mouth of criticism. The root of the evil is to be traced to the civilian paymaster-general, who from the peculiarity of his position was accountable to no one, and enjoyed total irresponsibility for full forty years. The King no doubt flattered himself that the men were regularly paid; the abuses took some time to attain to their height, and in the short reign of James the Second it is probable that his attention to military business did somewhat to improve matters. But while Charles was on the throne the paymaster-general did as he pleased. Though wages were nominally paid after each muster, they were often withheld for months, and even for years. Finally, when payment was at last made, it was discharged not in cash but in tallies or debentures which could only be sold at a discount; while the colonels' agents seized the opportunity to deduct a percentage in consideration of the trouble to which they had been subjected to obtain any payment whatever.

So the old foundations of fraud were renovated,

and on them was built during the next century and a 1660-1688. half a gigantic superstructure of rascality and corruption which is not yet wholly demolished. Let it not be thought that in the seventeenth century such malpractices were either new or confined to England. They were, as I have often repeated, as old almost as the art of war, and they were rampant all over Europe. The excuse of English officers for their dishonesty was always, "It is so in France," and in France, as the history of the French Revolution shows, the old evils endured and throve for another full century. But the sin and shame of England is, that though she had once put away the accursed thing from her, she returned to it again as the sow to her wallowing in the mire. In 1659 English soldiers were proud of their name and calling; in 1666 it had already become a scandal to be a Life Guardsman.[1] Recruits had been found without difficulty under the Commonwealth to make the military profession, as was the rule in those days, the business of their whole life; but after a very few years of the Stuarts the King was compelled to resort to the pressgang. The status of the soldier was lowered, and has never recovered itself to this day.

I turn from this melancholy tale of retrogression to contemplate the changes made in other departments of the service. Herein it will be most convenient to begin with the regimental organisation and equipment. First, then, let us glance at the cavalry, which at the Restoration appears definitely to have taken precedence as the senior service. The reader will remember that in the New Model the fixed strength of a regiment was six troops of one hundred men, which was reduced in time of peace to an establishment of sixty men. Setting aside the Life Guards, which were independent troops of two hundred gentlemen apiece, the regiment which first occupies our attention is the Blues, which began life with eight troops, each of sixty men. So far there was practically no change, but in 1680 the strength of the

[1] *Cal. S. P., Dom.* (30th June 1666), p. 478.

1660-1688. Blues was diminished to fifty men in a troop; and in 1687 the newly raised regiments were established at an initial strength of six or seven troops of forty men only. Finally, as shall presently be seen in the campaigns that lie before us in Flanders, the establishment of a troop for war sinks to fifty men, and the establishment for peace to thirty-six. Here, therefore, is Cromwell's excellent system overthrown. The troop of cavalry is so far weakened as to be not worth assorting into three divisions, one to each of the three officers, and the seeds of enforced idleness are sown, to bear fruit an hundredfold. Hardly less significant is the appointment, in 1661, of regimental adjutants to help the majors in the duties which they had hitherto discharged without assistance.

The equipment of the Horse was likewise altered. The trooper retained the iron head-piece[1] and cuirass, the pistols and the sword of the New Model, but he was now further supplied with a carbine, which was slung at his back, and with a cartridge box for his ammunition. The new equipment was served out to the household troops in 1663, and to other regiments of Horse in 1677. It marks a new birth of the futile practice of firing from the saddle, which has wasted untold ammunition with infinitesimal results. As regards horses it was still the rule, which had been little modified during the Civil War, that the trooper should bring with him his own horse; if he had none the King supplied him with one, at an average price, and the money was stopped, if necessary, from the trooper's pay.

The drill still bore marks of Cromwell's influence, for the men were drawn up in three ranks only; and though the attack was opened by the discharge of carbines and pistols, yet it was distinctly laid down that when the fire-arms were empty, there must be no thought of reloading, but immediate resort to the sword.

[1] Which, however, was soon discarded for the hat, with or without an iron skull-piece beneath it.

Moreover, although the front was still increased or diminished by the doubling of ranks or files, there were already signs of the manœuvre by small divisions that was to displace it. 1660-1688.

Passing next to the dragoons, the reader will have noticed that this arm was not represented in the original Army formed by Charles the Second. Notwithstanding the high reputation which dragoons had enjoyed during the Civil War, it was not until 1672 that a regiment of them was raised, and then only to be disbanded after a brief existence of two years. The Tangier Horse, now called the First Royal Dragoons, was converted into a regiment of dragoons on its return from foreign service in 1684; and four years later there was added to the establishment a Scotch regiment which bears a famous name. It was made up in 1681 of three independent troops that had been raised three years before, and was completed by three additional troops, under the name of the Royal Regiment of Dragoons of Scotland. It now ranks as the Second regiment of the Cavalry of the Line, and is known to all the world as the Scots Greys.

Dragoons still preserved their original character of mounted infantry. Twelve men of each troop besides the non-commissioned officers were armed with the halberd and a pair of pistols, while the remainder were equipped with matchlock muskets, bandoliers, and, after 1672, with bayonets. In 1687 this equipment was improved by the substitution of flintlocks for matchlocks, of cartridge boxes for bandoliers, and of buckets, in addition to the old slings, for the carriage of muskets. The tactical unit of the dragoons was still called the company, though at the close of the Civil War often denominated the troop; but the tendency of dragoons to assimilate themselves to horse is seen in the substitution of cornet for ensign as the title of the junior subaltern. This tendency was perhaps the stranger, since the companies of dragoons, eighty men strong, must have presented a favourable contrast to the weak and attenuated troops of horse.

1660-1688. A new description of mounted soldier appeared in 1683,[1] in the shape of the Horse-grenadier. I shall have more to say presently of grenadiers, when treating of the infantry, so it is sufficient to state here that Horse-grenadiers were practically only mounted men of that particular arm, who as a rule linked their horses for action and fought on foot like the dragoons. There were in all three troops of Horse-grenadiers, which were attached to the three troops of Life Guards. Their peculiarity was that the two junior officers of each troop were both lieutenants, instead of lieutenant and cornet.

The infantry, like the cavalry, suffered an alteration in the regimental establishments after the Restoration. The old strength of one hundred and twenty to a company was reduced to one hundred, and in time of peace sank to eighty, sixty, and even fifty men. The number of companies to a battalion was also altered. The First Guards began life with twelve companies; and though for a time the Coldstreamers and newly raised regiments retained the original number of ten, yet twelve gradually became the usual, and after the accession of James the Second, the accepted, strength of a battalion. It must be noted that after 1672 a battalion and a regiment of foot cease to be synonymous terms, the First Guards being in that year increased to twenty-four companies and two battalions, a precedent which was soon extended to sundry other regiments.

On the accession of James there was added to the twelve companies of every regiment an additional company of grenadiers. These were established first in 1678, and took their name from the grenade,[2] the new weapon with which they were armed. The hand grenade was simply a small shell of from one to two inches in diameter, kindled by a fuse and thrown by

[1] Some say in 1678, but no sign of them appears in the Army Lists or Commission Registers till 1683.

[2] Spanish *granada*, a pomegranate. Grenadiers were established in France in 1667.

the hand. Hence it was entrusted to the tallest and finest men in the regiment, who might reasonably be expected to throw it farthest. The white plume, supposed to be symbolic of the white smoke of the fuse, was not apparently used at first as the distinctive mark of grenadiers. They, and the fusiliers likewise, wore caps instead of broad-brimmed hats, to enable them to sling their firelocks over both shoulders with ease. These caps, which were at first of fur, were soon made of cloth, and assumed the shape of the mitre which Hogarth has handed down to us. Another peculiarity of grenadiers was that they were always armed with firelocks and with hatchets,[1] and that both of their subaltern officers were lieutenants. 1660-1688.

Another new branch of the infantry was the regiment of Fusiliers, so called from the fusil or flintlock, as opposed to the matchlock, with which they were armed. They were, in fact, simply an expansion of the companies of firelocks which formed part of the New Model in the department of the Train; they were borne for duty with the artillery specially, and therefore included one company of miners. Miner-companies were armed with long carbines and hammer-hatchets peculiar to themselves, and they had but one subaltern officer, a lieutenant. Like the grenadiers, the fusiliers did not recognise the rank of ensign, and their junior subalterns were therefore called second lieutenants.[2]

It is somewhat remarkable that so much should have been made of a weapon so familiar as the firelock. Men who, like Gustavus Adolphus, saw that the whole future of warfare turned on the fire of musketry, had long accepted its superiority to the matchlock; and George Monk, on marching into London in 1660, had at once ordered the Coldstreamers to return their

[1] The hatchet was issued for the hewing down of the palisades at the attack of a fortified place. This is one reason why the grenadiers were nearly always told off for the assault of a fortress.

[2] But this rank was not confined to them. The Royal Scots at this period possessed second lieutenants in addition to ensigns.

1660-1688. matchlocks into store and to draw firelocks in their stead. Nor was this preference confined solely to military reformers, for we find the Assemblies of Barbados and Jamaica, remote islands in which old fashions might have been expected to die their hardest, uncompromisingly rejecting the matchlocks prescribed for them by the English Government and insisting on arming themselves with "fusees."[1] At home, however, jobbery and corruption were doubtless at work, for the Coldstream Guards reverted to the matchlock in 1665. Finally, after many compromises, the Guards were in 1683 armed exclusively with firelocks, while the other regiments carried a fixed proportion, probably not less than one-half, of the superior weapon among their matchlocks.

Correspondingly we find throughout these reigns a steady diminution in the use of the pike. In companies of grenadiers and regiments of fusiliers they were utterly abolished; in other corps the proportion, which had once been one-half, had already sunk at the Restoration to one-third, whence it speedily declined to one-fourth and one-fifth.[2] We find them, however, still in use during the wars of William the Third, and we shall see that they did not want advocates even at the close of the Seven Years' War, to say nothing of the part that they played in the French Revolution.[3] As a weapon for officers it survived for many generations under the form of the half-pike or spontoon,[4] even as the halberd prolonged its life as the peculiar weapon of sergeants. To the officers also was assigned by a singular coincidence the preservation of the memory of

[1] *Cal. S. P., Col.* (1677-1680), Nos. 397, 1141.

[2] The allowance in 1692 is fourteen per company.

[3] For the reluctance of the French to part with pikes see Belhomme, *L'Armée Française en* 1690, pp. 24, 25. The word *piquet* descends from the time when the pikemen were but a small body in the centre of the battalion, *ibid.*, p. 42.

[4] Thus General Cadogan, when virtually commander-in-chief, carried a half-pike at a review of the Guards in June 1722. *Flying Post*, 14th June 1722 (Marlborough died 16th June 1722).

the armour which had once been worn by all pikemen; 1660-1688.
and the gorget survived as a badge of rank on their breasts long after corslet and tassets had vanished from the world.[1]

None the less the pike had received its death-blow through the invention of the bayonet. This new and revolutionary weapon had been invented in 1640, when it consisted of a double-edged blade, like a pike-head, mounted on two or three inches of wooden haft, which could be thrust into the barrel of the musket. In this form the bayonet was issued first to the Tangier regiment[2] alone in 1663, and to all the infantry and dragoons in 1673, but only to be withdrawn, until in 1686 it was finally reissued to the Foot Guards. It was not until after the Revolution that bayonets were served out to the whole of the infantry.

In the matter of drill there was little or no change. The front was still increased or diminished by the doubling of ranks and of files, and the file still consisted of six men. The reduction of the numbers of pikemen, however, greatly increased the homogeneity of the infantry and contributed not a little to simplify its movements. Moreover, although the file might consist of six men, it is not likely, considering how far the musket and bayonet had superseded the pike, that the formation for action was greater than three ranks in depth. The platoon is not mentioned in the drill books, the probable reason being that it was not favoured by the French School, in which Charles and James had both of them received their training. But for this, there is every reason to suppose that the army encamped on Hounslow Heath would not have been found behind the times in the matter of exercise and equipment if it could have been transported without change to the field of Blenheim.

1 The pikemen of the Guardes Suisses in France, however, clung to the defensive armour for years after it had been discarded by others, a curious survival of the old glory of the Swiss.

2 2nd Queen's.

1660-1688. Of the artillery there is still little to be said. Until 1682 gunners seem to have enjoyed their original distribution into small, independent bodies, in charge of the various scattered garrisons. Even such small organisation as appeared in the New Model seems to have been lost, and field-guns appear to have been told off to battalions of infantry, or to have been worked by such of the escort of fusiliers as had been trained by the few expert gunners. The artilleryman had long looked upon himself as a superior mortal,[1] but in 1682 he was brought under the Ordnance, subjected to military discipline, and regularly exercised at his duty. The time was not far distant when the organisation of the gunners was to be improved. Of engineers I can say no more than the few details already given when describing the Ordnance Office and the fusiliers.

A word remains to be said of the foundation of Chelsea Hospital. It has been told that Queen Mary was the first of our sovereigns who showed any care for old soldiers, and that Elizabeth was intolerably impatient of such miserable creatures. Two generations, however, had bred a softer heart in English sovereigns, and when Charles the Second had been twenty years on the throne, and England was again thronged with maimed and infirm soldiers who had served their time in Tangier, in the West Indies, or in the Low Countries, it was felt to be a reproach that faithful fighting-men should be left to starve or to beg their bread. Kilmainham Hospital in Dublin was the first-fruit of this sentiment, and was founded in 1680; Chelsea followed it in the succeeding year. Sir Stephen Fox, the paymaster-general, was the man who was foremost in the work, and it is to his credit that, having made so much money out of the private soldier, he should have

[1] No better instance of this can be found than in Georg von Frundsberg, the famous landsknecht-leader, who once, being in supreme command of an army, took the linstock from a gunner and aimed and fired a gun himself. The "officer commanding artillery" at once came up, cashiered the gunner, and bade Georg look after his men and not meddle with other people's guns.

chosen this method of repaying him. The scheme of the hospital was submitted to the King, who was asked to grant a piece of land for a building. Charles, always gracious, readily complied, and offered the site of St. James's College, Chelsea. "But odso!" he added, "I now recollect that I have already given that land to Mistress Nell here." Whereupon, so runs the story, whether true or untrue, Nell gracefully forewent her grant for so good a purpose; and Chelsea Hospital is the British soldier's to this day. It is painful to have to add that the officials of the pay-office seem to have begun at once to steal part of the money contributed by the Army to its maintenance, though the fact will astonish no reader who has followed me through this chapter. But the friends of the Army have always been few, and the best of them in former times, strange conjunction, were a queen and a harlot. Had they endowed a fund for supplying African negroes with Bibles, or even with mass-books, much would be forgiven them in England; but they thought more of saving old soldiers from want, so Mary Tudor is still Bloody Mary, and Eleanor Gwyn the unspeakable Nell. 1660-1688.

AUTHORITIES.—The reader will find the fullest of references for the details in this chapter in Clifford Walton's *History of the British Standing Army*, with an index which will enable him to trace them without difficulty. Having myself perused the War Office books and papers in the Record Office, and the Calendars of the Domestic and Treasury State Papers independently, I can answer for the care and accuracy of the author in the preparation of this vast store of information, and gladly acknowledge my debt to it. The defect of the work is, of course, that it begins abruptly at the year 1660. Mr. Dalton's *Army Lists and Commission Registers* are also of great value, and claim the gratitude of all workers in the field of English military history. Sir Sibbald Scott's *British Army* is worth consulting occasionally for a few details, but is superseded by Hewitt's *Ancient Armour* on one side, and by Colonel Clifford Walton on the other. Mackinnon's *Coldstream Guards* contains a very valuable appendix of ancient documents. Sir F. Hamilton's *History of the Grenadier Guards* should be used only with extreme caution. The drill and exercise of the period may be studied in Venn's *Military Observations*, 1672.

BOOK V

CHAPTER I

Seldom has a man been confronted with such difficulties as those that beset William of Orange when the Revolution was fairly accomplished. So long as his success was still uncertain he stood in his favourite position of a military commander doing his worst against the power of France, while to the English nation he was a champion and a deliverer. Once seated on the throne he found that he had to do with a disorganised administration and a demoralised people. Forty years of revolution, interrupted by twenty-five of corrupt government, had done their work; and chaos reigned alike in the minds of private men and in all departments of the public service. Finally, as if this were not sufficient, there was a war in Ireland, a war in Flanders, and the practical certainty of an insurrection in Scotland.

His first trouble came quickly enough. Amid the general rejoicing over the overthrow of King James the English Army stood apart, surly and silent. The regiments felt that they had been fooled. They had been concentrated to resist foreign invasion, but had been withdrawn without any attempt to strike a blow. During his advance, and after his arrival in London, William had detailed the British regiments in the Dutch service for all duties which, if entrusted to foreigners, might have offended national sentiment; but his prudence could not reconcile the Army. The troops felt their disgrace keenly, and the burden of their dishonour was aggravated by the taunts of the foreigners. Moreover, the discipline of the Dutch had been so admirable that English folk had not failed to draw invidious comparisons between the well-conducted strangers and their own red-coats. Needless to say, they never reflected that Parliament, by withholding powers to enforce discipline, was chiefly responsible for the delinquencies of the English soldier. Discontent spread fast among the troops, and before the new king had been proclaimed a month, found vent in open mutiny.

1689. On the news of William's expedition to England, France had declared war against the States-General; and England, pursuant to obligations of treaty, was called upon to furnish her contingent of troops for their defence. On the 8th of March accordingly Lieutenant-General Lord Marlborough was ordered to ship four battalions of Guards and six of the Line[1] for Holland. Among these battalions was the Royal Scots, to which regiment William, doubtless with the best intentions, had lately appointed the Duke of Schomberg to be colonel. Schomberg was by repute one of the first soldiers in Europe. He had held a marshal's bâton in France and had sacrificed it to the cause of the Protestant religion. He had even fought by the side of the Royal Scots in more than one great action. But he was not a Scotsman, and the Scots had known no colonel yet but a Mackay, a Hepburn, or a Douglas. Moreover, the Parliament at Westminster, though not a Scottish Assembly, had, without consulting the regiment, coolly transferred its allegiance from James Stuart to William of Nassau.

With much grumbling the Scots marched as far as Ipswich on their way to their port of embarkation, and then, at a signal from some Jacobite officers, they broke into mutiny, seized four cannon, and, turning northward, advanced by forced marches towards Scotland. The alarm in London was great. "If you let this evil spread," said Colonel Birch, an old officer of Cromwell's

[1] 1st Battalion Royal Scots, Buffs, 7th, 21st, Collier's, Fitzpatrick's.

day, "you will have an army upon you in a few days." 1689.
William at once detached Ginkell, one of his best officers, with a large force in pursuit; the mutineers were overtaken near Sleaford, and, finding resistance hopeless, laid down their arms. William, selecting a few of the ringleaders only for punishment, ordered the rest of the regiment to return to its duty, and the Royal Scots sailed quietly away to the Maas. There the men deserted by scores, and even by hundreds,[1] but recruits were found, as good as they, to uphold the ancient reputation of the regiment.

Meanwhile good came out of evil, for the mutiny frightened the House of Commons not only into paying the expenses of William's expedition, but into passing the first Mutiny Act. It is true that the Act was passed for six months only, and that it provided for no more than the punishment of mutiny and desertion; but it recognised at least that military crime cannot be adequately checked by civil law, and it gave the Army more or less of a statutory right to exist. But readers should be warned once for all against the common fallacy that the existence of the Army ever depended on the passing of the annual Mutiny Act. The statute simply empowered the King to deal with certain military crimes for which the civil law made no provision. It made a great parade of the statement that the raising or keeping of a standing army in time of peace is against law, but the standing army was in existence for nearly thirty years before the Mutiny Act was passed, and continued to exist, as will be seen, for two short but distinct periods between 1689 and 1701 without the help of any Mutiny Act whatever. If, therefore, the keeping of a standing army in time of peace be against the law, it can only be said that during those periods Parliament deliberately voted money for the violation of the law, as indeed it is always prepared to do when convenient to itself. The Mutiny Act was not a protection to liberty; Parliament for the present

[1] *Cal. S. P., Dom.*, 23rd May 1689.

1689. reserved for itself no check on the military code that might be framed by the King; and the Act was therefore rather a powerful weapon placed in the hands of the sovereign. Nevertheless, the passing of the Mutiny Act remains always an incident of the first importance in the history of the Army, and the story of its origin is typical of the attitude of Parliament towards that long-suffering body. Every concession, nay, every commonest requirement, must be wrung from it by the pressure of fear.

It might have been thought that the news which came from Ireland a few days before the mutiny would have stirred the House of Commons to take some such measure in hand. Tyrconnel had already called the Irish to arms for King James, and on the 14th of March James himself, having obtained aid from the French king, had landed at Cork with some hundreds of officers to organise the Irish levies. The regular troops in the Irish establishment, already manipulated by Tyrconnel before the Revolution, were ready to join him. Some regiments went over to him entire; others split themselves up into Catholics and Protestants, and ranged themselves on opposite sides. It was evident that no less a task than the reconquest of Ireland lay before the English Government; and considering that several regiments had already been detached to Flanders, it was equally evident that the Army must be increased. Estimates were therefore prepared of the cost of six regiments of horse, two of dragoons, and twenty-five of foot, sixteen of which last were to be newly raised, for the coming campaign.

Of the new regiments a few lay ready to William's hand. The first was Lord Forbes's regiment, one of the many Irish corps brought over to England by King James in 1688, and the only one which, being made up entirely of Protestants, was not disbanded by William at his accession. It is still with us as the Eighteenth Royal Irish. The next three were corps which had been raised for the support of the Protestant cause at the Revolution. The first of them was a regiment of horse raised by the Earl of Devonshire among his tenantry in Derbyshire, which, long known by the name of the Black Horse, now bears the title of the Seventh Dragoon Guards. The second was a regiment of foot that had been formed at Exeter to join the Prince of Orange on his march from Torbay, and is still known as the Twentieth East Devon; and the third also remains with us as the Nineteenth of the Line. Three more regiments date their birth from March 1689—one raised by the Duke of Norfolk, one enlisted in the Welsh Marches, and a third which was recruited in Ireland but almost immediately brought over to England. These are now the Twenty-second, Twenty-third, and Twenty-fourth of the Line. Six more regiments of infantry which were raised in the same year, but disbanded at the close of the war, were Drogheda's, Lisburn's, Kingston's, Ingoldsby's, Roscommon's, and Bolton's. Of these, curiously enough, no fewer than three were dressed in blue instead of scarlet coats, possibly in flattering imitation of King William's famous Blue Guards. Thus, with ten thousand men to be enlisted, drilled, trained, and equipped, there was no lack of work for the recruiting officer, or for the Office of Ordnance, in the spring of 1689.

1689.

It was not long before William and Schomberg made the discovery that the old regiments would require as much watching as the new. There were significant symptoms of rottenness in the whole military system; and discontented spirits were already spreading false and calumnious reports as to the treatment of the English regiments in Flanders, with the evident design of kindling a mutiny. Moreover, there were loud complaints from citizens of oppression by the soldiery, from soldiers of the fraudulent withholding of their pay, and from every honest officer, not, alas! a very numerous body, of false musters, embezzlement, fraud, and every description of abuse. The King lost no time in appointing

1689. May 10. nine commissioners, with Schomberg at their head, to make the tour of the quarters in England, to inquire into the true state of the case, and if possible to restore order and discipline.[1]

Still more disquieting news came from the Prince of Waldeck, who commanded the confederate army in Flanders. The English regiments were far below the strength assigned to them on paper, their officers were ill-paid, and many of them, even the colonels, ill-conducted; the men were sickly, listless,[2] undisciplined, and disorderly; their shoes were bad, their clothing miserable, their very arms defective. William, whose eyes always rested by preference on the eastern side of the German Ocean, lost no time in sending his best officer to Flanders; but even the Earl of Marlborough had much ado to reduce these unruly elements to order. Nevertheless he persevered; and in the one serious action wherein the British were engaged during the

August $\frac{15}{25}$. campaign, that against Marshal d'Humières at Walcourt, Marlborough opened the eyes of Waldeck to the qualities of his men and to his own capacity. This was Marlborough's first brush with a Marshal of France; and it would seem that it was never forgotten by William. With this we may dismiss the campaign in Flanders for 1689.

Meanwhile another soldier of remarkable talent, and an old comrade of William, had rushed into rebellion in Scotland. The dragoons with which Dundee had harried the Covenanters and earned the name of "Bloody Claver'se" were still ready to his hand, and to these, by fanning the undying flame of tribal feud, he presently added an array of Highland clans. The flight of Dundee from Edinburgh on his errand of insurrection warned the city to take speedy measures for its defence. Lord Leven caused the drums to beat, and within two hours, it is said, had raised eight hundred men; but

[1] *Cal. S. P., Dom.*, 10th May 1689.

[2] "Nonchalants" is Waldeck's expression. See *Cal. S. P., Dom.*, 1st June, 28th June, 18th Sept., 23rd Sept.

the work of these two hours has lasted for two centuries, 1689.
for the regiment thus hastily enlisted is still alive as the Twenty-fifth of the Line. Shortly after, William sent up three Scotch regiments of the Dutch service under a veteran officer, Mackay; and the Highland war began in earnest. Skilful, however, as Mackay might be on the familiar battle-grounds of Flanders, he was helpless in the Highlands, where one week with George Monk would have helped him more than all the campaigns of Turenne. He crawled over the country conscientiously enough in pursuit of an enemy that he could never overtake, without further result than to exhaust the strength of both horses and men. It was not until one stage of a desultory campaign had been ended and a new one begun, that he at last met his enemy at Killiecrankie.

There is no need for me to repeat the story told July 27.
once for all by Lord Macaulay, of that romantic action; but it is worth while to glance at some few of its peculiarities. Mackay's force consisted of five battalions —the three Scottish regiments already mentioned, Hastings', now the Thirteenth Light Infantry, and the newly raised Twenty-fifth, together with two troops of horse. Of these the Scottish battalions, trained in the Dutch School by competent officers, should unquestionably have been the most efficient; yet all three of them broke before the charge of the Highlanders, threw down their arms, and would not be rallied. The two troops of horse took to their heels and disappeared; the Twenty-fifth broke like the other Scottish regiments, as was pardonable in such young soldiers, though they made some effort to rally. The only regiment that stood firm was the Thirteenth, which kept up a murderous fire to the end, and retired with perfect coolness and good order. Yet this was their first action, and Hastings, their colonel, was one of the most unscrupulous scoundrels, even in those days of universal robbery, that ever robbed a regiment.[1] Thus the troops

[1] He was cashiered for dressing his regiment in the cast clothes of another regiment.

1689. which should have done best did worst, and those that might have been expected to do worst did best; and the moral would seem to be that inexperienced troops are sometimes safer than troops trained in civilised warfare for the rough-and-ready fighting of a savage campaign.

A still more curious example of the same peculiarity was seen before the close of the war. At the end of the first stage of Mackay's campaign it was found necessary to raise fresh troops; and it was hoped that the Covenanters of Western Scotland, who of all men had most reason to detest bloody Claverhouse, might be willing to furnish recruits. But the Covenanters had scruples about joining the army of King William, wherein they might be set shoulder to shoulder with the immoral and, even worse, with the unorthodox. Even Mackay, a man of extreme piety,[1] was suspected by them. They held a tumultuous meeting, wherein the majority, little knowing probably how terribly true their words then were of the British Army, declared that military service was a sinful association. Nevertheless there was still a minority from which the Earl of Angus formed a body of infantry, twelve hundred strong, which, though now numbered Twenty-sixth of the Line, is still best known by its first name of the Cameronians. Their ideas of military organisation were peculiar. They desired that each company should furnish an elder, who with the chaplain should constitute a court for the suppression of immorality and heresy; and though the elders were never appointed, and the officers bore the usual titles of captain, lieutenant, and ensign, yet the chaplain, a noted hill-preacher, supplied in his own person fanaticism for all. So in spite of the ravings of the majority a true Puritan regiment once more donned the red coat, under the youngest colonel—for Angus was no more than eighteen—that had led such men since Henry Cromwell.

August 21. Within four months they were engaged against four times their number of Highlanders at Dunkeld. They

[1] "The piousest man I ever knew." *Burnet.*

were still imperfectly disciplined, still somewhat of a congregation that preferred elders to officers. They would not be satisfied that their mounted officers would not gallop away, until the lieutenant-colonel and major offered to shoot their horses before their eyes. Then they braced themselves, and fought such a fight as has seldom fallen to the lot of a regiment of recruits. The battle was fought amid the roar of a burning town. Angus was not present—short though his time was to be, it was not yet come—and his place was taken by Lieutenant-Colonel Cleland. The action was hardly opened before Cleland fell dead. The major stepped forward to his place, and a minute after was pierced by three mortal wounds. The men too fell fast; the musketry crackled round them, and the flames roared behind them; but still they fought on. Ammunition failed them at last; everything conspired to make the trial too hard for a young regiment to endure; but nothing could break the spirit of these men. At last, after four long hours, the Highlanders rolled back in disorder. The Cameronians had won their first battle and ended the Highland war. 1689.

But that war brought something more to the British Army even than two famous Scottish regiments. For Mackay had noticed that at Killiecrankie his Scotsmen had not had time to fix the clumsy plug-bayonets into the muzzles of their muskets, and had consequently been unable to meet the Highland charge. He therefore ordered bayonets to be made so that they could be screwed on to the outside of the barrel, thus enabling the men to fire with bayonets fixed. So finally was accomplished the blending of pike and musket into a single weapon, a great era in the history of the art of war.[1]

But while recruiting officers were beating their drums through the market towns of England, and Mackay was toiling in pursuit of the Highlanders, Protestant Ireland was standing desperately at bay

[1] The French had introduced this improvement some time before.

1689. against King James at Londonderry and Enniskillen. There is no need for me to recall the triumph of the unconquerable defenders of Derry; and it would be pleasanter, were it possible, to pass over the somewhat discreditable behaviour of the Army in relation to their relief. Five days, indeed, before the city was invested two English regiments, the Ninth and Seventeenth Foot, had arrived in the bay, but had been persuaded by the treacherous governor, Lundy, to return and to leave Derry to its fate. Colonels Cunningham and Richards, who commanded these corps, were both of them superseded on their arrival in England; but no further help came until on the 15th of June General Kirke sailed into Lough Foyle with the Second, Ninth, and Eleventh Foot. Even then he would not stir for six whole weeks, when he received positive orders from home to relieve the city.

Meanwhile all operations of the Irish Protestants that were not wholly defensive were directed from Enniskillen, which was filled with refugees from Munster and Connaught. With extraordinary energy these Protestants organised a body of horse and another of foot, with which they kept up an incessant harassing warfare against the insurgent Irish. On Kirke's arrival they applied to him for reinforcements. These he refused to give; but he sent them arms and he sent them officers, one of whom, Colonel Wolseley, equalled
July 31. at Newtown Butler Dundee's feat of Killiecrankie, of beating trained soldiers with raw but enthusiastic levies. After this action the force of the Enniskilleners was reorganised into two regiments of dragoons and three of foot, which are represented among us to this day by the Fifth Royal Irish Dragoons, now Lancers, the Sixth Enniskillen Dragoons, and the Twenty-seventh Enniskillen regiment of the infantry of Line.

The time was now come when the great English expedition for the reconquest of Ireland should set sail. The untrained Irish Protestant had played his part gallantly, and it was the turn of the English soldier. For months great preparations had been going forward; the new regiments had been raised; and on paper at any rate there were not only horse, foot, and dragoons, but a respectable train of artillery and of transport. Moreover, the failure of Cunningham and Richards had led Parliament to inquire into the conduct of that expedition; and it had been discovered that the supply of transport-ships had been so insufficient that the men had not had space even to lie down, while the biscuit provided for them had been mouldy and uneatable, and the beer so foul and putrid that they preferred to drink salt water. These shortcomings had occurred in the dispatch of a couple of battalions only; it remained to be seen how the military departments could cope with the transport and maintenance of an entire army. The total force to be employed in Ireland was close on nineteen thousand men, of which about one-fourth was already on the spot. 1689.

William had chosen Marshal Schomberg to command the expedition. Though past fourscore, the veteran was still active and fit for duty; and in reputation there was no better officer in Europe. On the 13th of August he landed with his army at Bangor and detached twelve regiments to besiege Carrickfergus. The garrison held out for a week, and was then permitted to capitulate and to march away to Newry. But that week was sufficient to open Schomberg's eyes. The new regiments proved to be mobs of undisciplined boys. Their officers were ignorant, negligent, and useless. The arms served out from the Tower were so ill-made, and the men so careless in the handling of them, that nearly every regiment required to be re-armed. The officers of artillery were not only ignorant and lazy, but even cowardly,[1] while their guns were so defective that a week of easy work had sufficed to render most of them unserviceable.[2] Senior officers were as deficient as August 13.

[1] *Cal. S. P., Dom.*, Schomberg to the King, 27th August 1689.

[2] But this was nothing uncommon in all the armies of Europe. French ordnance would break down in the same way, and many

1689. junior: there was not one qualified to command a brigade; and the commissary, in spite of reports that he had made all needful provision, had failed to supply sufficient stores. Lastly, in spite of the warning given by the experience of Cunningham and Richards, the transport across St. George's Channel was so shamefully conducted that one regiment of horse, that now known as the Queen's Bays, lost every charger and troop-horse in the passage.[1] The result was that all was confusion, and that every detail in every department required the personal supervision of the Commander-in-Chief.

Fortunately James's Irish were so far demoralised by previous failures that his officer at Belfast thought it prudent to evacuate that town. Schomberg therefore threw a garrison into it, and marched with his whole force upon Newry. The Duke of Berwick, who was guarding the road, fell back on his approach to Drogheda, where James had collected twenty thousand men; and Schomberg, advancing through a wasted and deserted country, halted, and entrenched himself at Dundalk. James struggled forward to within a league of him to try and tempt him to an action, but Schomberg was not to be entrapped; and by the second week in September the campaign was over.

The fact was that a month's service in the field had completely broken the English Army down. By the time when it reached Dundalk it was on the brink of starvation. The Commissary-General, one Shales, was a man of experience, for he had been purveyor to King James's camp at Hounslow; and he had accumulated stores—bad stores, it is true, but nevertheless stores—at the base, Belfast. But he had made no provision for carrying any part of them with the Army. He had bought up large numbers of horses in Cheshire, but,

of the guns at Carrickfergus were Dutch. See Belhomme, *L'Armée Française en* 1690, p. 131; and *Commons Journals*, 19th March 1706-7.

[1] *Cal. S. P., Dom.*, 12th September 1689.

instead of transporting them to Ireland, had let them 1689.
out to the farmers of the district for the harvest, and pocketed their hire.[1] Again, the artillery could not be moved because the Ordnance Department looked to Shales to provide horses, while Shales declared the artillery to be no business of his. Moreover, had the horses been on the spot, there was not a shoe ready for their feet.[2] No measures had been taken, in spite of Schomberg's representations, to victual the troops by sea, though Cromwell had shown forty years before, in Scotland, how readily the work could be done. But indeed the expedition would have been better managed than it was by following the guidance of so old a master as King Edward the Third.[3] Never was there a more signal example of English ignorance, neglect, and sloth in respect of military administration.

By the 18th of September victuals at Dundalk were at famine price, and the men began to perish by scores and by hundreds. It was hardly surprising, for they were not only unfed but unclothed; there was not so much as a greatcoat in the whole of the English infantry; the cavalry were without cloaks, boots, and belts, and almost the entire force wanted shoes. Moreover, the English were shiftless; when ordered to build themselves huts they could not be at the pains to obey, even with the example of their Dutch and Huguenot comrades before them. Sickness spread rapidly among them, and there was no hospital; and had there been a hospital there were no medicines. Finally, the behaviour of the officers was utterly shameful. "The lions in Africa," wrote one who was on the spot, "are not more barbarous than some of our officers are to the sick."[4] "I never saw officers more wicked and more interested," wrote Schomberg almost on the same day.[5] The

1 Authorities in Macaulay.
2 *Cal. S. P., Dom.*, Schomberg to the King, 3rd October 1689.
3 See Rymer's *Fœdera, anno* 1346.
4 Harbord's letter, *Cal. S. P., Dom.*, 18th September 1689.
5 Schomberg's letter, *ibid.* 20th September 1689.

1689. Commander-in-Chief did his best to interpose on behalf of the men, but his hands were already overfull. The colonels were perhaps the worst of all the officers; they understood pillage better than the payment of their men, and filled their empty ranks with worthless Irish recruits, simply because these were more easily cheated than English.[1] It cost Schomberg a week's work to ensure that the pay of the soldiers went into their own and not into their captains' pockets.

Yet on the whole it was not the military officers that were chiefly to blame. The constant complaint of Schomberg was that he could get no money; and for this the Treasurer of the Army was responsible. This functionary, William Harbord, a civilian and a member of the House of Commons, appears to have been on the whole the most shameless of all the officials in Ireland. By some jobbery he had contrived to obtain an independent troop of cavalry, for which he drew pay as though it were complete, though the troop in reality consisted of himself, two clerks whom he put down as officers, and a standard which he kept in his bedroom.[2] This was the only corps which was regularly paid. The other regiments he turned equally to his own advantage by sending home false muster-rolls[3] in order to draw the pay of the vacancies; but whenever the question of payment of the men was raised, he evaded it and went to England, pleading the necessity of attending to his duties in the House of Commons. It was Harbord again who was responsible for the failure of the hospital. He admitted, indeed, that if he had known as much about hospitals at the beginning as at the end of the campaign, he might have saved two-thirds of the men; but the truth was that he would never at any time supply a penny for it.[4] By Christmas Schomberg began to relent towards his officers, for he discovered that they

[1] Schomberg's letters, *Cal. S. P., Dom.*, 12th Oct., 26th December.
[2] Schomberg, 26th December 1689, *ibid.*
[3] Do., 30th December 1689, *ibid.*
[4] Harbord, 23rd October 1689, 9th January 1690, *ibid.*

were penniless, not having received a farthing of pay for four months.[1] Meanwhile civilians were growing fat. Shales was buying salt at ninepence a pound and selling it at four shillings;[2] and junior commissaries were acting as regimental agents and advancing money to the unhappy officers at exorbitant interest.[3] 1689.

In such a state of affairs Schomberg, rightly or wrongly, considered himself powerless. William ordered him from time to time to advance on Dublin; and Harbord, with incredible impertinence, urged him to march against the enemy.[4] Schomberg answered William by a plain statement of his condition, and Harbord by a surly and contemptuous growl. In truth his Dutch and Huguenot regiments, which alone were well clad and well looked after by their officers, were the only troops on which he could rely. The English continued to die like flies. Schomberg wisely endeavoured to distract their thoughts from their own misery by keeping them at drill. He found that not one in four had the slightest idea how to load or fire his musket, while the muskets themselves fell to pieces in the handling. Pestilence increased, and with it callousness and insubordination. The men used the corpses of their comrades to stop the draughts under their tent-walls, and robbed any man whose appearance promised hope of gain. Nor was this indiscipline confined to Dundalk. The Enniskilleners, who have generally been represented as superior to the English, were quite as fond of plunder, and robbed William Harbord himself, despite his protestations, in broad daylight.[5] Happily for Schomberg, James's forces were in as ill condition as his own, so that he was able to retire into winter quarters from Dundalk without molestation. Of fourteeen thousand men in the camp, upwards of six thousand had perished.[6] Nov. 5.

1 Schomberg, 24th December 1689, *Cal. S. P., Dom.*

2 Do. 16th October 1689, *ibid.*

3 Do. 26th December 1689, *ibid.*

4 Harbord, 23rd October 1689, *ibid.*

5 Schomberg, 30th December 1689, *ibid.*

6 Further details as to this Irish campaign will be found, with

1689. Gradually and painfully the winter wore away, but without abatement in the mortality of the troops. Meanwhile the House of Commons, awaking to the terrible state of things in Ireland, addressed the King for the arrest of Shales. William replied that he had already put him under arrest; and the name of Shales was accordingly constantly before the House in the course of the next few months, but without any result. He seems to have escaped scot-free; and indeed there was no lack of men as corrupt as he in the House of Commons and in all places of trust. William then took the extraordinary step of asking the House to appoint seven members to superintend the preparations for the next campaign; but this it very wisely declined to do. It appointed a Committee, however, to examine into the expenses of the war,[1] and finally passed a Mutiny Act with new clauses against false musters and other abuses—clauses which were as old as King Edward the Sixth, and for all practical purposes as dead. It was not legislation that was wanted, but enforcement of existing laws. William, however, appears early to have abandoned in despair the hope of finding an honest man in England.

1690. And now, with the experience of 1689 before them, the King and Schomberg began to arrange their plans for the campaign of 1690. In the matter of troops Schomberg was vehement against further employment of regiments of miserable English and Irish boys;[2] and it was therefore decided to transport twenty-seven thousand seasoned men, seventeen thousand of them British and the remainder Dutch and Danish, from England and Holland. Artillery and small arms were imported from Holland, since the Office of Ordnance had been found wanting; and as a daring experiment,

all authorities, in Clifford Walton's *History of the Standing Army*, pp. 70 *sqq*. Some details are also in Macaulay. Several of Schomberg's letters are printed complete in Dalrymple's *Memoirs*.

[1] *Commons Journals*, 8th November 1689.

[2] Schomberg, 10th February 1690, *Cal. S. P., Dom.*

which proved to be a total failure, the King took the clothing of several regiments out of their colonels' hands into his own.[1] Finally care was taken for the proper organisation of the transport-service. The plan of campaign in its broad lines was mapped out by a civilian, Sir Robert Southwell,[2] the secretary for Ireland. The country, he said, must be attacked simultaneously from north and south, for while the ports of Munster were open France could always pour in reinforcements and supplies. While, therefore, Schomberg advanced from the north, a descent should be made on the south, and Cork should be the objective. Finally, Southwell or some other sensible man did what William should have done the year before, and drew out a succinct account of the principles followed in Ireland with such signal success by that forgotten General, Oliver Cromwell.[3] 1690.

I shall not dwell further on the Irish campaigns of 1690 and 1691. There is little of importance to the History of the Army to be found in them; and the reader will more readily follow Lord Macaulay than myself over this familiar ground. The battle of the Boyne was won without great credit to William's skill, and paid for rather dearly by the death of gallant old Schomberg. The troops learned something of active service, and something, though not nearly so much as they should have learnt, of discipline. The lesson of Cromwell was not taken to heart; and the Protestant Irish were allowed to set an example of plunder which was but too readily followed by the English. Ginkell's final campaign of 1691 was more successful, more brilliant, and more satisfactory in every respect, inasmuch as the Irish fought with distinguished gallantry. For the rest, the English showed at Aghrim and at Athlone their usual desperate valour; succeeding, even when experienced commanders, like St. Ruth, confessed with

1 Carmarthen to the King, February 1691, *Cal. S. P., Dom.*

2 Southwell, January 1690, *ibid.*

3 See the very remarkable memorandum in *Cal. S. P., Dom.* (1691), pp. 398-400.

1690. admiration that they had thought their success impossible. But in the matter of skill the quiet and unostentatious captures of Cork and Kinsale in 1690 were far the most brilliant achievements of the war; and these were the work of John, Earl of Marlborough.[1]

[1] The Irish campaigns are treated with great fulness by Colonel Clifford Walton, and Marlborough's part in them in particular in Lord Wolseley's *Life of Marlborough*.

CHAPTER II

I PASS now to Flanders, which is about to become for the second time the training ground of the British Army. The judicious help sent by Lewis the Fourteenth to Ireland had practically diverted the entire strength of William to that quarter for two whole campaigns; and though, as has been seen, there were English in Flanders in 1689 and 1690, the contingents which they furnished were too small and the operations too trifling to warrant description in detail. After the battle of the Boyne the case was somewhat altered, for, though a large force was still required in Ireland for Ginkell's final pacification of 1691, William was none the less at liberty to take the field in Flanders in person. Moreover, Parliament with great 1690,
good-will had voted seventy thousand men for the ensuing October.
year, of which fully fifty thousand were British,[1] so that England was about to put forth her strength in Europe on a scale unknown since the loss of Calais.

But first a short space must be devoted to the theatre of war, where England was to meet and break down the overweening power of France. Few studies are more difficult, even to the professed student, than that of the old campaigns in Flanders, and still fewer more hopeless of simplification to the ordinary reader. Nevertheless, however desperate the task, an effort must be made once for all to give a broad idea of the scene of innumerable great actions.

Taking his stand on the northern frontier of France

[1] Four troops of life guards, ten regiments of horse, five of dragoons, forty-seven battalions of foot.

1690. and looking northward, the reader will note three great rivers running through the country before him in, roughly speaking, three parallel semicircles, from south-east to north-west. These are, from east to west, the Moselle, which is merged in the Rhine at Coblentz, the Meuse, and the Scheldt, all three of which discharge themselves into the great delta whereof the southern key is Antwerp. But for the present let the reader narrow the field from the Meuse in the east to the sea in the west, and let him devote his attention first to the Meuse. He will see that, a little to the north of the French frontier, it picks up a large tributary from the south-west, the Sambre, which runs past Maubeuge and Charleroi and joins the Meuse at Namur. Thence the united rivers flow on past the fortified towns of Huy, Liège, and Maestricht to the sea. But let the reader's northern boundary on the Meuse for the present be Maestricht, and let him note another river which rises a little to the west of Maestricht and runs almost due west past Arschot and Mechlin to the sea at Antwerp. Let this river, the Demer, be his northern, and the Meuse from Maestricht to Namur his eastern, boundary.

Returning to the south, let him note a river rising immediately to the west of Charleroi, the Haine, which joins the Scheldt at Tournay, and let him draw a line from Tournay westward through Lille and Ypres to the sea at Dunkirk. Let this line from Dunkirk to Charleroi be carried eastward to Namur ; and there is his southern boundary. His western boundary, is, of course, the sea. Within this quadrilateral, Antwerp (or more strictly speaking the mouth of the Scheldt), Dunkirk, Namur, and Maestricht, lies the most famous fighting-ground of Europe.

Glancing at it on the map, the reader will see that this quadrilateral is cut by a number of rivers running parallel to each other from south to north, and flowing into the main streams of the Demer and the Scheldt. The first of these, beginning from the east, are the Great and Little Geete, which become one before they join the main stream. It is worth while to pause for a 1690. moment over this little slip of land between the Geete and the Meuse. We shall see much of Namur, Huy, Liège, and Maestricht, which command the navigation of the greater river, but we shall see still more of the Geete, and of two smaller streams, the Jaar and the Mehaigne, which rise almost in the same table-land with it. On the Lower Jaar, close to Maestricht, stands the village of Lauffeld, which shall be better known to us fifty years hence. On the Little Geete, just above its junction with its greater namesake, are the villages of Neerwinden and Landen. In the small space between the heads of the Geete and the Mehaigne lies the village of Ramillies. For this network of streams is the protection against an enemy that would threaten the navigation of the Meuse from the north and west, and the barrier of Spanish Flanders against invasion from the east; and the ground is rich with the corpses and fat with the blood of men.

The next stream to westward is the Dyle, which flows past Louvain to the Demer, and gives its name, after the junction, to that river. The next in order is the Senne, which flows past Park and Hal and Brussels to the same main stream. At the head of the Senne stands the village of Steenkirk; midway between the Dyle and Senne are the forest of Soignies and the field of Waterloo.

Here the tributaries of the Demer come to an end, but the row of parallel streams is continued by the tributaries of another system, that of the Scheldt. Easternmost of these, and next in order to the Senne, is the Dender, which rises near Leuse and flows past Ath and Alost to the Scheldt at Dendermond. Next comes the Scheldt itself, with the Scarpe and the Haine, its tributaries, which it carries past Tournay and Oudenarde to Ghent, and to the sea at Antwerp. Westernmost of all, the Lys runs past St. Venant, where in Cromwell's time we saw Sir Thomas Morgan and his immortal six thousand, past Menin and Courtrai, and is merged in the Scheldt at Ghent.

1690. The whole extent of the quadrilateral is about one hundred miles long by fifty broad, with a great waterway to the west, a second to the east, and a third, whereof the key is Ghent, roughly speaking midway between them. The earth, fruitful by nature and enriched by art, bears food for man and beast, the waterways provide transport for stores and ammunition. It was a country where men could kill each other without being starved, and hence for centuries the cockpit of Europe.

A glance at any old map of Flanders shows how thickly studded was this country with walled towns of less or greater strength, and explains why a war in Flanders should generally have been a war of sieges. Every one of these little towns, of course, had its garrison; and the manœuvres of contending forces were governed very greatly by the effort on one side to release these garrisons for active service in the field, and on the other to keep them confined within their walls for as long as possible. Hence it is obvious that an invading army necessarily enjoyed a great advantage, since it menaced the fortresses of the enemy while its own were unthreatened. Thus ten thousand men on the Upper Lys could paralyse thrice their number in Ghent and Bruges and the adjacent towns. On the other hand, if an invading general contemplated the siege of an important town, he manœuvred to entice the garrison into the field before he laid siege in form. Still, once set down to a great siege, an army was stationary, and the bare fact was sufficient to liberate hostile garrisons all over the country; and hence arose the necessity of a second army to cover the besieging force. The skill and subtlety manifested by great generals to compass these different ends is unfortunately only to be apprehended by closer study than can be expected of any but the military student.

A second cause contributed not a little to increase the taste for a war of sieges, namely the example of France, then the first military nation in Europe.[1] The

[1] I had almost written that France was then, as always, the first

Court of Versailles was particularly fond of a siege, since it could attend the ceremony in state and take nominal charge of the operations with much glory and little discomfort or danger. The French passion for rule and formula also found a happy outlet in the conduct of a siege, for while there is no nation more brilliant or more original, particularly in military affairs, there is also none that is more conceited or pedantic. The craving for sieges among the French was so great that the King took pains, by the grant of extra pay and rations, to render this species of warfare popular with his soldiers.[1] 1690.

Again, it must be remembered that the object of a campaign in those days was not necessarily to seek out an enemy and beat him. There were two alternatives prescribed by the best authorities, namely, to fight at an advantage or to subsist comfortably.[2] Comfortable subsistence meant at its best subsistence at an enemy's expense. A campaign wherein an army lived on the enemy's country and destroyed all that it could not consume was eminently successful, even though not a shot was fired. To force an enemy to consume his own supplies was much, to compel him to supply his opponent was more, to take up winter-quarters in his territory was very much more. Thus to enter an enemy's borders and keep him marching backwards and forwards for weeks without giving him a chance of striking a blow, was in itself no small success, and success of a kind which galled inferior generals, such as William of Orange, to desperation and so to disaster. The tendency to these negative campaigns was heightened once more by French example. The French ministry of war interfered with its generals to an extent that was always dangerous, and eventually proved calamitous.

military nation; and though Prussia wrested the position from her under Frederick the Great and again in 1870, the lesson of history seems to teach that she is as truly the first military, as England is the first naval, nation.

[1] Belhomme, p. 153.

[2] Feuquières.

1690. Nominally the marshal commanding-in-chief in the field was supreme; but the intendant or head of the administrative service, though he received his orders from the marshal, was instructed by the King to forward those orders at once by special messenger to Louvois, and not to execute them without the royal authority. Great commanders such as Luxemburg had the strength from time to time to kick themselves free from this bondage, but the rest, embarrassed by the surveillance of an inferior officer, preferred to live as long as possible in an enemy's country without risking a general action. It was left to Marlborough to advance triumphant in one magnificent campaign from the Meuse to the sea.

Next, a glance must be thrown at the contending parties. The defenders of the Spanish Netherlands, for they cannot be called the assailants of France, were confederate allies from a number of independent states—England, Holland, Spain, the Empire, sundry states of Germany, and Denmark, all somewhat selfish, few very efficient, and none, except the first, very punctual. From such a heterogeneous collection swift, secret, and united action was not to be expected. King William held the command-in-chief, and, from his position as the soul of the alliance, was undoubtedly the fittest for the post. But though he had carefully studied the art of war, and though his phlegmatic temperament found its only genuine pleasure in the excitement of the battle-field, he was not a great general. He could form good plans, and up to a certain point could execute them, but up to a certain point only. It would seem that his physical weakness debarred him from steady and sustained effort. He was strangely incapable of conducting a campaign with equal ability throughout; he would manœuvre admirably for weeks, and forfeit all the advantage that he had gained by the carelessness of a single day. In a general action, of which he was fonder than most commanders of his day, he never shone except in virtue of conspicuous personal bravery. He lacked tactical instinct, and above all he lacked patience; in 1690. a word, to use a modern phrase, he was a very clever amateur.

France, on the other hand, possessed the finest and strongest army in Europe,—well equipped, well trained, well organised, and inured to work by countless campaigns. She had a single man in supreme control of affairs, King Lewis the Fourteenth; a great war-minister, Louvois; one really great general, Luxemburg; and one with flashes of genius, Boufflers. Moreover she possessed a line of posts in Spanish Flanders extending from Dunkirk to the Meuse. On the Lys she had Aire and Menin; on the Scarpe, Douay; on the Upper Scheldt, Cambray, Bouchain, Valenciennes, and Condé; on the Sambre, Maubeuge; between Sambre and Meuse, Philippeville and Marienburg; and on the Meuse, Dinant. Further, in the one space where the frontier was not covered by a friendly river, between the sea and the Scheldt, the French had constructed fortified lines from the sea to Menin and from thence to the Scheldt at Espierre. Thus with their frontier covered, with a place of arms on every river, with secrecy and with unity of purpose, the French enjoyed the approximate certainty of being able to take the field in every campaign before the Allies could be collected to oppose them.

The campaign of 1691 happily typifies the relative 1691. positions of the combatants in almost every respect. The French concentrated ten thousand men on the Lys. This was sufficient to paralyse all the garrisons of the Allies on and about the river. They posted another corps on the Moselle, which threatened the territory of Cleves. Now Cleves was the property of the Elector of Brandenburg, and it was not to be expected that he should allow his contingent of troops to join King William at the general rendezvous at Brussels, and suffer the French to play havoc among his possessions. Thus the Prussian contingent likewise was paralysed. So while William was still ordering his troops to con-

1691. centrate at Brussels, Boufflers, who had been making preparations all the winter, suddenly marched up from Maubeuge and, before William was aware that he was in motion, had besieged Mons. The fortress presently surrendered after a feeble resistance, and the line of the Allies' frontier between the Scheldt and Sambre was broken. William moved down from Brussels across the Sambre in the hope of recovering the lost town, outmanœuvred Luxemburg, who was opposed to him, and for three days held the recapture of Mons in the hollow of his hand. He wasted those three days in an aimless halt; Luxemburg recovered himself by an extraordinary march; and William, finding that there was no alternative before him but to retire to Brussels and remain inactive, handed over the command to an incompetent officer and returned to England. Luxemburg then closed the campaign by a brilliant action of cavalry, which scattered the horse of the Allies to the four winds. As no British troops except the Life Guards were present, and as they at any rate did not disgrace themselves, it is unnecessary to say more of the combat of Leuse. It, had however, one remarkable effect: it increased William's dread of the French cavalry, already morbidly strong, to such a pitch as to lead him subsequently to a disastrous military blunder.

The campaign of 1691 was therefore decidedly unfavourable to the Allies, but there was ground for hope that all might be set right in 1692. The Treasurer, Godolphin, was nervously apprehensive that Parliament might be unwilling to vote money for an English army in Flanders; but the Commons cheerfully voted a total of sixty-six thousand men, British and foreign; which, after deduction of garrisons for the safety of the British Isles, left forty thousand free to cross the German Ocean.

Of these, twenty-three thousand were British, the most important force that England had sent to the Continent since the days of King Henry the Eighth. The organisation was remarkably like that of the New Model. William was, of course, commander-in-chief, and under him a general of horse and a general of foot, with a due allowance of lieutenant-generals, major-generals, and brigadiers. There is, however, no sign of an officer in command of artillery or engineers, nor any of a commissary in charge of the transport.[1] The one strangely conspicuous functionary is the Secretary-at-War, who in this and the following campaigns for the last time accompanied the Commander-in-Chief on active service. But the most significant feature in the list of the staff is the omission of the name of Marlborough. Originally included among the generals for Flanders, he had been struck off the roll, and dismissed from all public employment, in disgrace, before the opening of the campaign. Though this dismissal did not want justification, it was perhaps of all William's blunders the greatest. 1692.

As usual, the French were beforehand with the Allies in opening the campaign. They had already broken the line of the defending fortresses by the capture of Mons; they now designed to make the breach still wider. All through the winter a vast siege-train was collecting on the Scheldt and Meuse, with Vauban, first of living engineers, in charge of it. In May all May.
was ready. Marshal Joyeuse, with one corps, was on the Moselle, as in the previous year, to hold the Brandenburgers in check. Boufflers, with eighteen thousand men, lay on the right bank of the Meuse, near Dinant; Luxemburg, with one hundred and fifteen thousand more, stood in rear of the river Haine. On May $\frac{10}{20}$.
the 20th of May, King Lewis in person reviewed the grand army; on the 23rd it marched for Namur; and May $\frac{13}{23}$.
on the 26th it had wound itself round two sides of the town, while Boufflers, moving up from Dinant, completed May $\frac{16}{26}$.
the circuit on the third side. Thus Namur was completely invested; unless William could save it, the line of the Sambre and one of the most important fortresses on the Meuse were lost to the Allies.

[1] That is to say, of land-transport. After the sad experience of the Irish war the marine transport was entrusted to an officer specially established for the purpose.—*Commons Journals.*

1692. William, to do him justice, had strained every nerve to spur his indolent allies to be first in the field. The contingents, awaked by the sudden stroke at Namur, came in fast to Brussels; but it was too late. The French had destroyed all forage and supplies on the direct route to Namur, and William's only way to the city lay across the Mehaigne. Behind the Mehaigne lay Luxemburg, the ablest of the French generals. The best of luck was essential to William's success, and instead of the best came the worst. Heavy rain swelled the narrow stream into a broad flood, and the building of bridges became impossible. There was beautiful fencing, skilful feint, and more skilful parry, between the two generals, but William could not get
May 26/June 5. under Luxemburg's guard. On the 5th of June, after a discreditably short defence, Namur fell, almost before William's eyes, into the hands of the French.

Then Luxemburg thought it time to draw the enemy away from the vicinity of the captured city; so recrossing the Sambre, and keeping Boufflers always between himself and that river, he marched for the Senne as if to threaten Brussels. William followed, as in duty bound; and French and Allies pursued a parallel course to the Senne, William on the north and
July 23/August 2. Luxemburg on the south. The 2nd of August found both armies across the Senne, William at Hal, facing west with the river in his rear, and Luxemburg some five miles south of him with his right at Steenkirk, and his centre between Hoves and Enghien, while Boufflers lay at Manny St. Jean, seven miles in his rear.

The terrible state of the roads owing to heavy rain had induced Luxemburg to leave most of his artillery at Mons, and as he had designed merely to tempt the Allies away from Namur, the principal object left to him was to take up a strong position wherein his worn and harassed army could watch the enemy without fear of attack. Such a position he thought that he had found at Steenkirk.[1] The country at this point is more

[1] I spell the village according to the popular fashion in England,

broken and rugged than is usual in Belgium. The camp lay on high ground, with its right resting on the river Sennette and its right front covered by a ravine, which gradually fades away northward into a high plateau of about a mile in extent. Beyond the ravine was a network of wooded defiles, through which Luxemburg seems to have hoped that no enemy could fall upon him in force unawares. It so happened, however, that one of his most useful spies was detected, in his true character, in William's camp at Hal; and this was an opportunity not to be lost. A pistol was held at the spy's head, and he was ordered to write a letter to Luxemburg, announcing that large bodies of the enemy would be in motion next morning, but that nothing more serious was contemplated than a foraging expedition. This done, William laid his plans to surprise his enemy on the morrow. 1692.

An hour before daybreak the advanced guard of William's army fell silently into its ranks, together with a strong force of pioneers to clear the way for a march through the woods. This force consisted of the First Guards, the Royal Scots, the Twenty-first, Fitzpatrick's regiment of Fusiliers, and two Danish regiments of great reputation, the whole under the command of the Duke of Würtemberg. Presently they moved away, and as the sun rose the whole army followed them in two columns, without sound of drum or trumpet, towards Steenkirk. French patrols scouring the country in the direction of Tubise saw the two long lines of scarlet and white and blue wind away into the woods, and reported what they had seen at headquarters; but Luxemburg, sickly of constitution, and, in spite of his occasional energy, indolent of temperament, rejoiced to think that, as his spy had told him, it was no more than a foraging party. Another patrol presently sent in another message that a large force of cavalry was July 23 / August 3.

and according to the Flemish pronunciation. So many names in Flanders seem to halt between the Flemish and the French that it is difficult to know how to set them down.

1692. July 23/August 3. advancing towards the Sennette. Once more Luxemburg lulled himself into security with the same comfort.

Meanwhile the allied army was trailing through narrow defiles and cramped close ground, till at last it emerged from the stifling woods into an open space. Here it halted, as the straitness of the ground demanded, in dense, heavy masses. But the advanced guard moved on steadily till it reached the woods over against Steenkirk, where Würtemberg disposed it for the coming attack. On his left the Bois de Feuilly covered a spur of the same plateau as that occupied by the French right, and there he stationed the English Guards and the two battalions of Danes. To the right of these, but separated from them by a ravine, he placed the three remaining British battalions in the Bois de Zoulmont. His guns he posted, some between the two woods, and the remainder on the right of his division. These dispositions complete, the advanced party awaited orders to open the attack.

It was now eleven o'clock. Luxemburg had left his bed and had ridden out to a commanding height on his extreme right, when a third message was brought to him that the Allies were certainly advancing in force. He read it, and looking to his front, saw the red coats of the Guards moving through the wood before him, while beyond them he caught a glimpse of the dense masses of the main body. Instantly he saw the danger, and divined that William's attack was designed against his right. His own camp was formed, according to rule, with the cavalry on the wings; and there was nothing in position to check the Allies but a single brigade of infantry, famous under the name of Bourbonnois, which was quartered in advance of the cavalry's camp on his extreme right. Moreover, nothing was ready, not a horse was bridled, not a man standing to his arms. He despatched a messenger to summon Boufflers to his aid, and in a few minutes was flying through the camp with his staff, energetic but perfectly self-possessed, to set his force in order of battle. The two battalions of Bourbonnois fell in hastily before their camp, with a battery of six guns before them. The dragoons of the right wing dismounted and hastened to seal up the space between Bourbonnois and the Sennette. The horse of the right was collected, and some of it sent off in hot speed to the left to bring the infantry up behind them on their horses' croups. All along the line the alarm was given, drums were beating, men snatching hastily at their arms and falling into their ranks ready to file away to the right. Such was the haste, that there was no time to think of regimental precedence, a very serious matter in the French army, and each successive brigade hurried into the place where it was most needed as it happened to come up.

1692. July 23 / August 3.

Meanwhile Würtemberg's batteries had opened fire, and a cunning officer of the Royal Scots was laying his guns with admirable precision. French batteries hastened into position to reply to them with as deadly an aim, and for an hour and a half the rival guns thundered against each other unceasingly. All this time the French battalions kept massing themselves thicker and thicker on Luxemburg's right, and the front line was working with desperate haste, felling trees, making breastworks, and lining the hedges and copses while yet they might. But still Würtemberg's division remained unsupported, and the precious minutes flew fast. William, or his staff for him, had made a serious blunder. Intent though he was on fighting a battle with his infantry only, he had put all the cavalry of one wing of his army before them on the march, so that there was no room for the infantry to pass. Fortunately six battalions had been intermixed with the squadrons of this wing, and these were now with some difficulty disentangled and sent forward. Cutts's, Mackay's, Lauder's, and the Twenty-sixth formed up on Würtemberg's right, with the Sixth and Twenty-fifth in support; and at last, at half-past twelve, Würtemberg gave the order to attack.

His little force shook itself up and pressed forward

1692. July 23 / August 3. with eagerness. The Guards and Danes on the extreme left, being on the same ridge with the enemy, were the first that came into action. Pushing on under a terrible fire at point-blank range from the French batteries, they fell upon Bourbonnois and the dragoons, beat them back, captured their guns, and turned them against themselves. On their right the Royal Scots, Twenty-first, and Fitzpatrick's plunged down into the ravine into closer and more difficult ground, past copses and hedges and thickets, until a single thick fence alone divided them from the enemy. Through this they fired at each other furiously for a time, till the Scots burst through the fence with their Colonel at their head and swept the French before them. Still further to the right, the remaining regiments came also into action; muzzle met muzzle among the branches, and the slaughter was terrible. Young Angus, still not yet of age, dropped dead at the head of the Cameronians, and the veteran Mackay found the death which he had missed at Killiecrankie. He had before the attack sent word to General Count Solmes, that the contemplated assault could lead only to waste of life, and had been answered with the order to advance. "God's will be done," he said calmly, and he was among the first that fell.

Still the British, in spite of all losses, pressed furiously on; and famous French regiments, spoiled children of victory, wavered and gave way before them. Bourbonnois, unable to face the Guards and Danes, doubled its left battalion in rear of its right; Chartres, which stood next to them, also gave way and doubled itself in rear of its neighbour Orléans. A wide gap was thus torn in the first French line, but not a regiment of the second line would step into it. The colonel of the brigade in rear of it ordered, entreated, implored his men to come forward, but they would not follow him into that terrible fire. Suddenly the wild voice ceased, and the gesticulating figure fell in a heap to the ground: the colonel had been shot dead, and the gap was still unfilled.

The first French line was broken; the second and third were dismayed and paralysed: a little more and the British would carry the French camp. Luxemburg perceived that this was a moment when only his best troops could save him. In the fourth line stood the flower of his infantry, the seven battalions of French and Swiss Guards. These were now ordered forward to the gap; the princes of the blood placed themselves at their head, and without firing a shot they charged down the slope upon the British and Danes. The English Guards, thinned to half their numbers, faced the huge columns of the Swiss and stood up to them undaunted, till by sheer weight they were slowly rolled back. On their right the Royal Scots also were forced back, fighting desperately from hedge to hedge and contesting every inch of ground. Once, the French made a dash through a fence and carried off one of their colours. The Colonel, Sir Robert Douglas, instantly turned back alone through the fence, recaptured the colour, and was returning with it when he was struck by a bullet. He flung the flag over to his men and fell to the ground dead.

1692. July 23 / August 3.

Slowly the twelve battalions retired, still fighting furiously at every step. So fierce had been their onslaught that five lines of infantry backed by two more of cavalry[1] had hardly sufficed to stop them, and with but a little support they might have won the day. But that support was not forthcoming. Message after message had been sent to the Dutch general, Count Solmes, for reinforcements, but there came not a man. The main body, as has been told, was all clubbed together a mile and a half from the scene of action, with the infantry in the rear; and Solmes, with almost criminal folly, instead of endeavouring to extricate the foot, had ordered forward the horse. William rectified the error as soon as he could, but the correction led to further delay and to the increased confusion which

[1] Fifty-three battalions of infantry and seven regiments of dragoons.—*Beaurain.*

1692. July 23/August 3. is the inevitable result of contradictory orders. The English infantry in rear, mad with impatience to rescue their comrades, ran forward in disorder, probably with loud curses on the Dutchman who had kept them back so long; and some time was lost before they could be re-formed. Discipline was evidently a little at fault. Solmes lost both his head and his temper. "Damn the English," he growled; "if they are so fond of fighting, let them have a bellyful"; and he sent forward not a man. Fortunately junior officers took matters into their own hands; and it was time, for Boufflers had now arrived on the field to throw additional weight into the French scale. The English Horse-grenadiers, the Fourth Dragoons, and a regiment of Dutch dragoons rode forward and, dismounting, covered the retreat of the Guards and Danes by a brilliant counter-attack. The Buffs and Tenth advanced farther to the right, and holding their fire till within point-blank range, poured in a volley which gave time for the rest of Würtemberg's division to withdraw. A demonstration against the French left made a further diversion, and the shattered fragments of the attacking force, grimed with sweat and smoke, fell back to the open ground in rear of the woods, repulsed but unbeaten, and furious with rage.

William, it is said, could not repress a cry of anguish when he saw them; but there was no time for emotion. Some Dutch and Danish infantry was sent forward to check further advance of the enemy, and preparations were made for immediate retreat. Once again the hardest of the work was entrusted to the British; and when the columns were formed, the grenadiers of the British regiments brought up the rear, halting and turning about continually, until failing light put an end to what was at worst but a half-hearted pursuit. The retreat was conducted with admirable order; but it was not until the chill, dead hour that precedes the dawn that the Allies regained their camp, worn out with the fatigue of four-and-twenty hours.

The action was set down at the time as the severest

ever fought by infantry, and the losses on both sides were very heavy. The Allies lost about three thousand killed and the same number wounded, besides thirteen hundred prisoners, nearly all of whom were wounded. Ten guns were abandoned, the horses being too weary to draw them; the English battalions lost two colours, and the foreign three or four more. The British, having borne the brunt of the action, suffered most heavily of all, the Guards, Cutts's, and the Sixth being terribly punished. The total French loss was about equal to that of the Allies, but the list of the officers that fell tells a more significant tale. On the side of the Allies four hundred and fifty officers were killed and wounded, no fewer than seventy lieutenants in the ten battalions of Churchill's British brigade being killed outright. The French on their side lost no less than six hundred and twenty officers killed and wounded, a noble testimony to their self-sacrifice, but sad evidence of their difficulty in making their men stand. In truth, with proper management William must have won a brilliant victory; but he was a general by book and not by instinct. Würtemberg's advanced guard could almost have done the work by itself but for the mistake of a long preliminary cannonade; his attack could have been supported earlier but for the pedantry that gave the horse precedence of the foot in the march to the field; the foot could have pierced the French position in a dozen different columns but for the pedantry which caused it to be first deployed. Finally, William's knowledge of the ground was imperfect, and Solmes, his general of foot, was incompetent. The plan was admirably designed and abominably executed. Nevertheless, British troops have never fought a finer action than Steenkirk. Luxemburg thought himself lucky to have escaped destruction; his troops were much shaken; and he crossed the Scheldt and marched away to his winter-quarters as quietly as possible. So ended the campaign of 1692.

1692. July 23 / August 3.

CHAPTER III

1692, November. In November the English Parliament met, heartened indeed by the naval victory of La Hogue, but not a little grieved over the failure of Steenkirk. Again, the financial aspect was extremely discouraging; and Sir Stephen Fox announced that there was not another day's subsistence for the Army in the treasury. The prevailing discontent found vent in furious denunciations of Count Solmes, and a cry that English soldiers ought to be commanded by English officers. The debate rose high. The hardest of hard words were used about the Dutch generals, and a vast deal of nonsense was talked about military matters. There were, however, a great number of officers in the House of Commons, many of whom had been present at the action. With great modesty and good sense they refused to join in the outcry against the Dutch, and contrived so to compose matters that the House committed itself to no very foolish resolution. The votes for the Army were passed; and no difficulty was made over the preparations for the next campaign. Finally, two new regiments of cavalry were raised—Lord Macclesfield's Horse, which was disbanded twenty years later; and Conyngham's Irish Dragoons, which still abides with us as the Eighth (King's Royal Irish) Hussars.

Meanwhile the French military system had suffered an irreparable loss in the death of Louvois, the source of woes unnumbered to France in the years that were soon to come. Nevertheless, the traditions of his rule were strong, and the French once more were first in the field, with, as usual, a vast siege-train massed on the Meuse and on the Scheldt. But a late spring and incessant rain delayed the beginning of operations till the beginning of May, when Luxemburg assembled seventy thousand men in rear of the Haine by Mons, and Boufflers forty-eight thousand more on the Scheldt at Tournay. The French king was with the troops in person ; and the original design was, as usual, to carry on a war of sieges on the Meuse, Boufflers reducing the fortresses while Luxemburg shielded him with a covering army. Lewis, however, finding that the towns which he had intended to invest were likely to make an inconveniently stubborn defence, presently returned home, and after detaching thirty thousand men to the war in Germany, left Luxemburg to do as he would. It had been better for William if the Grand Monarch had remained in Flanders. 1693.

The English king, on his side, assembled sixty thousand men at Brussels as soon as the French began to move, and led them with desperate haste to the Senne, where he took up an impregnable position at Park. Luxemburg marched up to a position over against him, and then came one of those deadlocks which were so common in those old campaigns. The two armies stood looking at each other for a whole month, neither venturing to move, neither daring to attack, both ill-supplied, both discontented, and as a natural consequence both losing scores, hundreds, and even thousands of men through desertion.

At last the position became insupportable, and on the 6th of July Luxemburg moved eastward as if to resume the original plan of operations on the Meuse. William thereupon resolved to create a diversion by detaching a force to attack the French lines of the Scheldt and Lys, a project which was brilliantly executed by Würtemberg, thanks not a little to three British regiments, the Tenth, Argyll's, and Castleton's, which formed part of his division. But meanwhile Luxemburg, quite ignorant of the diversion, advanced to the June 26/July 6.

1693. July. Meuse and laid siege to Huy, in the hope of forcing William to come to its relief. He judged rightly. William left his impregnable camp at Park and hurried to the rescue. But he came too late, and Huy fell after a trifling resistance. Luxemburg then made great seeming preparations for the siege of Liège, and William, trembling for the safety of that city and of Maestricht, detached eight thousand men to reinforce those garrisons, and then withdrew to the line of the Geete. Luxemburg watched the whole proceeding with grim delight. Würtemberg's success was no doubt annoying, but William had weakened his army by detaching this force to the Lys, and had been beguiled into weakening it still further by reinforcing the garrisons on the Meuse, which was exactly what he wanted. If he could bring the Allies to action forthwith he could reasonably hope for success.

The ground occupied by William was a triangular space enclosed between the Little Geete and a stream called the Landen Beck, which joins it at Leuw. The position was not without features of strength. The camp, which faced almost due south, was pitched on a gentle ridge rising out of a vast plain.[1] This ridge runs parallel to the Little Geete and has that river in its rear. The left flank was protected by marshy ground and by the Landen Beck itself, while the villages of Neerlanden and Rumsdorp, one on either side of the beck and the latter well forward on the plain, offered the further security of advanced posts. The right rested on a little stream which runs at right angles to the Geete and joins it at Elixheim, and on the villages of Laer and Neerwinden, which stand on its banks. From Neerlanden on the left to Neerwinden on the right the position measured close on four miles; and to guard this front, to say nothing of strong

[1] No battlefield can be taken in more readily at a glance than that of Landen. On the path alongside the railway from Landen Station is a mound formed of earth thrown out of a cutting, from the top of which the whole position can be seen.

garrisons for the villages, William had little more than fifty thousand men. Here then was one signal defect: the front was too long to permit troops to be readily moved from flank to flank, or to be withdrawn, without serious risk, from the centre. But this was not all. The depth of the position was less than half of its frontage, and thus allowed no space for the action of cavalry. This William ignored: he was afraid of the French horse, and was anxious that the action should be fought by infantry only. Finally, retreat was barred by the Geete, which was unfordable and insufficiently bridged, and therefore the forcing of the allied right must inevitably drive the whole army into a pinfold, as Leslie's had been driven at the battle of Dunbar. 1693. July.

Luxemburg, who knew every inch of the ground, was now anxious only lest William should retire before he could catch him. On the 28th of July, by a great effort and a magnificent march, he brought the whole of his army, eighty thousand strong, before William's position. He was now sure of his game, but he need not have been anxious, for William, charmed with the notion of excluding the French cavalry from all share in the action, was resolved to stand his ground. Many officers urged him to cross the Geete while yet he might, but he would not listen. Fifteen hundred men were told off to entrench the open ground between Neerwinden and Neerlanden. The hedges, mud-walls, and natural defences of Neerwinden and Laer were improved to the uttermost, and the ditches surrounding them were enlarged. Till late into the night the King rode backward and forward, ordering matters under his own eyes, and after a few hours' rest began very early in the morning to make his dispositions. July $\frac{18}{28}$.

The key of the position was the village of Neerwinden with the adjoining hamlet of Laer, and here accordingly he stationed the best of his troops. The defence of Laer was entrusted to Brigadier Ramsey with the Scots Brigade, namely, the Twenty-first, Twenty-fifth, Twenty-sixth, Mackay's and Lauder's

1693. regiments, reinforced by the Buffs and the Fourth Foot. Between Laer and Neerwinden stood six battalions of Brandenburgers, troops already of great and deserved reputation, of whom we shall see more in the years before us. Neerwinden itself was committed to the Hanoverians, the Dutch Guards, a battalion of the First and a battalion of the Scots Guards. Immediately to the north or left of the village the entrenchment was lined by the two remaining battalions of the First and Scots Guards, the Coldstream Guards, a battalion of the Royal Scots, and the Seventh Fusiliers. On the extreme left of the position Neerlanden was held by the other battalion of the Royal Scots, the Second Queen's, and two Danish regiments, while Rumsdorp was occupied by the Fourteenth, Sixteenth, Nineteenth, and Collingwood's regiments. In a word, every important post was committed to the British. The remainder of the infantry, with one hundred guns, was ranged along the entrenchment, and in rear of them stood the cavalry, powerless to act outside the trench, and too much cramped for space to manœuvre within it.

Luxemburg also was early astir, and was amazed to find how far the front of the position had been strengthened during the night. His centre he formed in eight lines over against the Allies' entrenchments between Oberwinden and Landen, every line except the second and fourth being composed of cavalry. For the attack on Neerlanden and Rumsdorp he detailed fifteen thousand foot and two thousand five hundred dismounted dragoons. For the principal assault on Neerwinden he told off eighteen thousand foot supported by a reserve of two thousand more and by eight thousand cavalry; while seventy guns were brought into position to answer the artillery of the Allies.

July $\frac{19}{29}$. Shortly after sunrise William's cannon opened fire against the heavy masses of the French centre; and at eight o'clock Luxemburg moved the whole of his left to the attack of Neerwinden. Six battalions, backed by dragoons and cavalry, were directed against Laer, and three columns, counting in all seven brigades, were launched against Neerwinden. The centre column, under the Duke of Berwick, was the first to come into action. Withholding their fire till they reached the village, the French carried the outer defences with a rush, and then meeting the Hanoverians and the First Guards, they began the fight in earnest. It was hedge-fighting, as at Steenkirk, muzzle to muzzle and hand to hand. Every step was contested; the combat swayed backwards and forwards within the village; and the carnage was frightful. The remaining French columns came up, met with the like resistance, and made little way. Fresh regiments were poured by the French into the fight, and at last the First Guards, completely broken by its losses, gave way. But it was only for a moment. They rallied on the Scots Guards; the Dutch and Hanoverians rallied behind them, and though the French had been again reinforced, they resumed the unequal fight, nine battalions against twenty-six, with unshaken tenacity. At Laer, on the extreme right, the fight was equally sharp. Ramsey for a time was driven out of the village, and the French cavalry actually forced its way into the Allies' position. There, however, it was charged in flank by the Elector of Bavaria, and driven out with great slaughter. Ramsey seized the moment to rally his brigade. The French columns, despite their success, still remained isolated and detached, and presented no united front. The King placed himself at the head of the Guards and Hanoverians, and with one charge British, Dutch, and Germans fell upon the Frenchmen and swept them out of both villages.

1693. July 19/29.

The first attack on Neerwinden had failed, and a similar attack on the allied left had been little more successful. At Neerlanden the First and Second Foot had successfully held their own against four French battalions until reinforcements enabled them to drive them back. At Rumsdorp the British, being but three thousand against thirteen thousand, were pushed out of the village, but being reinforced, recovered a part of it

1693. and stood successfully at bay. Luxemburg, however, was not easily discouraged. The broken troops in the
July 19/29. left were rallied, fresh regiments were brought forward, and a second effort was made to carry Neerwinden. Again French impetuosity bore all before it, and again the British and Germans, weakened and weary though they were, rallied when all seemed lost, and hurled the enemy back not merely repulsed but in confused and disorderly retreat.

On the failure of the second attack the majority of the French officers urged Luxemburg to retire; but the marshal was not to be turned from his purpose. The fourteen thousand men of the Allies in Laer and Neerwinden had lost more than a third of their numbers, while he himself had still a considerable force of infantry interlined with the cavalry in the centre. Twelve thousand of them, including the French and Swiss Guards, were now drawn off to the left for a third attack. When they were clear of the cavalry, the whole six lines of horse, which had stood heroically for hours motionless under a heavy fire, moved forward at a trot to the edge of the entrenchments;[1] but the demonstration, for such it seems to have been, cost them dear, for they were very roughly handled and compelled to retire. But now the French reinforcements supported by the defeated battalions drew near, and a third attack was delivered on Neerwinden. British and Dutch still made a gallant fight, but the odds against their weakened battalions were too great, and ammunition began to fail. They fought on indomitably till the last cartridge was expended before they gave way, but they were forced back, and Neerwinden was lost. Five French brigades then assailed the central entrenchment at its junction with Neerwinden, where stood the Coldstream Guards and the

[1] St. Simon. With the exception of one hollow, which might hold three or four squadrons in double rank in line, there is not the slightest shelter in the plain wherein the French horse could find protection.

1693. July 19/29.

Seventh Fusiliers. Wholly unmoved by the overwhelming numbers in their front and the fire from Neerwinden on their flank, the two regiments stood firm and drove their assailants back over the breastwork. Even when the French Household Cavalry came spurring through Neerwinden and fell upon their flank they fought on undismayed, and the Coldstreamers not only repelled the charge but captured a colour.

Such fighting, however, could not continue for long. William, on observing Luxemburg's preparations for the final assault, had ordered nine battalions from his left to reinforce his right. These never reached their destination. The Marquis of Feuquières, an officer even more celebrated for his acuteness as a military critic than for skill in the field, watched them as they moved and suddenly led his cavalry forward to the weakest point of the entrenchment. The battalions hesitated, halted, and then turned about to meet this new danger, but too late to save the forcing of the entrenchment. The battle was now virtually over. Neerwinden was carried, Ramsey after a superb defence had been driven out of Laer, the Brandenburgers had perforce retreated with him, the infantry that lined the centre of the entrenchment had forsaken it, and the French cavalry was pouring in and cutting down the fugitives by scores. William, who had galloped away in desperation to the left, now returned at headlong speed with six regiments of English cavalry,[1] which delivered charge after charge with splendid gallantry, to cover the retreat of the foot. On the left Tolmach and Bellasys by great exertion brought off their infantry in good order, but on the right the confusion was terrible. The rout was complete, the few bridges were choked by a heaving mass of guns, waggons, pack-animals, and men, and thousands of fugitives were cut down, drowned, or trampled to death. William did all that a gallant man could do to save the day, but in

[1] Life Guards, 1st, 3rd, 4th, 6th Dragoon Guards, Galway's Horse.

1693. vain. His troops had done heroic things to redeem his bad generalship; and against any living man but Marlborough or Luxemburg they would probably have held their own. It was the general not the soldiers that failed.

July 19/29.

The losses on both sides were very severe. That of the French was about eight thousand men; that of the Allies about twelve thousand, killed, wounded, and prisoners, and among the dead was Count Solmes, the hated Solmes of Steenkirk. The nineteen British battalions present lost one hundred and thirty-five officers killed, wounded, and taken. The French captured eighty guns and a vast quantity of colours, but the Allies, although beaten, could also show fifty-six French flags. And, indeed, though Luxemburg won, and deserved to win, a great victory, yet the action was not such as to make the allied troops afraid to meet the French. They had stood up, fifty thousand against eighty thousand, and if they were beaten they had at any rate dismayed every Frenchman on the field but Luxemburg. In another ten years their turn was to come, and they were to take a part of their revenge on the very ground over which many of them had fled.

The campaign closed with the surrender of Charleroi, and the gain by the French of the whole line of the Sambre. William came home to meet the House of Commons and recommend an augmentation of the Army by eight regiments of horse, four of dragoons, and twenty-five of foot. The House reduced this list by the whole of the regiments of horse, and fifteen of foot, but even so it brought the total establishment up to eighty-three thousand men. There is, however, but one new regiment of which note need be taken in the campaign of 1694, namely the Seventh Dragoons, now known as the Seventh Hussars, which, raised in 1689-90 in Scotland, now for the first time took its place on the English establishment and its turn of service in the war of Flanders.

1694. I shall not dwell on the campaign of 1694, which is memorable only for a marvellous march by which Luxemburg upset William's entire plan of campaign.

Nor shall I speak at length of the abortive descent on Brest, which is remembered mainly for the indelible stain which it has left on the memory of Marlborough. It is only necessary to say that the French, by Marlborough's information, though not on Marlborough's information only, had full warning of an expedition which had been planned as a surprise, and that Tolmach,[1] who was in command, unfortunately though most pardonably lacked the moral courage to abandon an attack which, unless executed as a surprise, was hopeless of success. He was repulsed with heavy loss, and died of wounds received in the action, a hard fate for a good soldier and a gallant man. But it is unjust to lay his death at Marlborough's door. For the failure of the expedition Marlborough was undoubtedly responsible, and that is quite bad enough; but Tolmach alone was to blame for attempting an enterprise which he knew to be hopeless. Marlborough cannot have calculated that he would deliberately essay to do impossibilities and perish in the effort, so cannot be held guilty of poor Tolmach's blunders. 1694.

Before the new campaign could be opened there had come changes of vital importance to France. The vast expense of the war had told heavily on the country, and the King's ministers were at their wit's end to raise money. Moreover, the War Department had deteriorated rapidly since the death of Louvois; and to this misfortune was now added the death of Luxemburg, a loss which was absolutely irreparable. Lastly, with the object of maintaining the position which they had won on the Sambre, the French had extended their system of fortified lines from Namur to the sea. Works so important could not be left unguarded, so that a considerable force was locked up behind these entrenchments, and was for all offensive purposes useless. We 1695. January.

[1] This is, of course, the Talmash of *Tristram Shandy* and of Macaulay's History. He signed his name, however, as I spell it here, and I use his own spelling the more readily since it is more easily identified with the Tollemache of to-day.

1695. shall see before long how a really great commander could laugh at these lines, and how in consequence it became an open question whether they were not rather an encumbrance than an advantage. The subject is one which is still of interest; and it is remarkable that the French still seem to cling to their old principles in the works which they have constructed for defence against a German invasion.

His enemy being practically restricted to the defensive, William did not neglect the opportunity of initiating aggressive operations. Masking his design by a series of feints, he marched swiftly to the Meuse and invested Namur. This fortress, more famous through its connection with the immortal Uncle Toby even than as the masterpiece of Cohorn carried to yet higher perfection by Vauban, stands at the junction of the Sambre and the Meuse, the citadel lying in the angle between the two rivers, and the town with its defences on the left bank of the Meuse. To the northward of the town outworks had been thrown up on the heights of Bouge by both of these famous engineers; and it was against these outworks that William directed his first attack.

Ground was broken on the 3rd of July, and three
June 23 / July 3. days later an assault was delivered on the lines of Bouge. As usual, the hardest of the work was given to the
June 26 / July 6. British, and the post of greatest danger was made over, as their high reputation demanded, to the Brigade of Guards. On this occasion the Guards surpassed themselves alike by the coolness of their valour and by the fire of their attack. They marched under a heavy fire up to the French palisades, thrust their muskets between them, poured in one terrible volley, the first shot that they had yet fired, and charged forthwith. In spite of a stout resistance, they swept the French out of the first work, pursued them to the second, swept them out of that, and gathering impetus with success, drove them from stronghold to stronghold, far beyond the original design of the engineers, and actually to the gates of the

town. In another quarter the Royal Scots and the 1695. Seventh Fusiliers gained not less brilliant success; and in fact it was the most creditable action that William had fought during the whole war. It cost the Allies two thousand men killed and wounded, the three battalions of Guards alone losing thirty-two officers. The British were to fight many such bloody combats during the next twenty years—combats forgotten since they were merely incidents in the history of a siege, and so frequent that they were hardly chronicled and are not to be restored to memory now. I mention this, the first of such actions, only as a type of many more to come.

The outworks captured, the trenches were opened against the town itself, and the next assault was directed against the counterguard of St. Nicholas gate. This again was carried by the British, with a loss of eight hundred men. Then came the famous attack on the counterscarp before the gate itself, where Captain Shandy received his memorable wound. This gave William the possession of the town. Then came the siege of the citadel, wherein the British had the honour of marching to the assault over half a mile of open ground, a trial which proved too much even for them. Nevertheless, it was they who eventually stormed a breach from which another of the assaulting columns had been repulsed, and ensured the surrender of the citadel a few days later. For their service on this occasion the Eighteenth Foot were made the Royal Irish; and a Latin inscription on their colours still records that this was the reward of their valour at Namur.

Thus William on his return to England could for the first time show his Parliament a solid success due to the British red-coats; and the House of Commons gladly voted once more a total force of eighty-seven thousand men. But the war need be followed no further. The campaign of 1696 was interrupted by a futile attempt of the French to invade England, and in 1697 France, reduced to utter exhaustion, gladly concluded the Peace of Ryswick. So ended, not without honour,

1697. the first stage of the great conflict with King Lewis the Fourteenth. The position of the two protagonists, England and France, was not wholly unlike that which they occupied a century later at the Peace of Amiens. The British, though they had not reaped great victories, had made their presence felt, and terribly felt, on the battlefield; and as the French in the Peninsula remembered that the British had fought them with a tenacity which they had not found in other nations, not only in Egypt but even earlier at Tournay and Lincelles, so, too, after Blenheim and Ramillies they looked back to the furious attack at Steenkirk and the indomitable defence of Neerwinden. "Without the concurrence of the valour and power of England," said William to the Parliament at the close of 1695, "it were impossible to put a stop to the ambition and greatness of France." So it was then, so it was a century later, and so it will be again, for though none know better the superlative qualities of the French as a fighting people, yet the English are the one nation that has never been afraid to meet them. With the Peace of Ryswick the 'prentice years of the standing Army are ended, and within five years the old spirit, which has carried it through the bitter schooling under King William, will break forth with overwhelming power under the guiding genius of Marlborough.

AUTHORITIES.—The leading authority for William's campaigns on the English side is D'Auvergne, and on the French side the compilation, with its superb series of maps, by Beaurain. Supplementary on one side are Tindal's History, Carleton's Memoirs, and Sterne's *Tristram Shandy*; and on the other the *Mémoires* of Berwick and St. Simon, Quincy's *Histoire Militaire de Louis XIV.*, and in particular the *Mémoires* of Feuquières. Many details as to Steenkirk, in particular as to the casualties, are drawn from *Present State of Europe, or Monthly Mercury*, August 1692; and as to Landen from the official relation of the battle, published by authority, 1693. Beautiful plans of both actions are in Beaurain, rougher plans in Quincy and Feuquières. All details as to the establishment voted are from the Journals of the House of Commons. Very elaborate details of the operations are given in Colonel Clifford Walton's *History of the British Standing Army*.

CHAPTER IV

Peace having been signed, there arose the momentous question, what should be done with the Army. To understand aright the attitude of Parliament towards it, a brief sketch must be given of its relations therewith apart from the mere question of voting supplies. It has been seen that the scandals of Schomberg's first campaign had opened the eyes of Parliament to the iniquities that were then going forward; but, though a scape-goat had been made of the Commissary-General, the matter had not been sifted to the bottom. 1697.

The primary and principal difficulty was, of course, lack of money. In the case of the Irish war this had been overcome by grants of the Irish estates which had been forfeited after the conquest, the mere expectation and hope of which had sufficed to set the minds of many creditors at rest. For the war in Flanders, however, there was no such resource. The treasury was empty, and the funds voted by Parliament were so remote that they could only be assigned to creditors in security for payment at some future time. Many of these creditors, however, were tradesmen who could not afford to wait until tallies should be issued in course of payment, and were therefore compelled to dispose of these securities at a ruinous discount. The mischief naturally did not end there. Capitalists soon discovered that to buy tallies at huge discount was a much more profitable business than to lend money direct to the State at the rate of seven per cent, and accordingly devoted all their money to it. Thus the "tally-traffic,"

1697. as it was called, grew so formidable that the Lord Treasurer, Godolphin, was obliged secretly to offer larger interest for loans than was authorised by Parliament.[1]

The result of this financial confusion was that the close of every campaign found the Army in Flanders in a miserable state, owing to the exhaustion of its money and its credit. When it is remembered that a large proportion of the pay of officers and men was kept on principle one year in arrear, that they had to pay discount for anticipation of its payment at the best of times, and that to this charge was now added the further discount on the tallies of the State, it will be seen that their loss became very serious. The incessant difficulties of all ranks from want of their pay and arrears gave rise to much discontent and frequently hampered active operations. Officers were obliged to sell the horses, which they had bought for purposes of transport, before the campaign opened, and were very often driven to supply not only themselves but their men out of their own pockets.

Of all this it is probable that the House of Commons knew little, and as in 1691 it had appointed Commissioners to inquire into the public accounts, it doubtless awaited their report before taking any active step. In 1694, however, the House was rudely surprised by certain revelations respecting a notorious crimp of London, named Tooley, who went so far in his zeal to procure recruits that he not only forced the King's shilling on them when they were drunk—a practice which was common in France and has not long been extinct in England—but resorted to kidnapping pure and simple.[2] Here was one gross infringement of the liberty of the subject; and this scandal was quickly followed by another. At the end of 1694 there came a petition from the inhabitants of Royston, complaining

[1] Godolphin to the King, 2nd February 1691, *S. P., Dom.*

[2] *Commons Journals*, 24th February, 5th March, 1693-1694. A full account will be found in Colonel Clifford Walton, p. 483.

that the troops quartered there were exacting subsistence from the townsfolk on a fixed scale. Inquiry proved the truth of the allegation: the troops were unpaid, and had taken their own measures to save themselves from starvation. Almost simultaneously the Commissioners of Public Accounts reported that their inquiries had been baffled by the refusal of several regimental agents to show their books; and they gave at the same time an unvarnished relation of the shameful extortion practised by agents towards officers and men, and of one case of glaring misconduct on the part of a colonel. The House brought the recalcitrant agents to their senses by committing them to custody, and addressed the King with an earnest prayer that he would put a stop to these iniquities.[1] The King accordingly cashiered the colonel[2] and promised amendment, which promise was discharged so far as orders could fulfil it. But the case demanded not new orders but execution of existing regulations. 1697.

There, however, the matter rested for the time, the Commons being occupied with the task of purging corruption from their own body, which was very inadequately performed by the expulsion of the Speaker. Nevertheless, to the end of the war fresh petitions continued to come in from towns, from widows of officers, and from private soldiers, all complaining of the dishonesty of officers and of agents; and the House thus established itself as in some sort a mediator between officers and men. Such a mediator, it must be confessed, was but too sadly needed, but in the interests of discipline it was a misfortune that the House should ever have accepted the position. The immediate result was to overwhelm the Commons with a vast amount of business which they were incompetent to transact, and to suggest an easy remedy for soldiers' grievances in the abolition of all soldiers.

William was not unaware of the danger, and had

[1] *Commons Journals*, 26th February 1693-1694.
[2] Hastings of the Thirteenth.

1697. taken measures to meet it. Before meeting Parliament in December 1697, he had already disbanded ten regiments, and having thrown this sop to English prejudice, he delivered it as his opinion in his speech from the throne that England could not be safe without a land-force. But agitators and pamphleteers had been before him. The old howl of "No Standing Army" had been raised, and reams of puerile and pedantic nonsense had been written to prove that the militia was amply sufficient for England's needs. The arguments on the other side were stated with consummate ability by Lord Somers; but the old cry was far too pleasant in the ears of the House to be easily silenced. Another reason which may well have swayed the House was that, though his English soldiers had fought for William as no other troops in the world, he had never succeeded
Dec. 11. in winning a victory. Be that as it may, within eight days the House, on the motion of Robert Harley, resolved that all forces raised since September 1680 should be disbanded.

The resolution, in the existing condition of European affairs, was a piece of malignant folly; but the accounts submitted two days later by the Paymaster-
Dec. 13. General probably did much to confirm it. The arrears of pay due to the Army since April 1692 amounted to twelve hundred thousand pounds, and the arrears of subsistence to a million more, while yet another hundred thousand was due to regiments on their transfer from the Irish to the English establishment.[1] To meet this debt there was eighty thousand pounds in tallies which no one would discount at any price, while to make matters worse, taxation voted by the House to produce three millions and a half had brought no more than two millions into the treasury. Attempts were made
1698. in January 1698 to rescind the resolution, but in vain. The Government yielded, and after struggling hard to

[1] That is to say, to meet the difference between English and Irish pay, the rate being lower in Ireland than in England owing to the greater cheapness of provisions.

obtain four hundred thousand pounds, was fain to accept fifty thousand pounds less than that sum for the service of the Army in the ensuing year. 1698.

The effect of the vote was immediate. The enemies of the Army were exultant, and heaped abuse and insult on the soldiers who for five years had spent their blood and their strength for a people that had not paid them so much as their just wages. All William's firmness was needed to restrain the exasperated officers from wreaking summary vengeance on the most malignant of these slanderers. It was the old story. Men who had grown fat on the "tally-traffic" could find nothing better than bad words for the poor broken lieutenant who borrowed eighteenpence from a comrade to buy a new scabbard for his sword, being ashamed to own that he wanted a dinner.[1] The distress in the Army soon became acute. Petitions poured in from the disbanded men for arrears, arrears, arrears. Bad soldiers tried to wreak a grudge against good officers, good soldiers to obtain justice from bad officers; all military men of whatever rank complained loudly of the agents.[2] Then came unpleasant reminders that the expenses of the Irish war were not yet paid. Colonel Mitchelburne, the heroic defender of Londonderry, claimed, and justly claimed, fifteen hundred pounds which had been owing to him since 1690.[3] The House strove vainly to stem the torrent by voting a gratuity of a fortnight's subsistence to every man, and half-pay as a retaining fee to every officer, until he should be paid in full. The claims of men and officers continued to flow in, and at last the Commons addressed the King to May 28.
appoint persons unconnected with the Army to examine and redress just grievances, and to punish men who complained without cause.

[1] See Farquhar's *Trip to the Jubilee.*

[2] See *C. J.* 19th, 25th March, 16th December 1696; 5th, 7th, 15th, 23rd January 1697; 3rd, 7th, 10th, 12th, 17th, 24th, 27th January; 7th, 9th, 14th, 15th, 16th February 1698.

[3] *C. J.* 8th June 1698.

1698. On the 7th of July the House was delivered from further importunities by a dissolution; and William returned to his native Holland. Before his departure he left certain instructions with his ministers concerning the Army. The actual number of soldiers to be maintained was not mentioned in the Act of Parliament, but was assumed, from the proportion of money granted, to be ten thousand men. William's orders were to keep sixteen thousand men, for he still had hopes that Parliament might reconsider the hasty votes of the previous session.[1] These expectations were not realised. The clamour against the Army had been strengthened by a revival of the old outcry against the Dutch, and against the grant of crown-lands in general, and to Dutchmen in particular. Moreover, the House had no longer the pressure of the war to unite it in useful and patriotic work. The inevitable reaction of peace after long hostilities was in full vigour. All the selfishness, the prejudice, and the conceit that had been restrained in the face of great national peril was now let loose; and the House, with a vague idea that there were many things to be done, but with no clear perception what these things might be, was ripe for any description of mischief.

Dec. 12. William's speech was tactful enough. Expressing it as his opinion that, if England was to hold her place in Europe, she must be secure from attack, he left the House to decide what land-force should be maintained, and only begged that, for its own honour, it would provide for payment of the debts incurred during the war. The speech was not ill-received; and William, despite the warnings of his ministers, was sanguine that
Dec. 17. all was well. Five days later a return of the troops was presented to the House, showing thirty thousand men divided equally between the English and Irish establishments. Then Harley, the mover of the foolish resolution of the previous year, proposed that the English establishment should be fixed at seven thousand men, all of them to be British subjects. This was con-

[1] Burnet.

firmed by the House on the following day, together 1698.
with an Irish establishment of twelve thousand men to be maintained at the expense of the sister island. The words of the Act that embodied this decision were peremptory; it declared that on the 26th March 1699 all regiments, saving certain to be excepted by proclamation, were actually disbanded. Finally, the Mutiny Act, which had expired in April 1698, was not renewed by the House, so that even in this pittance of an Army the officers had no powers of enforcing discipline.

There is no need to dilate further on this resolution, which for three years placed England practically at the mercy of France. It was an act of criminal imbecility, the most mischievous work of the most mischievous Parliament that has ever sat at Westminster. William was so deeply chagrined that he was only with difficulty dissuaded from abdication of the throne. Apart from the madness of such wholesale reduction of the Army, the clause restricting the nationality of the seven thousand was directly aimed at the King's favourite regiment, the Dutch Blue Guards. He submitted, however, with dignity enough, merely warning the House that he disclaimed all responsibility for any disaster that might follow. Just at that moment came a rare opportunity for undoing in part the evil work of the Commons. The death of the Electoral Prince of Bavaria brought the question of the succession to the Spanish throne to an acute stage; and the occasion was utilised to ask Parliament for the grant of a larger force. William, however, with an unwisdom which even his loyalty to his faithful troops cannot excuse, pleaded as a personal favour for the retention of his Dutch Guards. The request preferred on such grounds was refused, and a great opportunity was lost.

Nothing, therefore, remained but to make the most of the slender force that was authorised by the Act of Disbandment. The ministers with great adroitness contrived to extort from the Commons an additional three thousand men under the name of marines, for the

1698. collective wisdom of the nation will often give under one name what it refuses under another; but as regards the Army proper, the only expedient was to preserve the skeleton of a larger force. Thus finally was established the wasteful and extravagant system which has been followed even to the present day. The seven thousand troops for England were distributed into nineteen, and the twelve thousand for Ireland into twenty-six, distinct corps, with an average proportion of one officer to ten men.[1] In addition to these, three corps of cavalry and seven of infantry were maintained in Scotland, while the Seventh Fusiliers were retained apparently in the Dutch service, or at any rate in Holland. The Artillery was specially reserved on a new footing by the name of the regimental train, first germ of the Royal Regiment that was to come,

[1] The following was the strength and distribution of the corps :—

England.—Three troops of Life Guards, and one of Horse-Grenadier Guards, each 180 of all ranks. Two regiments of Horse (Blues, 1st D.G.), each of nine troops, 37 officers, 353 non-commissioned officers and men. Five regiments of Horse (3rd, 5th, 6th, 7th D.G., Macclesfield's), each of six troops, 24 officers, 244 non-commissioned officers and men. Three regiments of Dragoons (Royals, 3rd and 4th Hussars), each of six troops, 24 officers, 259 non-commissioned officers and men. First Guards and Coldstream Guards, each of fourteen companies, 139 officers, 1826 non-commissioned officers and men. 2nd, 3rd, and 4th Foot, each of ten companies, 34 officers, 411 men.

Ireland.—Two regiments of Horse (2nd D.G. and 4th D.G.). Three regiments of Dragoons (5th and 6th D., 8th H.). Twenty-one battalions of Foot, 1st Royals (2 battalions), 5th, 6th, 8th, 9th, 10th, 11th, 12th, 13th, 14th, 15th, 16th, 17th, 18th, 19th, 20th, 22nd, 23rd, 24th, 27th. The establishments were on much the same scale as in England.

Scotland.—One troop of Horse Guards. Two regiments of Dragoons (Greys and 7th H.). Scots Guards, Collier's, 21st, 25th, 26th, George Hamilton's, Strathnaver's.

I may add that I have found the greatest difficulty in the compilation of this note. The proclamation regarding England is to be found in the British Museum; that for Ireland is neither in the Museum nor the Record Office, but the list was after much searching disinterred from an Entry Book (*H.O. Mil. Entry Book*, vol. iii. pp. 374-386). The Scotch establishment I have made up as best I could from various sources, but I cannot vouch for its accuracy.

and contained four companies, each of thirty men, with 1698. the usual proportion of an officer to every ten men. To these were added ten officers of engineers.[1] Within the next two years the principle of a skeleton army was pushed still further, and in each of the regiments of dragoons thirty-three officers and thirty sergeants and corporals looked minutely to the training of two hundred and sixteen men. Large numbers of officers, who were retained for emergencies by the allowance of half-pay, also drew heavily on the niggardly funds granted by the Commons; and it was a current jest of the time that the English Army was an army of officers.[2]

The sins of Parliament soon found it out. Before 1699. it had sat a month petitions from officers and men began to pour in, as during the previous sessions, with claims for arrears and with complaints of all kinds. As the Commons were the fountain of pay, it was natural and right that the clamour for wages should be directed at them; but the fashion had been set for soldiers to resort to them for redress of all grievances, and it would seem that men used the petition to Parliament as a means of openly threatening their officers.[3] Moreover, by some extraordinary blunder the grant of half-pay had been limited to such officers only as at the time of disbandment were serving in English regiments. This regulation naturally caused loud outcry from officers who, after long service in English regiments, had been transferred to Scottish corps on promotion. A prorogation at the end of April brought relief to the Commons for a time; but no sooner was it reassembled November. than the petitions streamed in with redoubled volume. The House thus found itself converted almost into a military tribunal. Appeal was made to it on sundry

[1] *H.O. Mil. Entry Book*, vol. iii. p. 327, May 1698.

[2] Burnet. Even prior to the disbandment one Irish regiment of horse numbered 103 commissioned officers in a total of 490 of all ranks.

[3] See the petition of men disbanded from Macclesfield's Horse. *Commons Journals*, 18th April, 3rd May 1699.

1699. points that were purely of military discipline, and private soldiers sought to further their complaints by alleging that their officers had spoken disrespectfully and disdainfully of the House itself.[1]

To do them justice, the Commons were woefully embarrassed by these multitudinous petitions. Once they interfered actively by taking up the cause of an officer, whom they knew, or should have known, to be a bad character,[2] and threatened his colonel with their vengeance unless the wrongs of the supposed sufferer were redressed. The reply of the colonel was so disconcerting as effectually to discourage further meddling of this kind. Nevertheless the grievances urged by the men must many of them have been just, while some of the allegations brought forward were most scandalous. In one of the disbanded regiments, Colonel Leigh's, it was roundly asserted that the officers had made all the men drunk, and then caused them to sign receipts in full for pay which had not been delivered to them.[3]
1700. Finally, in despair, a bill was introduced to erect a Court of Judicature to decide between officers and men. This measure, however, was speedily dropped, and the more prudent course was adopted of appointing Commissioners to inquire into the debt due to the Army.

But meanwhile another question had been raised, which brought matters into still greater confusion. A parliamentary inquiry as to the disposition of the Irish forfeited estates had revealed the fact that William had granted large shares of the same, not only in reward and compensation to deserving officers, which was just and right, but also to his discarded mistress, Elizabeth Villiers, and to his Dutch favourites, Portland and Albemarle. The King's conduct herein was the less defensible, inasmuch as the Irish government had

[1] Petition of Richard Nichols and others of the First Guards. *Commons Journals*, 6th December 1699.

[2] Petition of John Dorrell, *ibid.* 9th December 1699. The case had been investigated and dismissed in the previous Parliament.

[3] *Commons Journals*, 9th January 1699-1700.

counted upon these estates to defray the expenses, still unpaid, of the Irish war, and had thrown up its hands in despair when it found that this resource was to be withheld.[1] The House of Commons took up the question viciously, passed a sweeping and shameful bill resuming all property that had belonged to the Crown at the accession of James the Second, tacked it to a money-bill, and sent it up to the Lords. The Upper House, to save a revolution, yielded, after much protest, and passed the bill; and then none too soon William sent this most mischievous House of Commons about its business. 1700. April 11.

It was not until early in the following year that the King met the Parliament, more distinctly even than the last a Tory Parliament, which had been elected in the autumn. Once more he was obliged to remind it that, amid the all-important questions of the English succession and the Spanish succession, provision should be made for paying the debts incurred through the war. There could be no doubt about these debts, for the petitions which had formerly dropped in by scores, now, in consequence of the interference with the Irish grants, flowed in by hundreds. The Commons had flattered themselves that they had disposed of this disagreeable business by their appointment of commissioners, but they found that, owing to their own faulty instructions, the commissioners were powerless to deal with many of the cases presented to them. The complaints of officers against the Government became almost as numerous as those of men against officers, and every day came fresh evidence of confusion of military business worse confounded by the imbecility and mismanagement of the House.[2] 1701, February 14.

[1] *Cal. S. P., Dom.*, 1691, pp. 241, 393.

[2] Here is one instance. It was the rule that clothing should be provided for a regiment according to its establishment on paper, whether the muster-rolls were full or not; the allowance in payment for the same (which was deducted from the pay of the men) being granted to the colonels on the same basis at the close of the financial year. The colonels provided the clothing accordingly

1701. Where the matter would have ended, and whether it might not have led ultimately to a dangerous military riot, it is difficult to say. All, however, was cut short by the despatch of English troops to the Low Countries, and the evident approach of war; for the prospect of employment for every disbanded soldier and reduced officer sufficed in itself to quiet a movement which might easily have become formidable. Two more sessions such as those of 1698 and 1699 might have brought about a repetition of Cromwell's famous scene with the Long Parliament.

It is, however, impossible to leave these few stormy years of peace without taking notice of the apparent helplessness of the military administration. The War Office was in truth in a state of transition. The Secretary-at-War was still so exclusively the secretary to the Commander-in-Chief that he accompanied him on his campaigns; and it is difficult to say with whom, except with the Commander-in-Chief, rested the responsibility for the government of the Army. No ordinary standard should be used in judging of a man who was confronted with so many difficulties as King William the Third. His weak frame, the vast burden of his work in the department of foreign affairs, his failure to understand and his inability to sympathise with the English character, all these causes conspired to make the task of governing England and of commanding her Army too heavy for him. Still, making all

early in 1697. In December many regiments were disbanded, and all were much reduced by the Act of Disbandment, when, by the King's just order, all disbanded men were allowed to take away their clothing with them. In April 1698 the colonels applied for the allowance, but were told that the rule had been altered, and that no money would be issued to them except for men actually on the rolls at the time of reduction or disbandment. The colonels, thus defrauded of a large portion of their allowance, were unable to pay for the clothing, and were, of course, sued by the clothiers. It is added that the clothiers would accept in ready-money just half the price which they demanded in treasury-tallies. See the petition of the colonels to the House of Commons in *Journals*, 28th May and 4th June 1701.

possible allowance, and accepting as true Sterne's pictures of his popularity among the soldiers, it is difficult wholly to acquit him of blame for the misconduct of the military administration. His mind in truth was hardly well-suited for administrative detail. He could handle a great diplomatic combination with consummate skill and address, even as he could sketch the broad features of a movement or of a campaign; but he was a statesman rather than an administrator, a strategist rather than a general. In war his impatience guided him to a succession of crushing defeats, in peace his contempt for detail made his period of the command-in-chief one of the worst in our history. That, amid the corruption which he found in England, he should have despaired of finding an honest man is pardonable enough, but he took no pains to cure that corruption, preferring rather to conduct his business through his Dutch favourites than through the English official channels. Finally, his behaviour in the matter of the Irish forfeitures suggests that he was not averse to jobbery himself, nor over-severe towards the same weakness in others; and in truth the Dutch have no good reputation in the matter of corruption. Stern, hard, and cold, he had little feeling for England and Englishmen except as ministers to that hostility for France which was his ruling passion. Probably he felt more kindly towards the English soldier than towards any other Englishman; the iron nature melted at the sight of the shattered battalions at Steenkirk, and, if we are to believe Burnet, the cold heart warmed sufficiently towards the red-coat to prompt him to relieve the starving men, so shamefully neglected by Parliament, out of his own pocket. On the whole, it may be said that no commander was ever so well served by British troops, nor requited that service, whatever his good intent, so unworthily and so ill.

BOOK VI

CHAPTER I

A EUROPEAN quarrel over the succession to the Spanish throne,[1] on the death of the imbecile King Charles the Second, had long been foreseen by William, and had been provided against, as he hoped, by a Partition Treaty in the year 1698. The arrangement then made had been upset by the death of the Electoral Prince of Bavaria, and had been superseded by a second Partition Treaty in March 1700. In November of the same year King Charles the Second died, leaving a will wherein Philip, Duke of Anjou, and second son of the Dauphin, was named heir to the whole Empire of Spain. At this the second Partition Treaty went for naught. Lewis the Fourteenth, after a becoming interval of hesitation, accepted the Spanish crown for the Duke of Anjou under the title of King Philip the Fifth.

The Emperor at once entered a protest against the

[1] Philip III., d. 1621.

Philip IV., d. 1665.

Charles II., d. 1700.	Maria Theresa, m. Louis XIV.	Margaret, m. Leopold I.
	Louis, Dauphin, d. 1711.	Electress of Bavaria.
	Philip, Duke of Anjou (Philip V.).	Joseph, Electoral Prince, d. 1699.

1701. will, and Lewis prepared without delay for a campaign in Italy. William, however, for the present merely postponed his recognition of Philip the Fifth; and his example was followed by the United Provinces. Lewis, ever ready and prompt, at once took measures to quicken the States to a decision. Several towns[1] in Spanish Flanders were garrisoned, under previous treaties, by Dutch troops. Lewis by a swift movement surrounded the whole of them, and having thus secured fifteen thousand of the best men in the Dutch army, could dictate what terms he pleased. William expected that the House of Commons would be roused to indignation by this aggressive step, but the House was far too busy with its own factious quarrels. When, however, the States appealed to England for the ten thousand men, which under the treaty of 1677 she was bound to furnish, both Houses prepared faithfully to fulfil the obligation.

Then, as invariably happens in England, the work which Parliament had undone required to be done again. Twelve battalions were ordered to the Low Countries from Ireland, and directions were issued for the levying of ten thousand recruits in England to take their place. But, immediately after, came bad news from the West Indies, and it was thought necessary to despatch thither four more battalions from Ireland. Three regiments[2] were hastily brought up to a joint strength of two thousand men, and shipped off. Thus, within fifteen months of the disbandment of 1699, the garrison of Ireland had been depleted by fifteen battalions out of twenty-one; and four new battalions required to be raised immediately. Of these, two, namely Brudenell's and Mountjoy's, were afterwards disbanded, but two more, Lord Charlemont's and Lord Donegal's, are still with us as the Thirty-fifth and Thirty-sixth of the Line.

[1] Namur, Luxemburg, Mons, Charleroi, Ath, Oudenarde, Nieuport, Ostend.

[2] 12th, 22nd, 27th.

In June the twelve battalions[1] were shipped off to Holland, under the command of John, Earl of Marlborough, who since 1698 had been restored to the King's favour, and was to fill his place as head of the European coalition and General of the confederate armies in a fashion that no man had yet dreamed of. He was now fifty years of age; so long had the ablest man in Europe waited for work that was worthy of his powers; and now his time was come at last. His first duties, however, were diplomatic; and during the summer and autumn of 1701 he was engaged in negotiations with Sweden, Prussia, and the Empire for the formation of a Grand Alliance against France and Spain. Needless to say he brought all to a successful issue by his inexhaustible charm, patience, and tact. 1701.

Still the attitude of the English people towards the contest remained doubtful, until, on the death of King James the Second, Lewis made the fatal mistake of recognising and proclaiming his son as King of England. Then the smouldering animosity against France leaped instantly into flame. William seized the opportunity to dissolve Parliament, and was rewarded by the election of a House of Commons more nearly resembling that which had carried him through the first war to the Peace of Ryswick. He did not fail to rouse its patriotism and self-respect by a stirring speech from the throne, and obtained the ratification of his agreement with the Allies that England should furnish a contingent of forty thousand men, eighteen thousand of them to be British and the remainder foreigners. So the country was committed to the War of the Spanish Succession. September.

It was soon decided that all regiments in pay must be increased at once to war-strength, and that six more battalions, together with five regiments of horse and

[1] 1st batt. First Guards, 1st Royals (2 batts.), 8th, 9th, 10th, 13th, 15th, 16th, 17th, 18th, 23rd, 24th. The Guards had been substituted (after careful explanation to Parliament) by William's own direction in lieu of the 9th Foot.

1701. three of dragoons, should be sent to join the troops already in Holland. Then, as usual, there was a rush to do in a hurry what should have been done at leisure; and it is significant of the results of the late ill-treatment of the Army that, though the country was full of unemployed soldiers, it was necessary to offer three pounds, or thrice the usual amount of levy-money, to obtain recruits. The next step was to raise fifteen new regiments — Meredith's, Cootes', Huntingdon's, Farrington's, Gibson's, Lucas's, Mohun's, Temple's, and Stringer's of foot; Fox's, Saunderson's, Villiers', Shannon's, Mordaunt's and Holt's of marines. Of the foot Gibson's and Farrington's had been raised in 1694, but the officers of Farrington's, if not of both regiments, had been retained on half-pay, and, returning in a body, continued the life of the regiment without interruption. Both are still with us as the Twenty-eighth and Twenty-ninth of the Line. Huntingdon's and Lucas's also survive as the Thirty-third and Thirty-fourth, and Meredith's and Cootes', which were raised in Ireland, as the Thirty-seventh and Thirty-ninth, while the remainder were disbanded at the close of the war. Of the marines, Saunderson's had originally been raised in 1694, and eventually passed into the Line as the Thirtieth Foot, followed by Fox's and Villiers' as the Thirty-first and Thirty-second. Nothing now remained but to pass the Mutiny Act, which was speedily done; and on the 5th of May, just two months after the death of King William, the great work of his life was continued by a formal declaration of war.

The field of operations which will chiefly concern us is mainly the same as that wherein we followed the campaigns of King William. The eastern boundary of the cock-pit must for a time be extended from the Meuse to the Rhine, the northern from the Demer to the Waal, and the southern limit must be carried from Dunkirk beyond Namur to Bonn. But the reader should bear in mind that, in consequence of the Spanish alliance, Spanish Flanders was no longer hostile, but friendly, to France, so that the French frontier, for all practical purposes, extended to the boundary of Dutch Brabant. Moreover, the French, besides the seizure, already related, of the barrier-towns, had contrived to occupy every stronghold on the Meuse except Maestricht, from Namur to Venloo, so that practically they were masters so far of the whole line of the river.

A few leagues below Venloo stands the fortified town of Grave, and beyond Grave, on the parallel branch of the Rhine, stands the fortified city of Nimeguen. A little to the east of Nimeguen, at a point where the Rhine formerly forked into two streams, stood Fort Schenk, a stronghold famous in the wars of Morgan and of Vere. These three fortresses were the three eastern gates of the Dutch Netherlands, commanding the two great waterways, doubly important in those days of bad roads, which lead into the heart of the United Provinces.

It is here that we must watch the opening of the 1702.
campaign of 1702. There were detachments of the French and of the Allies opposed to each other on the Upper Rhine, on the Lower Rhine, and on the Lower Scheldt; but the French grand army of sixty thousand men was designed to operate on the Meuse, and the presence of a Prince of the blood, the Duke of Burgundy, with old Marshal Boufflers to instruct him, sufficiently showed that this was the quarter in which France designed to strike her grand blow. Marlborough being still kept from the field by other business, the command of the Allied army on the Meuse was entrusted to Lord Athlone, better known as that Ginkell who had completed the pacification of Ireland in 1691. His force consisted of twenty-five thousand men, with which he lay near Cleve, in the centre of the crescent formed by Grave, Nimeguen, and Fort Schenk, watching under shelter of these three fortresses the army of Boufflers, which was encamped some twenty miles to south-east of him at Uden and Xanten. On the 10th May 30
of June Boufflers made a sudden dash to cut off Athlone June 10.

1702. from Nimeguen and Grave, a catastrophe which Athlone barely averted by an almost discreditably precipitate retreat. Having reached Nimeguen Athlone withdrew to the north of the Waal, while all Holland trembled over the danger which had thus been so narrowly escaped.

Such was the position when Marlborough at last took the field, after long grappling at the Hague with the difficulties which were fated to dog him throughout the war. In England his position was comparatively easy, for though Prince George of Denmark, the consort of Queen Anne, was nominally generalissimo of all forces by sea and land, yet Marlborough was Captain-General of all the English forces at home and in Holland, and in addition Master-General of the Ordnance. But it was only after considerable dispute that he was appointed Commander-in-Chief of the allied forces, and then not without provoking much dissatisfaction among the Dutch generals, and much jealousy in the Prince of Nassau-Saarbrück and in Athlone, both of whom aspired to the office. These obstacles overcome, there came the question of the plan of campaign. Here again endless obstruction was raised. The Dutch, after their recent fright, were nervously apprehensive for the safety of Nimeguen, the King of Prussia was much disturbed over his territory of Cleve, and all parties who had not interests of their own to put forward made it their business to thwart the Commander-in-
June 21/July 2. Chief. With infinite patience Marlborough soothed them, and at last, on the 2nd of July, he left the Hague for Nimeguen, accompanied by two Dutch deputies, civilians, whose duty it was to see that he did nothing imprudent. Arrived there he concentrated sixty thousand men, of which twelve thousand were British,[1] recrossed the Waal and encamped at Ober-Hasselt over against Grave, within two leagues of the French. Then once more the obstruction of his colleagues caused

[1] Seven regiments of horse and dragoons, fourteen battalions of foot, fifty-six guns.

delay, and it was not until the 26th of July that he could cross to the left bank of the Meuse. "Now," he said to the Dutch deputies, as he pointed to the French camp, "I shall soon rid you of these troublesome neighbours." 1702. July 15/26.

Five swift marches due south brought his army over the Spanish frontier by Hamont. Boufflers thereupon in alarm broke up his camp, summoned Marshal Tallard from the Rhine to his assistance, crossed the Meuse with all haste at Venloo, and pushed on at nervous speed for the Demer. On the 2nd of August he lay between Peer and Bray, his camping-ground ill-chosen, and his army worn out by a week of desperate marching. Within easy striking distance, a mile or two to the northward, lay Marlborough, his army fresh, ready, and confident. He held the game in his hand; for an immediate attack would have dealt the French as rude a buffet as they were to receive later at Ramillies. But the Dutch deputies interposed; these Dogberries were content to thank God that they were rid of a rogue. So Boufflers was allowed to cross the Demer safely at Diest, and a first great opportunity was lost. July 22/August 2.

Marlborough, having drawn the French away from the Meuse, was now at liberty to add the garrison of Maestricht to his field-force, and to besiege the fortresses on the river. Boufflers, however, emboldened by his escape, again advanced north in the hope of cutting off a convoy of stores that was on its way to join the Allies. Marlborough therefore perforce moved back to Hamont and picked up his convoy; then, before Boufflers could divine his purpose he had moved swiftly south, and thrown himself across the line of the French retreat to the Demer. The French marshal hurried southward with all possible haste, and came blundering through the defiles before Hochtel on the road to Hasselt, only to find Marlborough waiting ready for him at Helchteren. Once again the game was in the Englishman's hand. The French were in great disorder, their left in particular being hopelessly entangled August 11/22.

1702. in marshy and difficult ground. Marlborough instantly gave the order to advance, and by three o'clock the artillery of the two armies was exchanging fire. At five Marlborough directed the whole of his right to fall on the French left; but to his surprise and dismay, the right did not move. A surly Dutchman, General Opdam, was in command of the troops in question and, for no greater object than to annoy the Commander-in-Chief, refused to execute his orders. So a second great opportunity was lost.

August 12/23. Still much might yet be won by a general attack on the next day; and for this accordingly Marlborough at once made his preparations. But when the time came the Dutch deputies interposed, entreating him to defer the attack till the morrow morning. "By to-morrow morning they will be gone," answered Marlborough; but all remonstrance was unavailing. The attack was perforce deferred, the French slipped away in the night, and though it was still possible to cut up their rear-guard with cavalry, a third great opportunity was lost.

Marlborough was deeply chagrined; but although with unconquerable patience and tact he excused Opdam's conduct in his public despatches, he could not deceive the troops, who were loud in their indignation against both deputies and generals. There was now nothing left but to reduce the fortresses on the Meuse, a part of the army being detached for the siege while the remainder covered the operations under the command of Marlborough. Even over their favourite pastime of a siege, however, the Dutch were dilatory beyond measure. "England is famous for negligence," wrote Marlborough, "but if Englishmen were half as negligent as the people here, they would be torn to pieces by Parliament."[1] Venloo was at length invested

August 18/29. on the 29th of August,[2] and after a siege of eighteen

[1] Coxe, vol. i. p. 182.

[2] So Quincy. Coxe gives August 25–September 5 as the date, but the difference depends merely on the interpretation of the word investment.

days compelled to capitulate. The English distinguished themselves after their own peculiar fashion. In the assault on the principal defence General Cutts, who from his love of a hot fire was known as the Salamander, gave orders that the attacking force, if it carried the covered way, should not stop there but rush forward and carry as much more as it could. It was a mad design, criminally so in the opinion of officers who took part in it,[1] but it was madly executed, with the result that the whole fort was captured out of hand. 1702.

The reduction of Stevenswaert, Maseyk, and Ruremond quickly followed; and the French now became alarmed lest Marlborough should transfer operations to the Rhine. Tallard was therefore sent back with a large force to Cologne and Bonn, while Boufflers, much weakened by this and by other detachments, lay helpless at Tongres. But the season was now far advanced, and Marlborough had no intention of leaving Boufflers for the winter in a position from which he might at any moment move out and bombard Maestricht. So no sooner were his troops released by the capture of Ruremond than he prepared to oust him. The French, according to their usual practice, had barred the eastern entrance to Brabant by fortified lines, which followed the line of the Geete to its head-waters, and were thence carried across to that of the Mehaigne. In his position at Tongres Boufflers lay midway between these lines and Liège, in the hope of covering both; but after the fall of so many fortresses on the Meuse he became specially anxious for Liège, and resolved to post himself under its walls. He accordingly examined the defences, selected his camping-ground, and on the 12th of October marched up with his army to occupy it. Quite unconscious of any danger he arrived within cannon-shot of his chosen position, and there stood Marlborough, calmly awaiting him with a superior force. For the fourth time Marlborough held his enemy within his grasp, but the Dutch deputies, as usual, Sept. 26 / October 7. Oct. $\frac{1}{12}$.

[1] See the description in Kane.

1702. interposed to forbid an attack; and Boufflers, a fourth time delivered, hurried away in the night to his lines at Landen. Had he thrown himself into Liège Marlborough would have made him equally uncomfortable by marching on the lines; as things were the French marshal perforce left the city to its fate.

The town of Liège, which was unfortified, at once opened its gates to the Allies; and within a week Marlborough's batteries were playing on the citadel. Oct. $\frac{12}{23}$. On the 23rd of October the citadel was stormed, the English being first in the breach, and a few days later Liège, with the whole line of the Meuse, had passed into the hands of the Allies. Thus brilliantly, in spite of four great opportunities marred by the Dutch, ended Marlborough's first campaign. Athlone, like an honest man, confessed that as second in command he had opposed every one of Marlborough's projects, and that the success was due entirely to his incomparable chief. He at any rate had an inkling that in Turenne's handsome Englishman there had arisen one of the great captains of all time.

Nevertheless the French had not been without their consolations in other quarters. Towards the end of the campaign the Elector of Bavaria had declared himself for France against the Empire, and, surprising the all-important position of Ulm on the Danube, had opened communication with the French force on the Upper Rhine. Villars, who commanded in that quarter, had seconded him by defeating his opponent, Prince Lewis of Baden, at Friedlingen, and had cleared the passages of the Black Forest; while Tallard had, almost without an effort, possessed himself of Treves and Trarbach on the Moselle. The rival competitors for the crown of Spain were France and the Empire, and the centre of the struggle, as no one saw more clearly than Marlborough, was for the present moving steadily towards the territory of the Empire.

While Marlborough was engaged in his operations on the Meuse, ten thousand English and Dutch, under the Duke of Ormonde and Admiral Sir George Rooke, had been despatched to make a descent upon Cadiz. The expedition was so complete a failure that there is no object in dwelling on it. Rooke would not support Ormonde, and Ormonde was not strong enough to master Rooke; landsmen quarrelled with seamen, and English with Dutch. No discipline was maintained, and after some weeks of feeble operations and shameful scenes of indiscipline and pillage, the commanders found that they could do no more than return to England. They were fortunate enough, however, on their way, to fall in with the plate-fleet at Vigo, of which they captured twenty-five galleons containing treasure worth a million sterling. Comforted by this good fortune Rooke and Ormonde sailed homeward, and dropped anchor safely in Portsmouth harbour. 1702.

Meanwhile a mishap, which Marlborough called an accident, had gone near to neutralise all the success of the past campaign. At the close of operations the Earl, together with the Dutch deputies, had taken ship down the Meuse, with a guard of twenty-five men on board and an escort of fifty horse on the bank. In the night the horse lost their way, and the boat was surprised and overpowered by a French partisan with a following of marauders. The Dutch deputies produced French passes, but Marlborough had none and was therefore a prisoner. Fortunately his servant slipped into his hand an old pass that had been made out for his brother Charles Churchill. With perfect serenity Marlborough presented it as genuine, and was allowed to go on his way, the French contenting themselves with the capture of the guard and the plunder of the vessel, and never dreaming of the prize that they had let slip. The news of his escape reached the Hague, where on his arrival rich and poor came out to welcome him, men and women weeping for joy over his safety. So deep was the fascination exerted on all of his kind by this extraordinary man.

A few days later he returned to England, where a

1702. new Parliament had already congratulated Queen Anne on the retrieving of England's honour by the success of his arms. The word retrieving was warmly resented, but though doubtless suggested by unworthy and factious animosity against the memory of William, it was strictly true. The nation felt that it was not in the fitness of things that Englishmen should be beaten by Frenchmen, and they rejoiced to see the wrong set right. Nevertheless party spirit found a still meaner level when Parliament extended to Rooke and Ormonde the same vote of thanks that they tendered to Marlborough. This precious pair owed even this honour to the wisdom and good sense of their far greater comrade, for they would have carried their quarrel over the expedition within the walls of Parliament, had not Marlborough told them gently that the whole of their operations were indefensible and that the less they called attention to themselves the better. The Queen, with more discernment, created Marlborough a Duke and settled on him a pension of £5000 a year. With the exaggerated bounty of a woman she wished Parliament to attach that sum forthwith permanently to the title, but this the Commons most properly refused to do. Moreover, the House was engaged just then on a work of greater utility to the Army than the granting of pensions even to such a man as Marlborough.

Nov. 11. On the 11th of November, the day before the public thanksgiving for the first campaign, the Committee of Public Accounts presented its report on the books of Lord Ranelagh, the paymaster-general. Ranelagh, according to their statement, had evinced great unwillingness to produce his accounts, and had met their inquiries with endless shuffling and evasion. In his office, too, an unusual epidemic of sudden illness, and an unprecedented multitude of pressing engagements, had rendered his clerks strangely inaccessible to examination. The commissioners, however, had persisted, and were now able to tell a long story of irregular book-keeping, false accounts, forged vouchers, and the clumsiest 1702. and most transparent methods of embezzlement and fraud.

Ranelagh defended himself against their charges not without spirit and efficiency, but the commissioners declined to discuss the matter with him. The Commons spent two days in examination of proofs, and then without hesitation voted that the Paymaster-General had been guilty of misappropriation of public money. It was thought by many at the time that Ranelagh was very hardly used; and it is certain that factious desire to discredit the late Government played a larger part than common honesty in this sudden zeal against corruption. Whig writers[1] assert without hesitation that there was no foundation whatever for the charges; and it is indubitable that many of the conclusions of the commissioners were strained and exaggerated. It is beyond question too that much of the financial confusion was due to the House of Commons, which had voted large sums without naming the sources from whence they should be raised, and where it had named the source had absurdly over-estimated the receipts. But it is none the less certain that Ranelagh's accounts were in disorder, and that, though his patrimony was small, he was reputed to have spent more money on buildings, gardens, and furniture than any man in England. Without attempting to calculate the measure of his guilt, it cannot be denied that his dismissal was for the good of the Army.

Had the House of Commons followed up this preliminary inquiry by further investigation much good might have been done, but its motives not being pure its actions could not be consistent. Ranelagh, for instance, had made one statement in self-defence which gravely inculpated the Secretary-at-War; but the House showed no alacrity to turn against that functionary. Very soon the question of the accounts degenerated into a wrangle with the House of Lords; and in March

[1] Burnet, Somerville, Tindall.

1702. 1704 the Commons were still debating what should be done with Ranelagh, while poor Mitchelburne of Londonderry, a prisoner in the Fleet for debt, was petitioning piteously for the arrears due to him since 1689.

It will, however, be convenient to anticipate matters a little, and to speak at once of the reforms that were brought about by this scandal in the paymaster's office. First, on the expulsion of Ranelagh the office was divided and two paymasters-general were appointed,
1705, May 10. one for the troops abroad, the other for those at home. Secondly, two new officers were established, with salaries of £1500 a year and the title of Controllers of the Accounts of the Army, Sir Joseph Tredenham and William Duncombe being the first holders of the office. Lastly, the Secretary-at-War definitely ceased to be mere secretary to the Commander-in-Chief, and became the civil head of the War Department. In William's time he had taken the field with the King, but from henceforth he stayed at home; while a secretary to the Commander-in-Chief, not yet a military secretary, accompanied the general on active service on a stipend of ten shillings a day. William Blathwayt, who had been Secretary-at-War since the days of Charles the
Commission dated April 20, 1704. Second, was got rid of, with no disadvantage to the service, and his place was taken by the brilliant but unprofitable Henry St. John.

CHAPTER II

THE force voted by Parliament for the campaign of 1703 consisted, as in the previous year, of eighteen thousand British and twenty-two thousand Germans. There had been much talk of an increase of the Army, and indeed Parliament had agreed to make an augmentation subject to certain conditions to be yielded by the Dutch ; but when the session closed no provision had been made for it, and the details required to be settled, as indeed such details generally were, by Marlborough himself. Four new British regiments formed part of the augmentation, and accordingly five new battalions were raised, which, as they were all disbanded subsequently, remain known to us only by the names of their colonels, Gorges, Pearce, Evans, Elliott, and Macartney. Finally, small contingents from a host of petty German states brought the total of mercenaries to twenty-eight thousand, which, added to twenty thousand British, made up a nominal total of fifty thousand men in the pay of England. But none of these additional troops could take the field until late in the campaign. 1703.

Such efforts were not confined to the side of the Allies. The French successes to the eastward of the Rhine had encouraged them to projects for a grand campaign, so their army too was increased, and every nerve was strained to make the preparations as complete as possible. The grand army under Villeroy and Boufflers, numbering fifty-four battalions and one hundred and three squadrons, was designed to recapture the strong places on the Meuse and to threaten the

1703. Dutch frontier. The frontiers towards Ostend and Antwerp were guarded by flying columns under the Marquis of Bedmar, Count de la Mothe, and the Spanish Count Tserclaes de Tilly. The entire force of the Bourbons in the Low Countries, including garrisons and field-army, included ninety thousand men in infantry alone.[1] With such a force to occupy the Allies in Flanders and with Marshal Tallard to hold Prince Lewis of Baden in check at Stollhofen on the Upper Rhine, Marshal Villars was to push through the Black Forest and join hands with the Elector of Bavaria. Finally, the joint forces of France and Savoy were to advance through the Tyrol to the valley of the Inn and combine with Villars and the Elector for a march on Vienna.

The design was grand enough in conception; but Marlborough too had formed plans for striking at the enemy in a vital part. A campaign of sieges was not to his mind, for he conceived that to bring his enemy to action and beat him was worth the capture of twenty petty fortresses; and accordingly on his arrival at the Hague he advocated immediate invasion of French Flanders and Brabant. But the project was too bold for the Dutch, whose commanders had changed and changed for the worse. Old Athlone was dead, and in his stead had risen up three new generals—Overkirk, who had few faults except mediocrity and age; Slangenberg, who combined ability with a villainous temper; and Opdam, who was alike cantankerous and incapable. Very reluctantly Marlborough was compelled to undertake the siege of Bonn, he himself commanding the besiegers, while Overkirk handled the covering army. Notwithstanding Dutch procrastination, Marlborough's energy had succeeded in bringing the Allies first into the field; and before Villeroy could strike a blow to hinder it, Bonn had capitulated, and Marlborough had rejoined Overkirk and was ready for active operations in the field.

March 6/17.

May 7/18.

[1] 180 battalions. At this period a battalion is generally taken at 500, and a squadron at 120 men.

The Duke now reverted to his original scheme of carrying the war into the heart of Brabant and West Flanders, and with this view ordered every preparation to be made for an attack on Antwerp. Cohorn, the famous engineer, was to distract the French by the capture of Ostend on the west side, a second force was to be concentrated under Opdam at Bergen-op-Zoom to the north, while Marlborough was to hold Villeroy in check in the east until all was ready. 1703.

The Duke's own share of the operations was conducted with his usual skill. Pressing back Villeroy into the space between the heads of the Jaar and the Mehaigne he kept him in continual suspense as to whether his design lay eastward or westward, against Huy or against Antwerp. Unfortunately, in an evil hour he imparted to Cohorn that he thought he might manage both.[1] The covetous old engineer had laid his own plans for filling his pockets; and no sooner did he hear of Marlborough's idea of attacking Huy than, fearful lest Villeroy should interrupt his private schemes for making money, he threw the capture of Ostend to the winds, and marched into West Flanders to levy contributions before it should be too late.

Still Marlborough was patient. He had hoped for Ostend first and Antwerp afterwards, but a reversal of the arrangement would serve. Cohorn having filled his pockets returned to the east of the Scheldt at Stabrock; Spaar, another Dutch general, took up his position at Hulst; Opdam remained at Bergen-op-Zoom; and thus the three armies lay in wait round the north and west of Antwerp, ready to move forward as soon as Marlborough should come up on the south-east. The Duke did not keep them long waiting. On the night of the 26th of June he suddenly broke up his camp, June 15/26. crossed the Jaar, and made for the bridge over the Demer at Hasselt. Villeroy, his eyes now thoroughly opened, hastened with all speed for Diest in order to be before him; and the two armies raced for Antwerp.

[1] Marlborough's *Despatches*, vol. i. p. 105.

1703. The Duke had hastened his army forward on its way by great exertions for six days, when the news reached him that Cohorn, unable to resist the temptation of making a little more money, had made a second raid into West Flanders, leaving Opdam in the air on the other side of the Scheldt. The Dutch were jubilant over Cohorn's supposed success, but Marlborough took a very different view. "If Opdam be not on his guard," he said, "he will be beaten before we can reach him"; and he despatched messengers instantly to give Opdam warning. As usual he was perfectly right. Villeroy hit the blot at once, and detached a force under Boufflers to take advantage of it. Opdam, in spite of Marlborough's warning, took no precautions, and finding himself surprised took to his heels, leaving Slangenberg to save his army. Thus the whole of Marlborough's combinations were broken up.[1]

The quarrels of the Dutch generals among them-

[1]

Order of Battle. Campaign of 1703.

Right Wing only.

Left. Right.

		Hamilton's Brigade.	Withers's Brigade.	Wood's Brigade.	Ross's Brigade.
1st Line.	Foreign Regiments.	8th Foot. 17th " 33rd " 20th " 13th "	1 Batt. 1st Guards. 1 " Royal Scots. 15th Foot. 24th " 23rd Royal Welsh. 9th Foot.	1st Dragoon Guards. 5th " " 7th " " 6th " " 3rd " "	1st Royal Dragoons. 5th Dragoons. Scots Greys. A Foreign Regiment.
2nd Line.	Foreign Regiments.		2nd Batt. Royal Scots. 16th Foot. 26th Cameronians. 21st Royal Scots Fusiliers. 10th Foot.	Foreign Cavalry.	

Daily Courant, June 2, 1703.

selves left no hope of success in further operations. 1703. Failing to persuade the Dutch to undertake anything but petty sieges he returned to the Meuse, and after the capture of Huy and Limburg closed the campaign. Thus a second year was wasted through the perversity of the Dutch.

Meanwhile things had gone ill with the Grand Alliance in other quarters. The King of Portugal had indeed been gained for the Austrian side and had offered troops for active operations in Spain, an event which will presently lead us to the Peninsula. The Duke of Savoy again had been detached from the French party, and the intended march over the Tyrol had been defeated by the valour of the Tyrolese; but elsewhere the French arms had been triumphant. Early in March Villars had seized the fort and bridge of Kehl on the Rhine, had traversed the Black Forest, joined hands with the Elector of Bavaria, and in spite of bitter quarrels with him had won in his company the victory of Hochstädt. Sept. $\frac{9}{20}$. Tallard too, though he took the field but late, had captured Old Brisach on the Upper Rhine, defeated the Prince of Hessen-Cassel at Spires, and recaptured Landau. The communications between the Rhine and the Danube were thus secured, and the march upon Vienna could be counted on for the next year. With her armies defeated in her front, and the Hungarian revolt eating at her vitals from within, the situation of the Empire was well-nigh desperate.

Marlborough, for his part, had made up his mind to resign the command, for he saw no prospect of success while his subordinates systematically disobeyed his orders. "Our want of success," he wrote, "is due to the want of discipline in the army, and until this is remedied I see no prospect of improvement."[1] Nevertheless a short stay in England seems to have restored him to a more contented frame of mind, while even before the close of the campaign he had begun to plan

[1] *Despatches*, vol. i. p. 198.

1703. a great stroke for the ensuing year, and to discuss it with the one able general in the Imperial service, Prince Eugene of Savoy. Frail and delicate in constitution, Eugene had originally been destined for the Church, and for a short time had been known as the Abbé of Savoy, but he had early shown a preference for the military profession and had offered his sword first to Lewis the Fourteenth. It was refused. Then Eugene turned to the Imperial Court, and after ten years of active service against Hungarians, Turks, and French, found himself at the age of thirty a field-marshal. At thirty-four he had won the great victory
[1697.] of Zenta against the Turks, and in the War of the Succession had made himself dreaded in Italy by the best of the French marshals. He was now forty years of age, having spent fully half of his life in war, and fully a quarter of it in high command. Marlborough was fifty-three, and until two years before had never commanded an army in chief.

Marlborough's design was nothing less than to commit the Low Countries to the protection of the Dutch, and, leaving the old seat of war with all its armies and fortresses in rear, to carry the campaign into the heart of Germany. The two great captains decided that it could and must be done; but it would be no easy task to persuade the timid States-General and a factious House of Commons to a plan which was bold almost to rashness.

Marlborough began his share of the work in England forthwith. Without dropping a hint of his great scheme he contrived to put some heart into the English ministers, and so into their supporters in Parliament. The Houses met on the 9th of November, and the Commons, after just criticism of the want of concert shown by the Allies, cheerfully voted money and men for the augmented force that had been proposed in the previous session. Then came a new difficulty which had been added to Marlborough's many troubles in the autumn. The treaty lately concluded with Portugal required the despatch of seven thousand troops to the Peninsula; and these it was decided to draw from the best British regiments in the Low Countries.[1] It was therefore necessary to raise one new regiment of dragoons and seven new battalions of foot,[2] a task which was no light one from the increasing difficulty of obtaining recruits. 1703.

But while the recruiting officers were busily beating their drums, and convicted felons were awaiting the decision which should send them either in a cart to Tyburn or in a transport to the Low Countries, the indefatigable Marlborough crossed the North Sea in the bitterest weather to see how the Dutch preparations were going forward. He found them in a state which caused him sad misgivings for the coming campaign, but he managed to stir up the authorities to increase supplies of men and money, and suggested operations on the Moselle for the next campaign. The same phrase, operations on the Moselle, was passed on to the King of Prussia and to other allies, and was repeated to the Queen and ministers on his return to England. Finally, early in April the Duke embarked for the Low Countries once more in company with his brother Charles, with general instructions in his pocket to concert measures with Holland for the relief of the Emperor. 1704. January 15.

Three weeks were then spent in gaining the consent of the States-General to operations on the Moselle, a consent which the Duke only extorted by threatening to march thither with the British troops alone, and in consultation with the solid but slow commander of the Imperial forces, Prince Lewis of Baden. To be quit of Dutch obstruction Marlborough asked only for the auxiliary troops in the pay of the Dutch, and obtained for his brother Charles the rank of General with the

[1] Royal Dragoons; 2nd, 9th, 11th, 13th, 17th, 33rd Foot.

[2] Erle's Dragoons. Rooke's, Paston's, Deloraine's, Inchiquin's, Ikerryn's, Dungannon's, and Orrery's Foot. All the foot, except the two first, were raised in Ireland.

1704. command of the British infantry. In the last week of April 24/May 5. April the British regiments began to stream out of their winter quarters to a bridge that had been thrown over May 7/18. the Meuse at Ruremonde, and a fortnight later sixteen thousand of them made rendezvous at Bedbourgh. Not a man of them knew whither he was bound, for it was only within the last fortnight that the Duke had so much as hinted his destination even to the Emperor or to Prince Lewis of Baden.

It is now time to glance at the enemy, who had entered on the campaign with the highest hopes of success. The dispositions of the French were little altered from those of the previous year. Villeroy with one army lay within the lines of the Mehaigne; Tallard with another army was in the vicinity of Strasburg, his passage of the Rhine secured by the possession of Landau and Old Brisach; and the Count of Coignies was stationed with ten thousand men on the Moselle, ready to act in Flanders or in Germany as occasion might demand. At Ulm lay the Elector of Bavaria and his French allies under Marsin, who had replaced Villars during the winter. The whole of this last force, forty-five thousand men in all, stood ready to march to the head-waters of the Danube, and there unite with the French that should be pushed through the Black Forest to meet it. The Elector, by the operations of the past campaign, had mastered the line of the Danube from its source to Linz within the Austrian frontier; he held also the keys of the country between the Iller and the Inn; and he asked only for a French reinforcement to enable him to march straight on Vienna.

To the passage of this reinforcement there was no obstacle but a weak Imperial force under Prince Lewis of Baden, which made shift to guard the country from Philipsburg southward to Lake Constance. The principal obstruction was certain fortified lines, of which the reader should take note, on the right bank of the Rhine, which ran from Stollhofen south-eastward to Bühl, and, since they covered the entrance into

Baden from the north-west, were naturally most 1704.
jealously guarded by Prince Lewis. From that point southward the most important points were held by weak detachments of regular troops, but a vast extent of the most difficult country was entrusted to raw militia and peasantry. To escort a reinforcement successfully through the defiles from Fribourg to Donaueschingen and to return with the escort in safety was no easy task, but it was adroitly accomplished by Tallard within the space of twelve days. The feat was lauded at the time with ridiculous extravagance, for, apart from the fact that Prince Lewis of Baden was remarkable neither for swiftness nor for vigilance, Tallard had hustled his unhappy recruits forward so unmercifully, along bad roads and in bad weather, that the greater part of them perished by the way.[1] Nevertheless the French had scored the first point of the game and were proportionately elated, while poor Tallard's head was, to his great misfortune, completely turned.

Marlborough meanwhile had begun his famous May $\frac{8}{19}$.
march, the direction lying up the Rhine towards Bonn. On the very day after he started he received urgent messages from Overkirk that Villeroy had crossed the Meuse and was menacing Huy, and from Prince Lewis that Tallard was threatening the lines of Stollhofen, both commanders of course entreating him to return to their assistance. Halting for one day to reassure them, the Duke told Overkirk that Villeroy had no designs against any but himself, and that the sooner reinforcements were sent to join the British the better. Prince Lewis he answered by giving him a rendezvous where his Hessians and Danes might also unite with his own army. This done he continued his march.

Marlborough's information was good. Villeroy had received strict orders to follow him to the Moselle, the French Court being convinced that he meditated opera-

[1] Quincy, vol. iv. p. 245. It is said that of seventeen battalions only 1500 men reached the Elector of Bavaria at Donaueschingen.

1704. tions in that quarter. The Duke stepped out of his
May 12/23. way to inspect Bonn in order to encourage this belief,
and then pushed on in all haste to Coblentz with his
cavalry only, leaving his brother to follow him with the
infantry, while the artillery and baggage was carried up
the Rhine to Mainz. Once again all his movements
seemed to point to operations on the Moselle, unless
indeed (for the French never knew what such a man
might do next) he designed to double back down the
river for operations near the sea. But wherever he
might be going he did not linger, but crossing the Rhine
and Moselle pushed constantly forward with his cavalry.
Starting always before dawn and bringing his men into
camp by noon he granted them no halt until he reached
May 18/29. the suburbs of Mainz at Cassel. Here he improved
his time by requesting the Landgrave of Hesse to send
the artillery, which he had prepared for a campaign on
the Moselle, to Mannheim. Again the French were
puzzled. Was Alsace, and not the Moselle, to be the
scene of the next campaign; and if not, why was the
English general bridging the Rhine at Philipsburg, and
May 21/June 1. why was his artillery moving up the river? Tallard
moved up to Kehl, crossed to the left bank of the
Rhine and took up a position on the Lauter, and
Villeroy sent to Flanders for reinforcements; but mean-
May 20/31. while Marlborough had crossed the Main, and still,
struggling on by rapid and distressing marches over
May 23/June 3. execrable roads, was within three more days across the
Neckar at Ladenburg and out of their reach.

His plans were now manifest enough, but it was too
late to catch him. He therefore halted two days by
Ladenburg to give orders for the concentration of the
troops that were on march to join him from the Rhine,
May 26/June 6. and then striking south-eastward across the great bend
of the Neckar, traversed the river for the second time
May 30/June 10. at Lauffen, and by the 10th of June was at Mondelheim.
Halting here for three days to allow his infantry to
come nearer to him, he was joined by Prince Eugene
whom he now met for the first time in the flesh. The

Prince inspected the English horse and was astonished 1704.
at the condition of the troops after their long and trying march. "I have heard much," he said, "of the English cavalry, and find it to be the best appointed and finest that I have ever seen. The spirit which I see in the looks of your men is an earnest of victory." Hither June $\frac{2}{13}$.
three days later came also a less welcome guest, Prince Lewis of Baden; and the three commanders discussed their plans for the future. Marlborough in vain tried to keep Eugene for his colleague, but it was ultimately decided that Eugene should take command in the lines of Stollhofen, to prevent the French if possible from crossing the Rhine, and to follow them at all hazards if they should succeed in crossing, while Baden should remain on the Danube and share the command of the allied army by alternate days with Marlborough.

Then the march was resumed south-eastward upon June $\frac{3}{14}$.
Ulm; and after one day's halt to perfect the arrangements for the junction with Prince Louis, the army reached the mountain-chain that bounded the valley of the Danube. The Pass of Geislingen, through which its road lay, could not in the most favourable circumstances be passed by any considerable number of troops in less than a day, and was now rendered almost impracticable by incessant heavy rain. To add to Marlborough's troubles the States-General, learning that Villeroy was astir, became frightened for their own safety and entreated for the return of their auxiliary troops. The Duke, to calm them, ordered boats to be ready to convey forces down the Rhine, and went quietly on with his own preparations, establishing magazines to the north of the Danube, and not forgetting to send a reinforcement of foreign troops to Eugene. At last the news came that Baden's army June $\frac{9}{20}$.
was come within reach; the British cavalry plunged into the defile, and two days later the junction of the June $\frac{11}{22}$.
two forces was effected at Ursprung.

The joint armies presently advanced to within eight June $\frac{14}{25}$.
miles of Ulm, whereupon the Elector of Bavaria with-

1704. drew to an entrenched camp further down the Danube between Lavingen and Dillingen. The Allies therefore turned northward to await the arrival of the British infantry at Gingen; for Charles Churchill, with the foot and the artillery, had found it difficult to march at great speed in the perpetual pouring rain. His troubles had begun from the moment when Marlborough had gone ahead with the cavalry from Coblentz. The ascent of a single hill in that mountainous country often cost the artillery[1] a whole day's work, and would have cost more but for the indefatigable exertions of the officers.[2] Marlborough's care for the comfort and discipline of these troops was incessant. A large supply of shoes, for instance, was ready at Heidelberg to make good defects, while constant injunctions in his letters to his brother testify to his anxiety that nothing should be omitted to lighten the burden of the march. Finally, anticipating Wellington in the Peninsula, he insisted that the men should pay honestly for everything that they took, and took care to provide money to enable them to do so. Such a thing had never been known in all the innumerable campaigns of Germany.

The joint armies after the arrival of Churchill amounted to ninety-six battalions, two hundred and two squadrons, and forty-eight guns; but a large contingent of Danish cavalry was still wanting, and not all Marlborough's entreaties could prevail with its commander, the Duke of Würtemberg, to hasten his march. Nevertheless it was necessary to move at once. Marlborough's objective had from the first been Donauwörth, which would give him at once a bridge over the Danube and a place of arms for the invasion of Bavaria. His move northward had revealed his intentions; and the Elector of Bavaria had detached Count d'Arco with ten thousand foot and twenty-five hundred

[1] Thirty-four English field-pieces and four howitzers took part in the famous march to the Danube. There were 2500 horses in all in the train.—*Postman*, 18th May.

[2] Hare's Journal.

horse to occupy the Schellenberg, a commanding height which covers Donauwörth on the north bank of the Danube. Marlborough pressed Baden hard to attack this detachment before it could be reinforced; and accordingly the army broke up from Gingen, and advancing parallel to the Danube encamped on the 1st of July at Amerdingen.

1704. June 18/29. June 20/July 1.

The next day was Marlborough's turn for command. It had not yet dawned when Quartermaster-General Cadogan was up and away with a party of cavalry, pioneers, and pontoons. At three o'clock marched six thousand men from the forty-five battalions of the left wing,[1] three regiments of Imperial Grenadiers, and thirty-five squadrons of horse. At five o'clock the rest of the army, excepting the artillery, followed in two columns along the main road towards a height that overhangs the river Wörnitz between Obermorgen and Wörnitzstein. By eight o'clock Cadogan was at Obermorgen, had driven back the enemy's picquets, and was engaged in marking out a camp; and at nine appeared the Duke himself, who, taking Cadogan's escort, went forward to reconnoitre the position.

June 21/July 2.

The Schellenberg, as its name implies, is a bell-shaped hill, some two miles in circumference at the base and with a flat top about half a mile wide, whereon was pitched the enemy's camp. On the south side, where the hill falls down to the Danube, the ascent is steeper than elsewhere; on the north-west the slope is gradual and about five hundred yards in length. To the south-west the hill joins the town of Donauwörth, from the outworks of which an entrenchment had been carried for nearly two miles round the summit to the river. This defence was strongest and most complete to the north-east, where a wood gave shelter for the formation of an attacking force; and at this point was

1 The British cavalry (seven regiments) formed the extreme left of the left wing in the line of battle, with ten British battalions immediately to their right. Four more British battalions formed the extreme left of the infantry of the second line. See p. 445.

1704. June 21/July 2. stationed a battery of cannon. To the north-west the works though incomplete were well advanced, and were strengthened by an old fort wherein the enemy had mounted guns. Marlborough, as he conned the position, could see that the enemy before him was so disposed as if expecting an attack on the northern and western sides. But looking to his right beyond Donauwörth, and across the Danube, he could see preparations of a more ominous kind, a camp with tents pitched on both wings and a blank space in the centre, sure sign that cavalry was already present and that infantry was expected. Closer and closer he drew to the hill, Prince Lewis and others presently joining him; and then puffs of white smoke began to shoot out from various points in the enemy's works as his batteries opened fire.

Finishing his survey undisturbed, Marlborough turned back to meet the advanced detachment of the army; for it was plain to him that the Schellenberg must be carried at once before more of the enemy's troops could reach it. So bad, however, was the state of the roads, that though the distance was but twelve miles, the detachment did not reach the Wörnitz until noon. It was then halted to give the men rest, for there were still three miles of bad road before them, and to allow the main body to come up. The cavalry was sent forward to cut fascines in the wood, pontoon bridges were thrown across the Wörnitz, and at three o'clock the advanced detachment passed the river. While this was going forward a letter arrived from Eugene that Villeroy and Tallard were preparing to send strong reinforcements to the Elector; and this intelligence decided Marlborough to take the work in hand forthwith. Without waiting for the rear of the main body to arrive he drew out sixteen battalions only, five of them British,[1] and led them and the advanced detachment straight on to the attack. The infantry

[1] These would appear to have been the 1st Guards, 1st Royals (2 batts.), 23rd, and perhaps the 37th.

of the detachment was formed in four lines, the English[1] being on the extreme left by the edge of the wood, and the cavalry was drawn up in two lines behind them. Eight battalions more were detailed to support the detachment or to deploy to its right if need should be, and yet eight more were held in reserve. 1704. June 21/July 2.

It was six o'clock in the evening before Marlborough gave the order to attack. Every foot-soldier took a fascine from the cavalry, and the columns, headed by two parties of grenadiers from the First Guards under Lord Mordaunt and Colonel Munden, marched steadily up the hill. The hostile batteries at once opened a cross-fire of round shot from the intrenchment and from the walls of Donauwörth, but the columns pressed on unheeding to within eighty yards of the intrenchment before they fired a shot. Then the enemy continued the fire with musketry and grape, and the slaughter became frightful. The grenadiers of the Guards fell down right and left, and very soon few of them were left. Still Mordaunt and Munden, the one with his skirts torn to shreds and the other with his hat riddled by bullets, stood up unhurt and kept cheering them on. General Goor, a gallant foreigner who commanded the attack, was shot dead, and many other officers fell with him under that terrible fire. The columns staggered, wavered, recovered, and went on. But now came an unlucky accident. In front of the intrenchment ran a hollow way worn in the hill by rain, into which the foremost men, mistaking it for the intrenchment, threw down their fascines, so that on reaching the actual lines they found themselves unable to cross them. Thus checked they suffered so heavily that they began to give way; and the enemy rushed out rejoicing to finish the defeat with the bayonet. But the English Guards, though they had suffered terribly, stood immovable as rocks, the Royal Scots and the Welshmen of the Twenty-

[1] Their strength would be 1820 men; 130 men from each of fourteen battalions.

1704. June 21 / July 2. third stood by them, and the counter-attack after desperate fighting was beaten back.

Meanwhile the enemy, finding the western face of the hill unthreatened, withdrew the whole of their force from thence to the point of assault. Their fire increased; the attacking columns wavered once more, and General Lumley was obliged to move up the entire first line of cavalry into the thick of the fire to support them. So the fight swayed for another half-hour, when the remainder of the Imperial army at last appeared on Marlborough's right, and finding the intrenchments deserted passed over them at once with trifling loss. Repulsing a charge of cavalry which was launched against them, they hurried on and came full on the flank of the French and Bavarians; yet even so this gallant enemy would not give way, and the allied infantry still failed to carry the intrenchment. Lumley now ordered the Scots Greys to dismount and attack on foot; but before they could advance the infantry by a final effort at last forced their way in. Then the Greys remounted with all haste and galloped forward to the pursuit, while Marlborough, halting the exhausted foot, sent the rest of the cavalry to join the Greys. The rout was now complete. Hundreds of men were cut off before they could reach Donauwörth, many were driven into the Danube, many more, flying to a temporary bridge to cross the river, broke it down by their weight and miserably perished. Of twelve thousand men not more than one-fourth rejoined the Elector's army.

The whole affair had lasted little more than an hour and a half, but the loss of the Allies in overcoming so gallant a defence cost them no fewer than fourteen hundred killed and three thousand eight hundred wounded. The losses of the British[1] were very heavy,

[1] 29 officers, 407 men killed; 86 officers, 1031 men wounded. Several details, with a full list of the casualties, will be found in the *Postman* of July 13, 1704. It is from this source that I draw the account of Mordaunt and Munden.

amounting to fifteen hundred of all ranks, or probably 1704.
more than a third of the numbers engaged. The June 21/July 2.
First Guards, Royal Scots, and the Twenty-third suffered most severely, every battalion of them having lost two hundred men or more, while the Guards at the close of the day could count but five officers unhurt out of seventeen. Of these five, wonderful to say, were Mordaunt and Munden, the one with three bullets through his clothes, the other with five through his hat, but neither of them scratched; but of eighty-two men whom they led to the assault only twenty-one returned. When it is remembered that the main body had been on foot fourteen hours, and the advanced detachment for sixteen hours, the exhaustion of the troops at the end of the day may be imagined. Nevertheless Donauwörth was taken and the enemy was not only beaten but demoralised.

The Elector of Bavaria on hearing the news broke down the bridge over the Lech, and entrenched himself at Augsburg. Marlborough on his part crossed the Danube, and set himself to cut off the Elector's supplies. The passage of the Danube he severed at Donauwörth, the road to the north by the capture of Rain, and that to the north-east by an advance south-eastward to Aichach, from which he presently moved on
to Friedberg, hemming his enemy tightly into his July 11/22.
entrenched camp. The Elector was at first inclined to come to terms, but hearing that the French were about to reinforce him he thought himself bound in honour to hold out. Marlborough was therefore compelled to put pressure on him by ravaging the country, a work which his letters show that he detested but felt obliged in duty to perform. The destruction was carried to the very walls of Munich; indeed, nothing but want of artillery, for which Prince Lewis of Baden was responsible, prevented an attack upon the city itself.[1] The prospect of the arrival of a French army gave the Duke little disquiet: if Bavaria were to

[1] *Despatches*, vol. i. p. 381.

1704. become the seat of war, so much the worse for Bavaria and for the cause of the Bourbons. So after sending thirty squadrons to reinforce Eugene, he prepared in the interim for the siege of Ingolstadt, which would give him command of the Danube from Ulm to Passau, and free access at all times into Bavaria. The Elector's country should feel the stress of war at any rate, and if fortune were propitious the French might feel it also. It is now time to return to the movements of those French.

CHAPTER III

We left Villeroy with his army in the Low Countries 1704.
endeavouring not very successfully to obey the orders which he had received, to watch Marlborough. On May 29/June 9. the 29th of May, when the Duke had already crossed the Neckar and fixed his quarters at Mondelheim, Villeroy was still at Landau waiting for him to repass the Rhine. On the following day, however, he took counsel with Tallard, with the result that, while Marl- June 21/July 2. borough was marching to the attack of the Schellenberg the French armies were streaming across the Rhine at Kehl. Tallard then moved south towards Fribourg, close to which he received intelligence of the Elector's defeat. Thereupon both he and Villeroy entered the defiles of the Black Forest, uniting at Horneberg, from which point Tallard pushed on eastward alone. Advancing to Villingen he wasted five precious days in an unsuccessful effort to take that town, a mistake which July 5-10/16-21. was not lost on Marlborough and Eugene. Called to his senses by an urgent message from the Elector, Tallard at last marched on by the south bank of the Danube, encamped before Augsburg on the 23rd of July 23/August 3. July, and three days later effected his junction with the Elector and Marsin a few miles to the north of the city.

Tallard was no sooner fairly on his way than Eugene, leaving a small garrison to hold the lines of Stollhofen, hurried on parallel with him along the north bank of the Danube, reaching Hochstädt on the day of the enemy's junction at Augsburg. Marlborough mean- July 26/August 6. while, at the news of Tallard's arrival, had fallen back

1704. northward in the direction of Neuburg on the Danube, and was lying at Schobenhausen some twelve miles to the south of the river. Hither came Eugene from Hochstädt to concert operations. The French and Bavarians were united to the south of the Danube; the Allies were divided on both sides of the river. If Marlborough fell back to Neuburg to join Eugene, the enemy could pass the Lech and enter Bavaria; if Eugene crossed the river to join Marlborough the enemy could pass to the north of the river and cut them off from Franconia, their only possible source of supplies. It was agreed that Prince Lewis of Baden should be detached with fifteen thousand men for the siege of Ingolstadt; and, as it was reported that the French were moving towards the Danube, Marlborough advanced closer to the river, so as to be able to cross it either at Neuburg or by the bridges which he had thrown over it by the mouth of the Lech at Merxheim.

July 29 / August 9. On the 9th of August Prince Lewis marched off to Ingolstadt, to the unspeakable relief of his colleagues, and Eugene took his leave. Two hours later, however, Eugene hurried back to report that the French were in full march to the bridge of Dillingen, evidently intending to cross the river and overwhelm his army. The Prince hastened back and withdrew his army eastward from Hochstädt to the Kessel. Marlborough, on his side, at midnight sent three thousand cavalry over the Danube to reinforce him, while twenty battalions under Churchill followed them as far as the bridge of Merxheim, with orders to halt on the south bank of the river. July 30 / August 10. Next morning the Duke brought the whole of the army up to Rain, within a league of the Danube, where he received fresh messages from Eugene urging him to hasten to his assistance. At midnight Churchill received his orders to pass the river and march for July 21 / August 11. the Kessel, and two hours later the whole army moved off in two columns, one to cross the Danube at Merxheim, the other to traverse the Lech at Rain and the Danube at Donauwörth. At five on the same after-

noon the whole of them were filing across the Wörnitz ; 1704.
by ten that night the junction was complete, and the united armies encamped on the Kessel, their right resting on Kessel-Ostheim, their left on the village of Munster and the Danube. Row's brigade of British was pushed forward to occupy Munster ; and then the wearied troops lay down to rest. The main body had been on foot for twenty hours, though it had covered no more than twenty-four miles. Both columns had passed the Danube and the Wörnitz, and the left column the Ach and the Lech in addition. It is easy to imagine how long and how trying such a march must have been ; it is less easy to appreciate the foresight and arrangement which enabled it to be performed at all.

The artillery, which had perforce been left to come August $\frac{1}{12}$.
up in the rear of the army, was by great exertions brought up at dawn on the following morning. A little later the Duke and Eugene rode forward with a strong escort to reconnoitre the ground before them, but perceiving the enemy's cavalry at a distance, ascended the church-tower of Tapfheim, from whence they descried the French quartermasters marking out a camp between Blenheim and Lutzingen, some three or four miles away. This was the very ground that they had designed to take up themselves, and it was with no small satisfaction that they perceived it to be occupied by the enemy. The French and Bavarian commanders had decided, after their junction on the Lech, that their best policy would be to cross the Danube, take up a strong position, and wait until want of supplies, by which Marlborough had already been greatly embarrassed, should compel the Allies to withdraw from the country. Tallard had no idea of offering battle; Marlborough indeed did not expect it of him, and had not dared to hope that the marshal would allow an action to be forced on him. But now that he had the chance, the Duke resolved not to let it slip. Men were not wanting to urge upon him the dangers of an attack

1704 on a superior force. "I know the difficulties," he answered, "but a battle is absolutely necessary, and I rely on the discipline of my troops."

The two camps lay some five miles apart, the ground between them consisting of a plain of varying breadth confined between a chain of woods and the Danube. This plain is cut by a succession of streams running down at right angles to the Danube, no fewer than three crossing the line of the march between the Kessel and the French position. The first of these, the Reichen, cuts a ravine through which the road passed close to the village of Dapfheim; and Marlborough, seeing that at this point the enemy could greatly embarrass his advance, sent forward pioneers to level the ravine, and occupied the village with two brigades of British and Hessian infantry.

Meanwhile the enemy entered their camp, Tallard taking up his quarters on the right, Marsin in the centre, and the Elector of Bavaria on the left. Tallard's force consisted of thirty-six battalions and forty-four squadrons of the best troops of France, his colleague's of forty-six battalions and one hundred and eight squadrons; yet notwithstanding this unequal distribution of the cavalry, the force was encamped not as one army but as two. The rule that infantry should be massed in the centre and the cavalry divided on each wing was followed, not for the entire host, but for each army independently. Thus the centre was made up of the cavalry of both armies without unity of command; the infantry was distributed on each flank of it; and on each flank of the infantry was yet another body of cavalry. Yet it was an axiom in those days that an army which ran the least risk of an engagement should be encamped as nearly as possible according to the probable disposition for action. This violation of rules was not unperceived by Marlborough.

The camp itself was situated at the top of an almost imperceptible slope, which descends for a mile, without affording the slightest cover, to a brook called the

Nebel. Its right rested on the village of Blenheim, little more than a furlong from the Danube; and here were Tallard's headquarters. The village having an extended front, and being covered by hedges and palisades, could easily be converted into a strong position. Half a mile above it a little boggy rivulet, called the Maulweyer, which was destined to play an important part in the next day's work, rises and flows down through the village to the Danube. About two miles up the Nebel from Blenheim, but on the opposite or left bank of the stream, stands the village of Unterglau; and a mile above this, on the same side of the stream as Blenheim, and about a hundred yards from the water, is another village called Oberglau. This Oberglau was the centre of the position, and Marsin's headquarters. A mile upward from Oberglau is another village, Lutzingen, resting on wooded country much broken by ravines. Here were the Elector's headquarters and the extreme left of the enemy's position. The Nebel, though no more than four yards broad at its mouth, was a troublesome obstacle, its borders being marshy, especially between Oberglau and Blenheim, and in many places impassable. Below Unterglau this swampy margin extended for a considerable breadth, while opposite Blenheim the stream parted in twain and flowed on each side of a small boggy islet. At the head of this islet was a stone bridge, over which ran the great road from Donauwörth to Dillingen. This had been broken down, or at least damaged, by Tallard; but herewith had ended his measures for obstructing the passage of the Nebel. 1704.

At two o'clock on the morrow morning, amid dense white mist, the army of the Allies broke up its camp, and passed the Kessel in eight columns, the two outermost on each flank consisting of cavalry, the four innermost of infantry. For this day the stereotyped formation was to be reversed; the cavalry was to form the centre and the infantry the wings. On reaching Tapfheim the army halted, and the two outlying August $\frac{2}{13}$.

1704. August 2/13. brigades, reinforced by eleven more battalions as well as by cavalry, formed a ninth column on the extreme left, to cover the march of the artillery along the great road and in due time to attack Blenheim. The new column was conspicuous from the red-coats of fourteen British battalions, with Cutts the Salamander at its head.

Then Marlborough, who commanded on the left, directed his generals to occupy the ground from the Danube to Oberglau, while Eugene's should prolong the line from Oberglau upwards to Lutzingen. The columns resumed the advance, spreading out like the sticks of a fan, wider and wider, as the Imperial troops streamed away to their appointed positions on the right. Fifty-two thousand men in all were tramping forward, and fifty-two guns groaning and creaking after them. Far in advance of all Marlborough and Eugene pushed on with a strong escort. At six o'clock they met and drove back the French advanced posts, and at seven they were on high ground within a mile of the Nebel and in full view of the enemy's camp.

Meanwhile Marshal Tallard was taking things at his ease, and had dispersed his cavalry to gather forage. Even while his vedettes were falling back before Marlborough's escort, he was calmly writing that the enemy had turned out early and was almost certainly on the march for Nördlingen. The morning was foggy, no uncommon thing on the banks of great and marshy rivers, and a dangerous enemy was within striking distance; yet no precautions had been taken against surprise. Then at seven o'clock the fog rolled away, and there, in great streaks of blue and white and scarlet, were the allied columns in full view, preparing to deploy on the other side of the Nebel. Presently the village of Unterglau and two mills farther down the stream burst into smoke and flame, and the outlying posts of the French came hurrying back across the stream. Then all was hurry and confusion in the French camp. Staff-officers flew off in all directions with orders, signal-guns brought the foragers galloping back, drums beat the assembly from end to end of the line, and the troops fell in hastily before their tents. 1704. August $\frac{2}{13}$.

Tallard's eyesight was very defective, but he had no difficulty in making out the red coats of Cutts's column, and he knew by this time that where the British were, there the heaviest fighting was to be expected. He therefore lost no time in occupying Blenheim. Four regiments of French dragoons trotted down to seal up the space between the village and the Danube, and presently almost the entire mass of the infantry faced to the right, and the white coats began striding away towards Blenheim itself. Eight squadrons of horse in scarlet, easily recognisable by Marlborough as the Gendarmerie, began Tallard's first line leftward from the village, and other squadrons presently prolonged it to Marsin's right wing. More cavalry supported these in a second line, together with nine battalions, which, being raw regiments, were not trusted to stand in the first line. Then the artillery came forward into position, ninety pieces in all, French and Bavarian. Four twenty-four pounders were posted before Blenheim, while a chain of batteries covered the line from end to end.

These dispositions completed, Tallard galloped off to the left, for Marsin had never yet commanded more than five hundred men in the field. Marsin's cavalry was already drawn up in two lines; his infantry and the Elector's was in rear of Oberglau and to the left of it, and the village itself was strongly occupied. Beyond this the left wing of cavalry stood in front of Lutzingen, and beyond them again a few battalions doubled back *en potence* protected the Elector's extreme left flank.

Marlborough on his side was equally busy. Blenheim and Oberglau were, as he saw, too far apart to cover the whole of the intervening ground with a cross-fire, and the French cavalry on the slope above were too remote to bar the passage of the Nebel. Officers were sent down to sound the stream, the stone bridge was repaired, and five pontoon bridges were laid, one

1704. August $\frac{2}{13}$. above Unterglau, the rest below it. Cutts formed his column into six lines, the first of Row's British brigade, the second of Hessians, the third of Ferguson's British brigade, and the fourth of Hanoverians, with two more lines in reserve. The four remaining columns of Marlborough's army were deployed between Wilheim and Oberglau in four lines, the first and fourth of infantry, with two lines of cavalry between them. The French esteemed this a "bizarre"[1] formation, but they understood its purport before the day was over.

At eight o'clock Tallard's batteries opened fire, though with little effect. Eugene thereupon took leave of Marlborough and hurried away to the right, while the Duke occupied himself with the posting of his artillery, every gun of which was stationed under his own eye. The chaplains came forward to the heads of the regiments and read prayers; and then the Duke mounted and rode down the whole length of his line. As he passed a round shot struck the ground under his horse and covered him with dust. For a moment every man held his breath, but in a few seconds the calm figure with the red coat and the broad blue ribbon reappeared, the horse moving slowly and quietly as before, and the handsome face unchangeably serene.

The inspection over, the Duke dismounted and waited till Eugene should be ready. The delay was long, and messenger after messenger was despatched to ask the cause. The answer came that the ground on the right was so much broken by wood and ravine that the columns had been compelled to make a long detour, and that formation had been hampered by the fire of the enemy's artillery as well as by the necessity for altering preconcerted dispositions. Marlborough waited with impatience, for, whether he hoped to carry Blenheim or not, every hour served to place it in a better state of defence. The French dragoons by the river had entrenched themselves behind a leaguer of waggons, and the infantry in the village had turned every wall

[1] Feuquières.

and hedge and house to good account. Moreover Marlborough had seen how strong the garrison of Blenheim was, having probably counted every one of the twenty-seven battalions into it, and identified them by their colours as the finest in the French army. 1704. August $\frac{2}{13}$.

At last, at half-past twelve, an aide-de-camp galloped up from Eugene to say that all was ready. Cutts was instantly ordered to attack Blenheim, while the Duke moved down towards the bridges over the Nebel. By one o'clock Cutts's two leading lines were crossing the stream by the ruins of the burnt mills under a heavy fire of grape. On reaching the other side they halted to reform under shelter of a slip of rising ground. There the Hessians remained in reserve; and the First Guards, Tenth, Twenty-first, Twenty-third, and Twenty-fourth, with Brigadier Row on foot at their head, advanced deliberately against Blenheim. They were received at thirty paces distance by a deadly fire from the French, but Row's orders were, that until he struck the palisades not a shot must be fired, and that the village must be carried with the steel. The British pressed resolutely on, Row struck his sword into the palisades, and the men pouring in their volley rushed forward, striving to drag down the pales by main strength in the vain endeavour to force an entrance. In a few minutes a third of the brigade had fallen, Row was mortally wounded, his lieutenant-colonel and major were killed in the attempt to bring him off, and the first line, shattered to pieces against a superior force in a very strong position, fell back in disorder. As they retired, three squadrons of the Gendarmerie swept down upon their flank and seized the colours of the Twenty-first, but pursuing their advantage too far were brought up by the Hessians, who repulsed them with great gallantry and recaptured the colours.

Cutts observing more of the Gendarmerie preparing to renew the attack asked for a reinforcement of cavalry to protect his flank, whereupon five English squadrons were ordered by General Lumley to cross the Nebel.

1704. August $\frac{2}{13}$. Floundering with the greatest difficulty through the swamp, these were immediately confronted by the Gendarmerie, who, however, with astonishing feebleness opened a fire of musketoons from the saddle. The English promptly charged them sword in hand and put them to flight, but pursuing as usual too far were galled by the flank fire from Blenheim and compelled to retire.

Cutts's two remaining lines now crossed the Nebel for a fresh attack on Blenheim. The enemy had by this time brought forward more artillery to sweep the fords with grape-shot, but the British made good their footing on the opposite bank and compelled the guns to retire. Then Ferguson's brigade advanced together with Row's against the village once more, carried the outskirts, but could penetrate no further in spite of several desperate attacks, and were finally obliged to fall back with very heavy loss. The subordinate generals would have thrown away more lives[1] had not Marlborough given orders that the regiments should take up a sheltered position and keep up a feigned attack by constant fire of platoons. Then, withdrawing the Hanoverian brigade to the infantry of the centre, the Duke turned the whole of his attention to that quarter.

During these futile attacks on Blenheim, the four lines of Marlborough's main army were struggling with much difficulty across the Nebel. The first line of infantry passed first, and drew up at intervals to cover the passage of the cavalry; while eleven battalions, under the Prince of Holstein-Beck, were detached to carry the village of Oberglau. Then the cavalry filed down to the stream, using fascines and every other means that they could devise to help them through the treacherous miry ground. The British cavalry had the hardest of the work, being on the extreme left, and therefore not only confronted with the worst of the ground, but exposed to the fire of the artillery at Blenheim. With immense difficulty the squadrons

[1] Kane.

extricated themselves and, with horses blown and heated, was forming up in front of the infantry, when the squadrons of the French right, fresh and favoured by the ground, came down full upon them. The first line of the British was borne back to the very edge of the stream, but the pursuit was checked by the fire of the infantry. Then the Prussian General Bothmar fell upon the disordered French with the second line of cavalry, and drove them in confusion behind the Maulweyer. Reinforced by additional squadrons he held the line of the rivulet and kept them penned in behind it, for the French could not cross it, and dared not pass round the head of it for fear of being charged in flank. It was not until two battalions had been sent from Blenheim to ply the allied squadrons with musketry that Bothmar retired, and some, but not all, of the French cavalry on this side was released. 1704. August $\frac{2}{13}$.

Meanwhile General Lumley had rallied his broken troops, and the squadrons further to the right had successfully crossed the Nebel. Still further up the water the Danish and Hanoverian cavalry had been put to the same trial as the British, being exposed to the fire from Oberglau and to the charges of Marsin's horse. While the combat was still swaying at this point the Prince of Holstein-Beck delivered his attack on Oberglau. He was instantly met by a fierce counter-attack from the Irish Brigade, which was stationed in the village. His two foremost battalions were cut to pieces, he himself was mortally wounded, and affairs would have gone ill had not Marlborough hastened up with fresh infantry and artillery, and forced the enemy back into Oberglau. Thus the passage for the central line of the allied cavalry was secured.

It was now three o'clock; and Marlborough sent an aide-de-camp to Eugene to ask how things fared with him. The Prince was holding his own and no more. His infantry had behaved admirably, but his horse had supported them but ill; and three consecutive attacks though brilliantly begun had ended in failure. The

1704. August $\frac{2}{13}$. fact was that the Elector, with better judgment than Tallard, had moved his troops down towards the water, and was straining every nerve to prevent his enemy from crossing. Meanwhile Marlborough, having at last brought the whole of his force across the Nebel, formed the cavalry in two grand lines for the final attack, the infantry being ranged at intervals to the left rear as rallying-points for any broken squadron. Tallard, on his side, brought forward the nine battalions of his centre from the second line to the first, a disposition which was met by Marlborough by the advance of three Hanoverian battalions and a battery of artillery. For a time these young French infantry stood firm against the rain of great and small shot, closing up their ranks as fast as they were broken; but the trial was too severe for them. Tallard strove hard to relieve them by a charge of the squadrons on their left, but his cavalry would not move; and Marlborough's horse crashed into the hapless battalions, cut them down by whole ranks, and swept them out of existence.

Then Tallard's sins found him out. The cavalry of Marsin's right, seeing their flank exposed, swerved back upon Marsin's centre; a wide gap was cut in the French line; and Tallard's army was left isolated and alone. The marshal sent urgent messages to Marsin for reinforcements, and to Blenheim for the withdrawal of the infantry; but Marsin could not spare a man, and the order reached Blenheim too late. Marlborough was riding along the ranks of his cavalry from right to left, and presently the trumpets sounded the charge, and the two long lines swept sword in hand up the slope. The French stood firm for a brief space, and then, after a feeble volley from the saddle, they broke, wheeled round upon their supports, and carried all away with them in confusion. Thirty squadrons fled wildly in rear of Blenheim towards the river. General Hompesch's division of horse by the Duke's order brought up their right shoulders and galloped after them; and the fugitives in panic madness plunged down the slope towards the Danube. The great river was before them, another stream and a swamp to their right; and there was no escape. Some dashed into the water and tried to swim away, others crept along the bank and over the morass towards Hochstädt, others again broke back over the slope towards Morselingen; but the relentless Hompesch left them no rest. Those that reached Hochstädt found themselves cut off, for another division of fugitives had fled thither straight from the field with Marlborough himself hard at their heels. Hundreds were drowned, hundreds were cut down, and a vast number taken prisoners. A few only preserving some semblance of order made good their retreat. 1704. August $\frac{2}{13}$.

Meanwhile Marsin and the Elector, seeing the collapse of Tallard's army, set fire to Oberglau and Lutzingen, and began their retreat, with Eugene in full march after them. Marlborough thereupon recalled Hompesch and prepared to break up this army also by a flank attack; but in the dusk Eugene's troops were mistaken for the enemy, so Marsin was permitted to escape, though with an army much shaken and demoralised. But there were still the French battalions in Blenheim, which Churchill, after the defeat of Tallard's cavalry, had made haste to envelope with his infantry and dragoons. Tallard had been captured while on his way to them, and the finest troops of France were locked up in the village without orders of any kind, helpless and inactive, and too much crowded together for effective action. At last they tried to break out to the rear of the village, but were headed back by the Scots Greys; they made another attempt on the other side, and were checked by the Irish Dragoons. Churchill was just about to attack them with infantry and artillery in overwhelming force, when the French proposed a parley. Churchill would hear of nothing but unconditional surrender. Regiment Navarre in shame and indignation burnt its colours rather than yield them, but there was no help for it;

1704. August $\frac{2}{13}$. and twenty-four battalions of infantry together with four regiments of dragoons laid down their arms, many of them not having fired a shot. The officers were stupefied by their misfortune, and could only ejaculate "Oh, que dira le Roi, que dira le Roi!" Seldom has harder fate overtaken brave men.

The day was closing when Marlborough borrowed a leaf from a commissary's pocket-book and wrote a note in pencil to his wife, the message and the handwriting both those of a man who is quite tired out.

"13*th August* 1704.

"I have not time to say more, but to beg you will give my duty to the queen, and let her know her army has had a glorious victory, Monsr. Tallard and two other generals are in my coach and I am following the rest. The bearer, my aide-de-camp, Colonel Parke, will give her an account of what has pass'd. I shall doe it in a day or two by another more at large.

"MARLBOROUGH."

So Colonel Parke galloped away with the news to England, and the broad Danube bore the same tale to the east as it rolled the white-coated corpses in silence towards the sea.

The total loss of the Allies amounted to four thousand five hundred killed and seven thousand five hundred wounded, of which the British numbered six hundred and seventy killed and over fifteen hundred wounded. No regimental list of the casualties seems to exist, but judging from their loss in officers the Tenth, Fifteenth, Sixteenth, Twenty-first, and Twenty-sixth regiments of Foot, and the Third, Sixth, and Seventh Dragoon Guards were the corps that suffered most severely—the Twenty-sixth in particular losing twenty officers, the Carabiniers ten officers and seventy-four horses, and the Seventh Dragoon Guards six officers and seventy-five horses. But most remarkable, and perhaps most splendid of all, is the record of the regiments which had been so terribly shattered at the Schellenberg.

The Guards lost their colonel and seven other officers; the two battalions of the Royal Scots lost twelve, and the Twenty-third nine officers, notwithstanding that the former had already lost thirty and the latter sixteen little more than a month before. Troops that will stand such punishment as this twice within a few weeks are not to be found in every army. 1704.

The losses of the French and their allies in killed, wounded, and prisoners, on the day of the battle and during the subsequent pursuit, fell little short of forty thousand men. Marlborough and Eugene divided eleven thousand prisoners, while the trophies included one hundred guns of various calibres, twenty-four mortars, one hundred and twenty-nine colours, one hundred and seventy-one standards and other less important items, together, of course, with the whole of the French camp.

The Allies lay on their arms on the field during the night after the battle, moved on for a short march on the morrow, and then halted for four days. The troops were very greatly fatigued, and Marlborough was much embarrassed by the multitude of his prisoners, so the pursuit, if pursuit it can be called, was left to the hussars of the Imperial Army. The Electar, however, needed no spur. On the night of the battle he crossed the Danube at Lavingen, and destroying the bridge behind him hurried back toward Ulm. Then, without pausing for a moment or attempting to obtain aid from Villeroy, he hastened on by forced marches, rather in flight than retreat, through the Black Forest to the Rhine. The sufferings of his troops were terrible. He had carried with him a thousand wounded officers and six thousand wounded men; and there was not a village on the line of march that had not its churchyard choked with the graves of those that had succumbed. The Imperial hussars too hung restlessly round his skirts, cutting off every straggler and bringing back multitudes of prisoners and deserters. Altogether it was a disastrous retreat. August $\frac{3}{14}$.

1704. August $\frac{8}{19}$. On the 19th of August Marlborough resumed his march up the Danube, having first recalled Prince Lewis of Baden from Ingolstadt, and occupied Augsburg. On arrival at Ulm a force was detached to besiege the town, while the main army marched back in three columns by the line of its original advance. By the August 28 / Sept. 8. 8th of September the whole force, strengthened by a reinforcement from Stollhofen, had crossed the Rhine and was concentrated at Philipsburg.

Villeroy, who with his own army and the remains of the Elector's had taken post in the Queich to cover Landau, now fell back without pausing to the Lauter, very much to the relief of Marlborough, who found it difficult to understand such feebleness even after such a Sept. $\frac{5}{16}$. defeat as that of Blenheim. Landau was accordingly invested by Prince Lewis of Baden, while Marlborough and Eugene covered the operations. The siege lasted long, and in October Marlborough, weary of such slow work, made a sudden spring upon Treves, gave orders for the siege of Trarbach, and so secured his winter Nov. $\frac{12}{23}$. quarters on the Moselle. The fall of Trarbach and the capture of Landau closed the campaign; and the occupation of Consaarbrück at the confluence of the Moselle and Saar showed what was to be the starting-point for the next year. A full week before the fall of Landau the English troops, so much weakened that their fourteen battalions had been temporarily reorganised into seven, were sent into winter quarters for the rest that they had earned so well.

Thus ended the famous campaign of Blenheim, a name which is rightly grouped with Creçy, Poitiers, Agincourt, and Waterloo. For well-nigh forty years the French arms had triumphed in every quarter of Europe, checked indeed by an occasional reverse, such as that of Namur, but by no failure that could be counted against the long succession of victories. But now an English general had rudely broken the chain of successes by a crushing defeat, with every circumstance of humiliation. First, the French marshals had been wholly outwitted by Marlborough's march to the Danube. Next, when they approached him it was without an idea of offering battle, but in full confidence that their manœuvres, added to their superior numbers, would compel him to withdraw. Yet to their astonishment the despised enemy had attacked them without hesitation, utterly destroyed one complete army and driven the relics of another in headlong flight to the Rhine. The dismay in Paris was profound; but mighty was the exultation in England, for the nation felt that the old traditions were right after all, and that the English were still better men than the French.[1] "Welcome to England, Sir," said an English butcher to Tallard, as the captured marshal was escorted with 1704.

[1] ORDER OF BATTLE CAMPAIGN OF 1704.

LEFT WING ONLY.

Left. — Right.

1st Line.

- {5th Royal Irish Dragoons. / Scots Greys, 1 squadron.}
- Four Foreign Squadrons.
- {1st ,, ,, 3 ,, / 5th ,, ,, 1 ,, / 7th ,, ,, 2 ,, / 6th ,, ,, 2 ,, / 3rd Dragoon Guards, 2 squadrons.}
- Thirty-two Foreign Squadrons in three Brigades.
- Hamilton's Brigade. {1 ,, 1st Guards. / 1 Batt. Royal Scots. / 16th ,, / 20th ,, / 8th Foot.}
- Row's Brigade. {3rd Buffs. / 21st Royal Scots Fusiliers. / 24th Foot. / 23rd Royal Welsh. / 10th Foot.}
- Foreign Battalions.

2nd Line

- Foreign Squadrons.
- Ferguson's Brigade. {2nd Batt. Royal Scots. / 26th Cameronians. / 37th ,, / 15th Foot.}
- Foreign Battalions.

From Dumont's *Histoire Militaire*.

1704. every mark of respect into Nottingham. "Welcome to England. I hope to see your master here next year." It was the revival of this feeling in all its old intensity, after a pause of nearly three centuries, that was to win for England her empire in East and West.

Yet amid all the noise of triumph and jubilation there were two men who preserved their modesty and tranquillity unmoved; and these were Marlborough and Eugene. Each quietly disclaimed credit for himself, each eagerly welcomed praise for the other. The French prisoners were comforted by Eugene's testimony to their gallant resistance to his own army, while even the unfortunate officers who had been swept into the net in the village of Blenheim found consolation in the thoughtful and generous courtesy of the great Duke.

CHAPTER IV

OUR attention is now claimed for a time by the Peninsula, where the War of the Spanish Succession was to be carried forward on Spanish soil. In January 1704 the Imperial claimant to the throne, the Archduke Charles of Austria, otherwise King Charles the Third of Spain, arrived in England, and was sent away with an English fleet and an English army to possess himself of his kingdom. Portugal had offered to help him with twenty-eight thousand men, to which the Dutch had added two thousand under General Fagel, and the British six thousand five hundred men,[1] under Mainhard, Duke of Schomberg, a son of the old marshal. The campaign of 1704 need not detain us. It was speedily found that the Portuguese army was ill-equipped and inefficient, the magazines empty, the fortresses in ruins, the transport not in existence. To add to these shortcomings, Schomberg and Fagel quarrelled so bitterly that they went off, each with his own troops, in two different directions. 1704.

The result might have been foreseen. King Philip, sometime Duke of Anjou, and the Duke of Berwick with twelve thousand French, marched down to the fortresses on the Portuguese frontier, and took them one after another without difficulty. So ready and eager were the Portuguese to surrender these strongholds that they made over not only themselves as prisoners of war, but also to their extreme indignation

[1] 2nd Dragoon Guards, Royal Dragoons, 2nd, 9th, 11th, 13th, 17th, 33rd Foot.

1704. two British regiments, the Ninth and Eleventh Foot, which had the misfortune to be in garrison with them. Marlborough, in all the press of his work on the Danube, was called upon to nominate a successor to the incompetent Schomberg and selected the Huguenot Ruvigny, Earl of Galway, for the post. With this appointment we may for the present take leave of the Peninsula.

July 26 / August 6. Meanwhile, however, the fleet under Sir George Rooke, and a handful of marines under Prince George of Hesse-Darmstadt, brought a new and unexpected possession to England by the surprise of Gibraltar, which, though captured for King Charles the Third, was kept for Queen Anne. The intrinsic value of the Rock in those days was small, and its value as a military position was little understood in England; but it was at any rate a capture and very soon it became a centre of sentiment.

After the surrender of Gibraltar the fleet sailed away, leaving Prince George with a good store of provisions and about two thousand men to hold it. These troops, though now numbered the Fourth, Thirty-first, and Thirty-second of the Line, were at that time Marines, a corps which, despite brilliant and incessant service by sea and land in all parts of the world, still contents itself with the outward record of a single name, Gibraltar. Prince George lost no time in repairing the fortifications, and with good reason, for at the end of August a Spanish force of eight thousand men marched down to

Sept. 23 / October 4. the isthmus, while a month later four thousand Frenchmen were disembarked at the head of the bay. These joint forces then began the siege of Gibraltar.

The operations were pushed forward with great vigour, and the besieged were soon hard beset. At the end of October Admiral Leake contrived to throw stores and a couple of hundred men on to the Rock, together with an officer of engineers, one Captain Joseph Bennett, whose energy and ability were of priceless value. The siege dragged on for another month,

the British repulsing an attack from the eastern side 1704.
with heavy loss; but by the end of November the garrison had dwindled to one thousand men, exhausted by the fatigue of incessant duty. At last, in the middle of December a stronger reinforcement of two December.
thousand men,[1] having first narrowly escaped capture by a French fleet, was successfully landed on the Rock; and then Prince George turned upon the besiegers, and by a succession of brilliant sorties almost paralysed further progress on their side.

In the middle of January, however, a reinforcement 1705.
of four thousand men reached the enemy's camp; their batteries renewed their fire, and a great breach was made in the Round Tower, which formed one of the principal defences on the western side. On the morning of the 27th an assault was delivered, and thirteen Jan. 27 / Feb. 7.
hundred men swarmed up to the attack of the Round Tower. They were met by a brave resistance by one-fifth of their number of British, but after a severe struggle they overpowered them, drove them out, and pressed on to gain possession of a gate leading into the main fortress. There, however, they were checked by 1705.
a handful of Seymour's Marines,[2] just seventeen men, under Captain Fisher. Few though they were, this gallant little band held its own, until the arrival of some of the Thirteenth and of the Coldstream Guards enabled them to force the enemy back and drive them headlong out of the Round Tower.

This brilliant little affair marked practically the close of the siege. Further reinforcements arrived for the garrison, and Marshal Tessé, who had taken command of the siege, fell back on the bombardment of the town, which was speedily laid in ruins. The advent of a French squadron seemed likely at one moment to hearten the besiegers to renewed efforts, but Bennett,

[1] Detachments of the 1st and Coldstream Guards, 13th and 35th of the Line.

[2] The 4th Foot. It had taken its marineship in exchange from another corps.

1705. who ever since his arrival had been the soul of the defence, had by that time constructed fresh batteries
March 10/21 and was fully prepared. Finally, in March Admiral Leake's fleet appeared on the scene, destroyed a third of the French squadron, and definitely relieved the fortress. By the middle of April the last of the Frenchmen had disappeared and Gibraltar was safe. Though the scale of the operations may seem small the siege had cost the enemy no fewer than twelve thousand men.

1704. Meanwhile Parliament had met on the 29th of the previous October, full of congratulations to the Queen on the triumphs of the past campaign. There were not wanting, of course, men who, in the madness of faction, doubted whether Blenheim were really a victory, for the very remarkable reason that Marlborough had won it, but they were soon silenced by the retort that the King of France at any rate had no doubts on the point.[1] The plans for the next campaign were designed on a large scale, and were likely to strain the resources of the Army to the uttermost. The West Indies demanded six battalions and Gibraltar three battalions for garrison; Portugal claimed ten thousand men, Flanders from twenty to twenty-five thousand; while besides this a design was on foot, as shall presently be seen, for the further relief of Portugal by a diversion
1705. in Catalonia. Five millions were cheerfully voted for the support of the war, and six new battalions were raised, namely, Wynne's, Bretton's, Lepell's, Soames's, Sir Charles Hotham's, and Lillingston's, the last of which alone has survived to our day with the rank of the Thirty-eighth of the Line.[2]

Marlborough's plan of campaign had been sufficiently

[1] St. Simon gives a curious account of Lewis's difficulty in arriving at the truth, owing to the general unwillingness to tell him bad news.

[2] It is stated in *Records and Badges of the Army* that Lillingston's was formed in 1702. But Narcissus Luttrell, Millar, and the Military Entry Books all give the date as 25th March (New Year's Day) 1705.

foreshadowed at the close of the previous year, namely, 1705. to advance on the line of the Moselle and carry the war into Lorraine. The Emperor and all the German Princes promised to be in the field early, the Dutch were with infinite difficulty persuaded to give their consent, and after much vexatious delay Marlborough joined his army at Treves on the 26th of May. Here May 15/26. he waited until the 17th of June for the arrival of the German and Imperial troops. Not a man nor a horse June 6/17. appeared. In deep chagrin he broke up his camp and returned to the Meuse, having lost, as he said, one of the fairest opportunities in the world through the faithlessness of his allies.[1]

His presence was sorely needed on the Meuse. Villeroy, who commanded the French in Flanders, finding no occasion for his presence on the Moselle, had moved out of his lines, captured Huy, and then May 21. marching on to Liège had invested the citadel. The States-General in a panic of fright urged Marlborough to return without delay, and Overkirk, who commanded the Dutch on the Meuse, added his entreaties to theirs. Marlborough, when once he had made up his mind to move, never moved slowly, and by the 25th of June he June 14/25. was at Düren, to the eastward of Aix-la-Chapelle. Here he was still the best part of forty miles from the Meuse, but that was too near for Villeroy, who at once abandoned Liège and fell back on Tongres. Marlborough, continuing his advance, crossed the Meuse at Visé on the 2nd of July, and on the same day united June 21/July 2. his army with Overkirk's at Haneff on the Upper Jaar. Villeroy thereupon retired ignominiously within his fortified lines.

These lines, which had been making during the past three years, were now complete. They started from the Meuse a little to the east of Namur, passed from thence to the Mehaigne and the Little Geete, followed the Little Geete along its left bank to Leuw and thence

1 Quincy's account of this portion of the campaign is, so far as concerns Marlborough, full of falsehoods.

1705. along the Great Geete to the Demer; from thence they ran up the Demer as far as Arschot, from which point a new line of entrenchments carried the barrier through Lierre to Antwerp. Near Antwerp Marlborough had already had to do with these lines in 1703, but hitherto he had made no attempt to force them. Villeroy and the Elector of Bavaria now lay before him with seventy thousand men, a force superior to his own, but necessarily spread over a wide front for the protection of the entrenchments. The marshal's headquarters were at Meerdorp, in the space between the Geete and the Mehaigne, which he probably regarded as a weak point. Marlborough posted himself over against him at Lens-les-Beguines, detaching a small force to recapture Huy while Overkirk with the Dutch army covered the siege from Vignamont. Thus, as if daring the French to take advantage of the dispersion of his army, he quietly laid his plans for forcing the lines.

The point that he selected was on the Little Geete between Elixheim and Neerhespen, exactly in rear of the battlefield of Landen. The abrupt and slippery banks of the river, which the English knew but too well, together with the entrenchments beyond it, presented extraordinary difficulties, but the lines were on that account the less likely to be well guarded at that particular point. Marlborough had already obtained the leave of the States-General for the project, but he had now the far more difficult task of gaining the consent of the Dutch generals at a Council of War. Slangenberg and others opposed the scheme vehemently, but were overruled; and the Duke was at length at liberty to fall to work.

June 30 / July 11. Huy fell on the 11th of July, but to the general surprise the besieging force was not recalled. Six days later Overkirk and the covering army crossed the
July. Mehaigne from Vignamont and pushed forward detachments to the very edge of the lines between Meffle and Namur. Villeroy fell into the trap, withdrew troops from all parts of the lines and concentrated

forty thousand men at Meerdorp. Marlborough then recalled the troops from Huy, and made them up to a total of about eight thousand men, both cavalry and infantry,[1] the whole being under the command of the Count of Noyelles. The utmost secrecy was observed in every particular. The corps composing the detachment knew nothing of each other, and nothing of the work before them; and, lest the sight of fascines should suggest an attack on entrenchments, these were dispensed with, the troopers only at the last moment receiving orders to carry each a truss of forage on the saddle before them. 1705.

At tattoo the detachment fell in silently before the camp of the right wing, and at nine o'clock moved off without a sound in two columns, the one upon Neerhespen, the other upon the Castle of Wange before Elixheim. An hour later the rest of the army followed, while at the same time Overkirk, under cover of the darkness, crossed the Mehaigne at Tourines and joined his van to the rear of Marlborough's army. The distance to be traversed was from ten to fifteen miles; the night though dry was dark; and the guides, frequently at fault, were fain to direct themselves by the trusses dropped on the way by the advanced detachment. Twelve years before to the very day a French army had toiled along the same route, wearied out and stifled by the sun, and only kept to its task by an ugly little hunch-backed man whom it had reverenced as Marshal Luxemburg. Now English and Dutch were blundering on to take revenge for Luxemburg's victory at the close of that march. The hours fled on, the light began to break, and the army found itself on the field of Landen, William's entrenchment grass-grown before it, Neerwinden and Laer lying silent to the left, and before the villages the mound that hid the corpses of the dead. Then some at least of the soldiers knew the work that lay before them.

July $\frac{6}{17}$.

July $\frac{6\text{-}7}{17\text{-}18}$.

[1] Four British regiments were of this detachment. Two battalions of the 1st Royals, the 3rd Buffs, and the 10th Foot.

1705. July $\frac{7}{18}$. At four o'clock the heads of the columns halted within a mile of the Geete, wrapped in a thick mist and hidden from the eye of the enemy. The advanced detachment quickly cleared the villages by the river, seized the bridge before the Castle of Wanghe, which had not been broken down, and drove out the garrison of the Castle itself. Then the pontoniers came forward to lay their bridges; but the infantry would not wait for them. They scrambled impatiently through hedges and over bogs, down one steep bank of the river and up the other, into the ditch beyond, and finally, breathless and dripping, over the rampart into the lines. So numerous were the hot-heads who thus broke in that they forced three regiments of French dragoons to retire before them without attempting resistance. Then the cavalry of the detachment began to file rapidly over the pontoon-bridges; but meanwhile the alarm had been given, and before the main army could cross, the French came down in force from the north, some twenty battalions and forty squadrons, in all close on fifteen thousand men, with a battery of eight guns.

The enemy advanced rapidly, their cavalry leading, until checked by a hollow way which lay between them and the Allies, where they halted to deploy. Marlborough took in the whole situation at a glance. Forming his thirty-eight squadrons into two lines, with the first line composed entirely of British, he led them across the hollow way and charged the French sword in hand. They answered by a feeble fire from the saddle and broke in confusion, but presently rallying fell in counter-attack upon the British and broke them in their turn. Marlborough, who was riding on the flank, was cut off and left isolated with his trumpeter and groom. A Frenchman galloped up and aimed at him so furious a blow that, failing to strike him, he fell from his horse and was captured by the trumpeter. Then the allied squadrons rallied, and charging the French once more broke them past all reforming and captured the guns. The French infantry now retired

very steadily in square, and the Duke sent urgent 1705.
messages for his own foot. But by some mistake the July 7/18.
battalions had been halted after crossing the Geete, so that the French were able to make good their retreat.

By this time Villeroy, who had spent the night in anxious expectation of an attack at Meerdorp, had hurried up with his cavalry, only to find that the Duke was master of the lines. Hastily giving orders for his scattered troops to pass the Geete at Judoigne he began his retreat upon Louvain. Presently up came Marlborough's infantry at an extraordinary pace, the men as fresh and lively after fifteen hours of fatigue as if they had just left camp. The Duke was anxious to follow up his success forthwith, a movement which the French had good reason to dread, but the Dutch generals opposed him, and Marlborough was reluctantly constrained to yield. The loss of the French seems to have been about two thousand men, most of them prisoners, a score of standards and colours, of which the Fifth Dragoon Guards claimed four as their own, and eighteen guns, eight of which were triple-barrelled and were sent across the Channel to be copied in England.[1]

The Allies halted for the night at Tirlemont, and July 8/19.
advancing next day upon Louvain struck against the rear of the French columns and captured fifteen hundred prisoners. That night they encamped within a mile to the east of Louvain, while the French, once again distributing their force along a wider front, lined the left bank of the Dyle from the Demer to the Yssche, with their centre at Louvain. Marlborough had hoped to push in at once, but he was stopped by heavy rains that rendered the Dyle impassable; and it was not until ten days later that, after infinite trouble with the Dutch, he was able to pursue his design.

The operations for the passage of the Dyle were July 18/29.
conducted in much the same way as in the forcing of the lines. An advanced detachment was pushed forward

[1] Narcissus Luttrell.

1705. from each wing of the army, that from the right or English[1] flank being appointed to cross the river under the Duke of Würtemberg at Corbeek Dyle, that from the left under General Heukelom to pass it at Neeryssche. The detachments fell in at five in the evening, reached their appointed destination at ten, and effected their passage with perfect success. The main bodies started at midnight, and went somewhat astray in the darkness, though by three o'clock the Dutch army was within supporting distance of its detachment and the British rapidly approaching it. The river had been in fact forced, when suddenly the Dutch generals halted their main body. Marlborough rode up to inquire the cause, and was at once taken aside by Slangenberg. "For God's sake, my Lord—" began the Dutchman vehemently, and continued to protest with violent gesticulations. No sooner was Marlborough's back turned than the Dutch generals, like a parcel of naughty schoolboys, recalled Heukelom's detachment. Thus the passage won with so much skill was for no cause whatever abandoned, without loss indeed, but also not without mischievous encouragement to the French, who boasted loudly that they had repulsed their redoubtable adversary.

Deeply hurt and annoyed though he was, the Duke, with miraculous patience, excused in his public despatches the treachery and imbecility which had thwarted him, and prepared to effect his purpose in another way. His movements were hastened by news that French reinforcements, set free by the culpable inaction of Prince Lewis of Baden, were on their way from Alsace. Unable to pass the Dyle he turned its head-waters at Genappe, and wheeling north towards the forest of Soignies encamped between La Hulpe and Braine l'Alleud.[2] The French at once took the alarm

August 5/16.

[1] It is worth noting that this was the first campaign in which Marlborough and the British took the post of honour at the extreme right of the Allied order of battle.

[2] His camp thus lay across the whole of Wellington's position

and posted themselves behind the river Yssche, with their left at Neeryssche, and their right at Overyssche resting on the forest of Soignies. Marlborough at once resolved to force the passage of the river. On the evening of the 17th of August he detached his brother Churchill with ten thousand foot and two thousand horse to advance through the forest and turn the French right; while he himself marched away at daybreak with the rest of the army and emerged into the plain between the Yssche and the Lasne. The Duke quickly found two assailable points, and choosing that of Overyssche, halted the army pending the arrival of the artillery. The guns were long in arriving, Slangenberg having insisted, despite the Duke's express instructions, on forcing his own baggage into the column for the express purpose of causing delay. At last about noon the artillery appeared, and Marlborough asked formal permission of the Dutch deputies to attack. To his surprise, although Overkirk had already consented, they claimed to consult their generals. Slangenberg with every mark of insolence condemned the project as murder and massacre, the rest solemnly debated the matter for another two hours, the auspicious moment passed away exactly as they intended, and another great opportunity was lost. The French reinforcements arrived, and having been the weaker became the stronger force. Nothing more could be done for the rest of the campaign, but to level the French lines from the Demer to the Mehaigne.

1705. August $\frac{6}{17}$. August $\frac{7}{19}$.

Thus for the third time a brilliant campaign was spoilt by the Dutch generals and deputies. Fortunately the public indignation both in England and in Holland was too strong for them, and Slangenberg, though not indeed hanged as he deserved, was deprived of all further command. Jealousy, timidity, ignorance, treachery, and flat imbecility seem to have been the motives that inspired these men, whose conduct has

at Waterloo, from east to west and considerably beyond it to westward, but fronted in the reverse direction.

1705. never been reprobated according to its demerit. It was they who were responsible for the prolongation of the war, for the burden that it laid on England, and for the untold misery that it wrought in France. Left to himself Marlborough would have forced the French to peace in three campaigns, and the war would not have been ended in shame and disgrace by the Treaty of Utrecht.[1]

Consolation for the disappointment in Flanders came from an unexpected quarter. In Portugal, indeed, comparatively little was done. An army was made up of about three thousand British[2] under Lord Galway, two thousand Dutch under General Fagel, and twelve thousand Portuguese under the Spanish General de Corsana; and to avoid friction it was arranged that these

[1] ORDER OF BATTLE. CAMPAIGN OF 1705.

RIGHT WING ONLY.

Left. — Right.

1st Line.

Foreign Troops.

- 15th "
- 24th "
- Evans's Foot.
- Macartney's Foot.
- 37th Foot.
- 21st Royal Scots Fusiliers.
- 3rd Buffs.

- 16th Foot.
- 26th Cameronians.
- Stringer's Foot.
- 28th Foot.
- 23rd Royal Welsh.
- 18th Royal Irish.
- 1 " Royal Scots.
- 1 Batt. 1st Guards.

- 3rd " " 2 "
- 6th " " 2 "
- 7th " " 2 "
- 5th " " 2 "
- 1st Dragoon Guards, 3 squadrons.

- 5th Dragoons, 3 "
- Scots Greys, 3 squadrons.

2nd Line.
Extreme Right of Centre.

- 8th "
- 29th Foot.
- Temple's Foot.
- 10th Foot.
- 2nd Batt. Royal Scots.

Foreign troops.

Newspaper.

[2] 2nd Dragoon Guards, 2nd, 9th (exchanged against the prisoners of Blenheim), 17th, 33rd, and Brudenell's Foot.

three generals should hold command alternately for a week at a time. In such circumstances it was surprising that they should even have accomplished the siege and capture of three weak fortresses, Valenza, Albuquerque, and Badajoz, with which achievements the campaign came to an end.[1] 1705.

But in Catalonia the operations were of a more brilliant kind. The Catalans were known to favour the Austrian side; and it was accordingly resolved in this year to send a fleet and an army to back them under Admiral Leake and Lord Peterborough, the latter to be joint admiral at sea as well as commander-in-chief ashore. The character of Peterborough is one of the riddles of history. He was now forty years of age, and had so far distinguished himself chiefly by general eccentricity, not always of a harmless kind, and, in common with most prominent men of his age, by remarkable pliancy of principle. His experience of active service was slight and had been gained afloat rather than ashore, and though he had long held the colonelcy of a regiment, he had never commanded in war nor in peace. His force consisted of six British[2] and four Dutch battalions, or about six thousand five hundred men in all. The expedition arrived at Lisbon early in June, when after some delay it was decided that June $\frac{9}{20}$. the fleet should proceed to Barcelona. Galway lent his two regiments of dragoons, the Royals and the Eighth; and with them Peterborough sailed to Gibraltar, where he picked up the eight battalions[3] of the garrison, leaving two of his own in their place, and proceeded to his destination. On the way up the Spanish coast a detachment was landed to capture Denia, and on the 23rd of August the main force was disembarked before August $\frac{12}{23}$.

[1] It is somewhat singular that the first regiment which signally distinguished itself in this first Peninsular War was the 33rd (Duke of Wellington's), which covered itself with honour at the storm of Valenza.

[2] 6th, 34th, 36th, Elliott's, J. Caulfield's (late Pearce's), Gorges's.

[3] Guards (mixed battalion of the 1st and Coldstream), 13th, 35th, Mountjoy's, and four of Marines.

1705. Barcelona and took up a position to the north-east of the town with its left flank resting on the sea.

The reports sent to England had represented Barcelona as ill-fortified and ill-garrisoned. Ill-fortified it may have been if compared with a creation of Vauban or Cohorn, but it was none the less a formidable fortress, well stocked with supplies and garrisoned by seven thousand troops under an energetic governor, by name Velasco. Peterborough, who grasped the situation, wished to abandon the project of a regular siege for operations of a livelier kind, but was prevailed upon to give it a trial for eighteen days, at the close of which he ordered the re-embarkation of the army. He was, however, again induced to change his mind, and then suddenly, on the evening of the 13th of
Sept. $\frac{2}{13}$. September, he produced an original scheme of his own.

About three-quarters of a mile to south-west of Barcelona stood the small fort of Montjuich, crowning a hill seven hundred feet above the fortress, strong by nature and strengthened still further by outworks, which though incomplete were none the less formidable. This Peterborough resolved to capture by escalade. Not a word was said to the men of the work before them. No further orders were issued than that twelve hundred English and two hundred Dutch should be ready in the afternoon to march towards Tarragona, while thirteen hundred men under Brigadier Stanhope were secretly detailed to cover the rear of the assaulting columns from any attack from Barcelona. At six o'clock the attacking force moved off under Lord Charlemont towards the north-west, continuing the march in this false direction for four hours, till Peterborough at last gave the order to turn about to southward. The night was dark, and much of the ground so rocky as to show no track, so that when the columns at length came up before Montjuich one complete body of two hundred was found to be missing, having evidently strayed away from the path of the remainder.

Sept. $\frac{3}{14}$. Half the force however was told off for simultaneous assault on the eastern and western extremities of the fort, Peterborough and Prince George of Hessen-Darmstadt accompanying the eastern column, which, since it was expected to meet with the sternest of the work, was made the stronger. The other moiety of the troops was held in reserve between the two columns. A little after daybreak the signal was given ; the storming parties dashed up the glacis under a heavy and destructive fire, and plunging in among the enemy drove them headlong from the outworks. Following the fugitives in hot pursuit Peterborough and Prince George captured the eastern bastion of the fort itself, threw up a barricade of loose stones in the gorge and entrenched themselves behind it. The western attack had met with equal success, and had likewise entrenched itself in a demi-bastion in that flank of the fort. Both parties being thus under cover the fire ceased, and Peterborough sent orders to Stanhope to bring up his reserve. 1705. Sept. $\frac{3}{14}$.

Meanwhile the Governor of Barcelona, being in communication with Montjuich, had at the sound of the firing despatched four hundred dragoons in all haste to reinforce the garrison. As they entered the fort they were received with loud shouts of welcome by the Spanish. Prince George, mistaking the sound for a cry of surrender, at once started up and advanced with all his men into the inner works. They were no sooner in the ditch than the Spaniards swept round them to cut them off. Two hundred were taken prisoners, Prince George fell mortally wounded, and the rest fell back in confusion. This was a severe blow ; but worse was to come. Peterborough hearing that fresh reinforcements were on their way to the enemy from Barcelona, rode out of the bastion to look for himself, and no sooner was he gone than the troops were seized with panic. Lord Charlemont was powerless to check it ; and in a few minutes the whole of the men, with Charlemont at their head, came running with unseemly haste out of the captured position.

1705. Sept. 3/14. They had not run far when up galloped Peterborough in a frenzy of rage. What he said no writer has dared to set down; but he snatched Charlemont's half-pike from his hand and waved the men back to the fort with a torrent of rebuke. Rallying instantly they regained their post without the loss of a man before the enemy had discovered their retreat; and the appearance of Stanhope with the reserve presently banished all further idea of panic. Meanwhile the Spanish reinforcements from Barcelona had met the English prisoners, and learning from them that Peterborough and Prince George were present in person before Montjuich, assumed that the British were attacking in overwhelming force. They therefore returned to Barcelona, leaving the fort to its fate. Three days of bombardment sufficed to overcome the resistance of the weakened garrison; and thus by a singular chapter of accidents Peterborough's design proved to be a success, and Montjuich was taken.

The siege of Barcelona was then pushed forward in form, aided by the guns of the fleet; and on the 9th of
Sept. 28/October 9. October the garrison capitulated with the honours of war. A fortnight later King Charles the Third made his public entry into the city; Peterborough scattered dollars with a liberal hand, and all was merriment and rejoicing. The picture would not be complete without the figure of a drunken English grenadier, whose vagaries afforded inexhaustible amusement to the populace;[1] but Peterborough was a disciplinarian, and the troops as a whole behaved remarkably well. Stanhope was at once sent home with the good news, and England awoke to the fact that she possessed a second officer who, though not to be named in the same breath with Marlborough, possessed a natural, if eccentric, genius for war.

The capture of Barcelona, and the subsequent reduction of Tarragona by the fleet, brought practically the whole of Catalonia to the side of King Charles.

[1] Carleton.

But now further operations were checked by lack of money and supplies. Peterborough, who saw the difficulty of supporting a large force in the field, was for dividing his little army into flying columns, and making good the deficiency of numbers by extreme mobility; but he could not gain acceptance for his views. He wrote piteous letters of his state of destitution, reviling, as his custom was, all his colleagues and subordinates with astonishing freedom. Very soon the troops in Barcelona became so sickly that he was compelled to distribute them in the fortresses of Catalonia, leaving further operations to the Catalan guerillas. By the exertions of these last the close of the year saw not only Catalonia but Valencia gained over, though on no very certain footing, to the side of King Charles. So ended the first serious campaign of the first Peninsular war. 1705.

CHAPTER V

1706. It is now time to revert to England and to the preparations for the campaign of 1706. Marlborough, as usual, directly that the military operations were concluded, had been deputed to visit the courts of Vienna and of sundry German states in order to keep the Allies up to the necessary pitch of unity and energy. These duties detained him in Germany and at the Hague until January 1706, when he was at last able to return to England. There he met with far less obstruction than in former years, but none the less with an increasing burden of work. The vast extension of operations in the Peninsula, and the general sickliness of the troops in that quarter, demanded the enlistment of an usually large number of recruits. One new regiment of dragoons and eleven new battalions of foot were formed in the course of the spring, to which it was necessary to add yet another battalion before the close of the year.[1] Again the epidemic sickness among the horses in Flanders had caused an extraordinary demand for horses. The Dutch, after their wonted manner, had actually taken pains to prevent the supply of horses to the British,[2] though, even if they had not, the Duke had a prejudice in favour of English horses, as of English men, as superior to any other. Finally, the stores of the Ordnance were unequal to the constant

[1] Peterborough's Dragoons; Mark Kerr's, Stanwix's, Lovelace's, Townsend's, Tunbridge's, Bradshaw's, Sybourg's, Price's Foot. Sybourg's was made up of Huguenots.

[2] Marlborough's *Despatches*, vol. ii. p. 262.

drain of small arms, and it was necessary to make good 1706. the deficiency by purchases from abroad. All these difficulties and a thousand more were of course referred for solution to Marlborough.

When in April he crossed once more to the Hague April $\frac{14}{25}$. he found a most discouraging state of affairs. The Dutch were backward in their preparations; Prussia and Hanover were recalcitrant over the furnishing of their contingents; Prince Lewis of Baden was sulking within his lines, refusing to communicate a word of his intentions to any one; and everybody was ready with a separate plan of campaign. The Emperor of course desired further operations in the Moselle for his own relief; but after the experience of the last campaign the Duke had wisely resolved never again to move eastward to co-operate with the forces of the Empire. The Dutch for their part wished to keep Marlborough in Flanders, where he should be under the control of their deputies; but the imbecile caprice of these worthies was little more to his taste than the sullen jealousy of Baden. Marlborough himself was anxious to lead a force to the help of Eugene in Italy, a scheme which, if executed, would have carried the British to a great fighting ground with which they are unfamiliar, the plains of Lombardy. He had almost persuaded the States-General to approve of this plan, when all was changed by Marshal Villars, who surprised Prince Lewis of Baden in his lines on the Motter, and captured two important magazines. The Dutch at once took fright and, in their anxiety to keep Marlborough for their own defence, agreed to appoint deputies who should receive rather than issue orders. So to the Duke's great disappointment it was settled that the main theatre of war should once again be Flanders.

Villeroy meanwhile lay safely entrenched in his position of the preceding year behind the Dyle, from which Marlborough saw little hope of enticing him. It is said that an agent was employed to rouse Villeroy by telling him that the Duke, knowing that the French

1706. were afraid to leave their entrenchments, would take advantage of their inaction to capture Namur.[1] Be that as it may, Villeroy resolved to quit the Dyle. He knew that the Prussian and Hanoverian contingents had not yet joined Marlborough, and that the Danish cavalry had refused to march to him until their wages were paid; so that interest as well as injured pride

May $\frac{8}{19}$. prompted the hazard of a general action. On the 19th of May, therefore, he left his lines for Tirlemont on the Great Geete. Marlborough, who was at Maestricht, saw with delight that the end, for which he had not dared to hope, was accomplished. Hastily making arrangements for the payment of the Danish troops, he

May $\frac{9}{20}$. concentrated the Dutch and British at Bilsen on the Upper Demer, and moved southward to Borchloen. Here the arrival of the Danes raised his total force to

May $\frac{11}{22}$. sixty thousand men, a number but little inferior to that of the enemy. On the very same day came the intelligence that Villeroy had crossed the Great Geete and was moving on Judoigne. The Duke resolved to advance forthwith and attack him there.

May $\frac{12}{23}$. At one o'clock in the morning, of Whitsunday the 23rd of May, Quartermaster-General Cadogan rode forward from the headquarters at Corswarem with six hundred horse and the camp-colours towards the head of the Great Geete, to mark out a camp by the village of Ramillies. The morning was wet and foggy, and it was not until eight o'clock that, on ascending the heights of Merdorp, they dimly descried troops in motion on the rolling ground before them. The allied army had not marched until two hours later than Cadogan, but Marlborough, who had ridden on in advance of it, presently came up and pushed the cavalry forward through the mist. Then at ten o'clock the clouds rolled away, revealing the whole of the French army in full march towards them.

Villeroy's eyes were rudely opened, for he had not expected Marlborough before the following day; but

[1] This is the story told in Lamberti.

he knew the ground well, for he had been over it before 1706.
with Luxemburg, and he proceeded to take up a position May $\frac{12}{23}$.
which he had seen Luxemburg deliberately reject. The table-land whereon he stood is the highest point in the plains of Brabant. To his right flowed the Mehaigne; in his rear ran the Great Geete; across his centre and left the Little Geete rose and crept away sluggishly in marsh and swamp.[1] In his front lay four villages: Taviers on the Mehaigne to his right, Ramillies, less advanced than Taviers, on the source of the Little Geete to his right centre, Offus parallel to Ramillies but lower down the stream to his left centre, Autréglise or Anderkirch between two branches of the Little Geete and parallel to Taviers to his left. Along the concave line formed by these villages Villeroy drew up his army in two lines facing due east.

The Mehaigne, on which his right rested, is at ordinary times a rapid stream little more than twelve feet wide, with a muddy bottom, but is bordered by swampy meadows on both sides, which are flooded after heavy rain. From this stream the ground rises northward in a steady wave for about half a mile, sinks gradually and rises into a higher wave at Ramillies, sinks once more to northward of that village and rolls downward in a gentler undulation to Autréglise. Between the Mehaigne and Ramillies, a distance of about a mile and a half, the ground east and west is broken by sundry hollows of sufficient inclination to offer decided advantage or disadvantage in a combat of cavalry. A single high knoll rises in the midst of these hollows, offering a place of vantage from which Marlborough must almost certainly have reconnoitred the disposition of the French right. The access to Ramillies itself is steep and broken both to north and south, but on the eastern front the ground rises to it for half a mile in a gentle, unbroken slope, which modern rifles would make impassable by the bravest troops. In rear, or to westward of the French position, the table-land is

[1] The ground, though now drained, is still very wet.

1706. May $\frac{12}{23}$. clear and unbroken, and to the right rear or south-west stands a mound or barrow called the tomb of Ottomond, still conspicuous and still valuable as a key to the actions of the day.[1] The full extent of the French front from Taviers to Autréglise covered something over four miles.

Having chosen his position, Villeroy lost no time in setting his troops in order. His left, consisting of infantry backed by cavalry,[2] extended from Autréglise to Offus, both of which villages were strongly occupied. His centre from Offus to Ramillies was likewise composed of infantry. On his right, in the expanse of sound ground which stretches for a mile and a half from the marshes of the Geete at Ramillies to those of the Mehaigne, were massed more than one hundred and twenty squadrons of cavalry with some battalions of infantry interlined with them, the famous French Household Cavalry (Maison du Roi), being in the first line. The left flank of this expanse was covered by the village of Ramillies, which was surrounded by a ditch and defended by twenty battalions and twenty-four guns. On the right flank not only Taviers but Franquinay, a village still further in advance, were occupied by detachments of infantry, while Taviers was further defended by cannon.

Marlborough quickly perceived the defects of Villeroy's dispositions, which were not unlike those of Tallard at Blenheim. Taviers was too remote from Ramillies for the maintenance of a cross-fire of artillery. Again, the cavalry of the French left was doubtless secure against attack behind the marshes of the Geete, but for this very reason it was incapable of aggressive action. The French right could therefore be turned, provided that it were not further reinforced ; and

1 I have described the field at some length, since the map given by Coxe is most misleading.

2 Coxe, by a singular error, makes the left consist exclusively of infantry, in face of Quincy, Feuquières, the *London Gazette* and other authorities, thereby missing almost unaccountably an important feature in the action.

accordingly the Duke opened his manœuvres by a demonstration against the French left. 1706. May $\frac{12}{23}$.

Presently the infantry of the allied right moved forward in two lines towards Offus and Autréglise, marching in all the pomp and circumstance of war, Dutch, Germans, and British, with the red coats conspicuous on the extreme right flank. Striding forward to the river they halted and seemed to be very busy in laying their pontoons. Villeroy marked the mass of scarlet, and remembering its usual place in the battlefield, instantly began to withdraw several battalions from his right and centre to his left. Marlborough watched the white coats streaming away to their new positions, and after a time ordered the infantry of his right to fall back to some heights in their rear. The two lines faced about and retired accordingly over the height until the first line was out of sight. Then the second line halted and faced about once more, crowning the ascent with the well-known scarlet, while the first marched away with all speed, under cover of the hill and unseen by the French, to the opposite flank. Many British battalions[1] stood on that height all day without moving a step or firing a shot, but none the less paralysing the French left wing.

About half past one the guns of both armies opened fire, and shortly afterwards four Dutch battalions were ordered forward to carry Franquinay and Taviers, and twelve more to attack Ramillies, while Overkirk advanced slowly on the left with the cavalry. Franquinay was soon cleared; Taviers resisted stoutly for a time but was carried, and a strong reinforcement on its way to the village was intercepted and cut to pieces. Then Overkirk, his left flank being now cleared, pushed forward his horse and charged. The Dutch routed the first French line, but were driven back in confusion by the second; and the victorious French were only

[1] Apparently the whole of Meredith's brigade, viz.: 1st, 18th, 29th, 37th, 24th, and 10th regiments. The place is still easily identifiable.

1706. May 12/23. checked by the advance of fresh squadrons under Marlborough himself. Even so the Allies were at a decided disadvantage; and Marlborough, after despatching messengers to bring up every squadron, except the British, to the left, plunged into the thick of the melée to rally the broken horse. He was recognised by some French dragoons, who left their ranks to surround him, and in the general confusion he was borne to the ground and in imminent danger of capture. His aide-de-camp, Captain Molesworth, dismounted at once, and giving him his own horse enabled him to escape. The cavalry, however, encouraged by the Duke's example, recovered themselves, and Marlborough took the opportunity to shift from Molesworth's horse to his own. Colonel Bringfield, his equerry, held the stirrup while he mounted, but Marlborough was hardly in the saddle before the hand that held the stirrup relaxed its hold, and the equerry fell to the ground, his head carried away by a round shot.[1]

Meanwhile the attack of the infantry on Ramillies was fully developed, and relieved the horse from the fire of the village. Twenty fresh squadrons came galloping up at the top of their speed and ranged themselves in rear of the reforming lines. But before they could come into action the Duke of Würtemberg pushed his Danish horse along the Mehaigne upon the right flank of the French, and the Dutch guards advancing still further fell upon their rear. These now emerged upon the table-land by the tomb of Ottomond, and the rest of the Allied horse dashed themselves once against the French front. The famous Maison du Roi after a hard fight was cut to pieces, and the whole of the French horse, despite Villeroy's efforts to stay them, were driven in headlong flight across the rear of their line of battle, leaving the battalions of infantry helpless and alone to be ridden over and trampled out of existence.

Villeroy made frantic efforts to bring forward the

[1] Molesworth escaped and was rewarded four years later, at the age of twenty-two, with a regiment of foot.

cavalry of his left to cover their retreat, but the ground was encumbered by his baggage, which he had carelessly posted too close in his rear. The French troops in Ramillies now gave way, and Marlborough ordered the whole of the infantry that was massed before the village to advance across the morass upon Offus, with the Third and Sixth Dragoon Guards in support. The French broke and fled at their approach; and meanwhile the Buffs and Twenty-first, which had so far remained inactive on the right, forced their way through the swamps before them, and taking Autréglise in rear swept away the last vestige of the French line on the left. Five British squadrons followed them up and captured the entire King's Regiment (Regiment du Roi). The Third and Sixth Dragoon Guards also pressed on, and coming upon the Spanish and Bavarian horse-guards, who were striving to cover the retreat of the French artillery, charged them and swept them away, only narrowly missing the capture of the Elector himself, who was at their head.[1] On this the whole French

1706. May 12/23.

[1] Order of Battle. Ramillies, 12th - 23rd May 1706.

Left. — Right Wing only. — Right.

1st Line.

Foreign Infantry.

- 15th Foot.
- Stringer's Foot.
- Macartney's Foot.
- Evans's Foot.
- 21st Royal Scots Fusiliers.
- 3rd Buffs.
- 8th Foot.
- 23rd Royal Welsh.
- 28th Foot.
- 26th Cameronians.
- 16th Foot.
- 1 ,, Royal Scots.
- 1 Batt. 1st Guards.
- Eighteen Dutch Squadrons.
- 3rd ,, ,,
- 6th ,, ,,
- 7th ,, ,,
- 5th ,, ,,
- 1st Dragoon Guards.
- 5th Royal Irish Dragoons.
- Scots Greys.

2nd Line.

Foreign Infantry.

- 10th ,,
- 24th ,,
- 37th ,,
- 29th Foot.
- 18th Royal Irish.
- 2nd Batt. Royal Scots.

Foreign Cavalry.

From Kane's *Campaigns*.

1706. May 12/23. army, which so far had struggled to effect an orderly retreat, broke up in panic and fled in all directions.

The mass of the fugitives made for Judoigne, but the ways were blocked by broken-down baggage-waggons and abandoned guns, and the crush and confusion was appalling. The British cavalry, being quite fresh, quickly took up the pursuit over the table-land. The guns and baggage fell an easy prey, but these were left to others, while the red-coated troopers, not without memories of Landen, pressed on, like hounds running for blood, after the beaten enemy. The chase lay northwards to Judoigne and beyond it towards the refuge of Louvain. Not until two o'clock in the morning did the cavalry pause, having by that time reached Meldert, fifteen miles from the battlefield; nay, even then Lord Orkney with some few squadrons spurred on to Louvain itself, rekindled the panic and set the unhappy French once more in flight across the Dyle.

May 13/24. Nor was the main army far behind the horse. Marching far into the night, the men slept under arms for two or three hours, started again at three o'clock, and before the next noon had also reached Meldert and were preparing to force the passage of the Dyle. Marlborough, who had been in the saddle with little intermission for nearly twenty-eight hours, here wrote to the Queen that he intended to march again that same night, but, through the desertion of the lines of the Dyle by the French, the army gained some respite. May 14/25. The next day he crossed the Dyle at Louvain and encamped at Betlehem, the next he advanced to May 15/26. Dieghem, a few miles north of Brussels, the next he May 16/27. passed the Senne at Vilvorde and encamped at Grimberghen, and here at last, after six days of incessant marching, the Duke granted his weary troops a halt, while the French, hopelessly beaten and demoralised, retired with all haste to Ghent.

So ended the fight and pursuit of Ramillies, which effectually disposed of the taunt levelled at Marlborough after Blenheim, that he did not know how to improve a victory. The loss of the French in killed, wounded, and prisoners was thirteen thousand men, swelled by desertion during the pursuit to full two thousand more. The trophies of the victors were eighty standards and colours, fifty guns, and a vast quantity of baggage. The loss of the Allies was from four to five thousand killed and wounded, which fell almost entirely on the Dutch and Danes, the British, owing to their position on the extreme right, being but little engaged until the close of the day. The chief service of the British, therefore, was rendered in the pursuit, which they carried forward with relentless thoroughness and vigour. The Dutch were delighted that their troops should have done the heaviest of the work in such an action, and the British could console themselves with the performance of their cavalry, and above all, with the reflection that the whole of the success was due to their incomparable chief. 1706.

The effect of the victory and of the rapid advance that followed it was instantaneous. Louvain and the whole line of Dyle fell into Marlborough's hands on the day after the battle; Brussels, Malines, and Lierre surrendered before the first halt, and gave him the line of the Senne and the key of the French entrenchments about Antwerp; and one day later, the surrender of Alost delivered to him one of the strongholds on the Dender. Never pausing for a moment, he sent forward a party to lay bridges on the Scheldt below Oudenarde in order to cut off the French retreat into France, a movement which obliged Villeroy forthwith to abandon the lines about Ghent and to retire up the Lys to Courtrai. Ghent, Bruges, and Damme thereupon surrendered on the spot; Oudenarde followed them, and after a few days Antwerp itself. Thus within a fortnight after the victory the whole of Flanders and Brabant, with the exception of Dendermond and one or two places of minor importance, had succumbed to the Allies, and the French had fallen back to their own frontier. May-June.

Nor was even this all. A contribution of two

1706. June. million livres levied in French Flanders brought home to the Grand Monarch that the war was now knocking at his own gates. Villars, with the greater part of his army, was recalled from the Rhine to the Lys, and a number of French troops were withdrawn to the same quarter from Italy. Baden had thus the game in his own hand on the Rhine, and though he was too sulky and incapable to turn the advantage to account, yet his inaction was no fault of Marlborough's. We are hardly surprised to find that in the middle of this fortnight the Duke made urgent request for fresh stores of champagne; he may well have needed the stimulant amid such pressure of work and fatigue.[1]

He now detached Overkirk to besiege Ostend and another party to blockade Dendermond, at the same time sending off five British battalions, which we shall presently meet again, for a descent on the Charente which was then contemplated in England. This done he took post with the rest of the Army at Rouslers, to westward of the Lys, whence he could at once cover the siege of Ostend and menace Menin and Ypres. The operations at Ostend were delayed for some time through want of artillery and the necessity of waiting for the co-operation of the Fleet; but the trenches June $\frac{6}{17}$. were finally opened on the 17th of June, and a few weeks later the town surrendered.

June 27 / July 8. Three days after this the army was reassembled for the siege of Menin. This fortress was of peculiar strength, being esteemed one of Vauban's masterpieces, and was garrisoned by five thousand men. Moreover, the French, being in command of the upper sluices of the Lys, were able greatly to impede the operations by cutting off the water from the lower stream, and thus rendering it less useful for purposes of transport. But all this availed it little; for three weeks after the opening of the trenches Menin surrendered. The Aug. $\frac{11}{22}$. British battalions[2] which had been kept inactive at

[1] *Despatches*, vol. ii. p. 554.

[2] The British regiments regularly employed in the besieging

Ramillies took a leading share in the work, and some 1706.
of them suffered very heavily, but had the satisfaction of recapturing four of the British guns that had been taken at Landen.

A few days later Dendermond was attacked in Aug. 25/Sept. 5. earnest and was likewise taken, after which Marlborough fell back across the Scheldt to secure the whole Sept. 12/23. line of the Dender by the capture of Ath. Ten days sufficed for the work, after which Ath also fell into the Sept. 21/Oct. 2. hands of the Allies. The apathy of the French throughout these operations sufficiently show their discouragement. Owing to the supineness of Prince Lewis of Baden Villars had been able to bring up thirty-five thousand men to the assistance of Marshal Vendôme, who had now superseded Villeroy, but even with this reinforcement the two commanders only looked on helplessly while Marlborough reduced fortress after fortress before their eyes. They were, indeed, more anxious to strengthen the defences of Mons and Charleroi, lest the Duke should break into France by that line, than to approach him in the field. Nor were they not wholly unreasonable in their anxiety, for Marlborough's next move was upon the Sambre; but incessant rain and tempestuous weather forbade any further operations, so that Ath proved to be the last conquest of the year. Thus ended the campaign of Ramillies, one of the most brilliant in the annals of war, wherein Marlborough in a single month carried his arms triumphant from the Meuse to the sea.

army were the 8th, 10th, and 18th, and Evans's Foot; the Scots Greys, 3rd and 6th Dragoon Guards. The total loss of the Allies was 32 officers and 551 men killed, 83 officers and 1941 men wounded. The 18th Royal Irish lost 15 officers alone, and in one attack over 100 men in half an hour.

CHAPTER VI

1706. FROM Flanders it is necessary to return to the Peninsula, where we left Peterborough bewailing his enforced inaction. Nothing is more remarkable in the story of these Peninsular campaigns than the utter want of unity in design between the forces of the Allies in Catalonia and in Portugal. Even in England the British troops in these two quarters were treated, for purposes of administration, as two distinct establishments, which might have been divided by the whole breadth of the Atlantic instead of by twice the breadth of England. Yet the fault could hardly be attributed to any English functionary, civil or military. Galway was as anxious as Peterborough to advance to Madrid; but the Portuguese were terrified at the prospect of moving far from their frontier, while the eyes of King Charles ever rested anxiously on the passes by which French reinforcements might advance into Catalonia. In such circumstances it was not easy to accomplish an effective campaign.

The Spaniards of the Austrian party, as has been told, had by the winter of 1705 gained a precarious hold on the whole province of Valencia. Just before the close of the year came intelligence that the Spanish General de las Torres had crossed the northern frontier from Arragon into Valencia and had laid siege to San Mateo. The town was important, inasmuch as it commanded the communications between Catalonia and Valencia, but it was held by no stronger garrison than thirty of the Royal Dragoons and a thousand Spanish irregular infantry under Colonel Jones. This officer defended himself as well as he could, but at once begged urgently for reinforcements. King Charles thereupon appealed for help to Peterborough, who forthwith ordered General Killigrew to march with his garrison from Tortosa and cross the Ebro, while he himself, riding night and day from Barcelona, caught up the column at the close of the first day's march. King Charles had represented the force of Las Torres as but two thousand strong, and had added that thousands of peasants were up in arms against it. Peterborough now discovered that the Spaniards numbered four thousand foot and three thousand horse, while the thousands of armed peasantry were wholly imaginary. His own force consisted of three weak British battalions, the Thirteenth, Thirty-fifth, and Mountjoy's Foot, together with one hundred and seventy of the Royal Dragoons, in all thirteen hundred men. With such a handful his only hope of success must lie in stratagem.

1706.

Dec. 26, 1705. / Jan. 6, 1706.

Advancing southward with all speed he split up his minute army into a number of small detachments, and pushing them forward by different routes arrived early in the morning, unseen and unsuspected, at Traguera, within six miles of the enemy's camp. That same day a spy was captured by the enemy and brought before Las Torres. On him was found a letter from Peterborough to Colonel Jones, written in the frankest and easiest style. "I am at Traguera," so it ran in effect, "with six thousand men and artillery. You may wonder how I collected them; but for transport and secrecy nothing equals the sea. Now, be ready to pursue Las Torres over the plain. It is his only line of retreat, for I have occupied all the passes over the hills. You will see us on the hill-tops between nine and ten. Prove yourself a true dragoon, and have your miquelets (irregulars) ready for their favourite plunder and chase." The spy, being threatened with death, offered to betray another messenger of Peterborough's who was lying concealed in the hills. This

Dec. 28, 1705. / Jan. 8, 1706.

1706. second spy was captured, and a duplicate of the same letter was found on him. The pair of them were questioned, when the first protested that he knew nothing of the strength of Peterborough's force, while the other declared that the despatch spoke truth. Suddenly came intelligence from the Spanish outposts that the enemy was advancing in force in several columns, and presently the red-coats appeared at different points on the hill-tops, making a brave show against the sky. Las Torres became uneasy. His depression was increased by the accidental explosion of one of his own mines before San Mateo ; and he hastily ordered an immediate retreat. Whereupon out came Jones with his garrison, and turned the retreat into something greatly resembling a flight ; while Peterborough with his thirteen hundred men walked quietly into San Mateo and took possession of the whole of the enemy's camp and material of war. The trick, for the whole incident of the captured spies had been carefully preconcerted, had proved a brilliant success.

Las Torres, though disagreeably shaken, was recovering his equanimity when, on the second day of the retreat, a friendly spy came to warn him that an English force was marching parallel to his left flank, was already in advance of him, and was likely to cut off his retreat by seizing the passes into the plain of Valencia. The warning was scouted as ridiculous, but the spy offered, if two or three officers would accompany him, to prove that he was right. Two officers, disguised as peasants, were accordingly guided to a point already indicated by the spy, where they were promptly captured by a picquet of ten of the Royal Dragoons. The spy, however, undertook to produce liquor, the dragoons succumbed or seemed to succumb to their national failing, and the three captives slipped out, took three of the dragoons' horses and galloped back with all speed to Las Torres to confirm the spy's story. Their escape did not prompt them to make the least of their adventure ; the housings of the horses testified incontestably to the actual presence of English dragoons; and Las Torres broke up his camp on the spot and hurried away once more. Once again the tricks of the eccentric Englishman had been successful; for the friendly spy was in reality a Spanish officer in his own army; and though there were undoubtedly ten English dragoons, who had been specially sent for the purpose, in advance of Las Torres at that particular moment, yet there were no more English within twenty miles of them. 1706.

Las Torres was still retreating southward by the coast-road, and Peterborough was making a show of pursuit by marching wide on his right flank, when a pressing message reached him from King Charles. A French force of eight thousand men was advancing into Catalonia from Roussillon; a second force of four or five thousand men under Count Tserclaes de Tilly was threatening Lerida, and a third under Marshal Tessé was marching through Arragon upon Tortosa. Seeing that the King was urgent for help in Catalonia, but intent on pursuing his own design in Valencia, Peterborough resolved to send his infantry to the coast at Vinaroz, to be transported if necessary by sea. The men, though ragged, shoeless, and much distressed by long marches through the wintry days, left him very unwillingly. Then summoning the garrison of Lerida[1] and a reinforcement of Spaniards to follow him to Valencia, Peterborough resumed the pursuit of Las Torres with one hundred and fifty dragoons. Jan. $\frac{1}{12}$.

He was too late to save Villa Real, which Las Torres took by treachery, and having taken massacred the entire male population; but while always concealing his own weakness he contrived by incessant harassing of the enemy's rear to inflict considerable loss and annoyance. Thus in due time he reached Nules, three days' march from the city of Valencia, a town of considerable strength, where Las Torres had left arms sufficient to equip a thousand of the townsmen. Peterborough January.

[1] 8th Dragoons (now Hussars), 30th and 34th Foot; two Dutch and two Neapolitan battalions.

1706. marched straight up to the gate with his handful of dragoons. The townspeople manned the walls and opened fire, but were speedily checked by a message from Peterborough, bidding them send out a priest or a magistrate instantly on pain of having their walls battered down and every soul put to the sword, in revenge for Villa Real. Some priests who knew him at once came out to him. "I give you six minutes," said Peterborough to the trembling cassocks. "Open your gates or I spare not a soul of you." The gates were quickly opened, and the General, riding in at the head of his tattered dragoons, demanded immediate provision of rations and forage for several thousand men.

The news soon reached Las Torres, who was little more than an hour ahead, and for the third time his unfortunate army was hurried out of camp and condemned to a weary retreat from an imaginary enemy. Peterborough, however, after taking two hundred horses from Nules, left the town to ponder over its fright and retired to Castallon de la Plana. Having there raised yet another hundred horses he ordered the Thirteenth Foot to march from Vinaroz to Oropesa and went thither himself to inspect them. The men marched in but four hundred strong, with red coats ragged and rusty, yellow facings in tatters, yellow breeches faded and torn, shoes and stockings in holes or more often altogether wanting. "I wish," said Peterborough when the inspection was over, "that I had horses and accoutrements for you, to try if you would keep up your good reputation as dragoons." The men doubtless glanced at their sore and unshod feet, and silently agreed. Presently they were marched up to the brow of a neighbouring hill, where to their amazement they found four hundred horses awaiting them, all fully equipped. The officers received commissions according to their rank in the mounted service, two or three only being detached to raise a new battalion in England; and thus within an hour Barrymore's Foot became Pearce's Dragoons.

Peterborough now called in such additional weak battalions of British as he could, and having collected a total force of three thousand men, one-third of it mounted, prepared to outwit a new general, the Duke of Los Arcos, who had superseded Las Torres. The relief of Valencia was Peterborough's first object, but to effect this he had first to gain possession of Murviedro, which lay on his road and was occupied by the enemy, and that, too, in such a way that Los Arcos should not move out against him in the open plain and crush him by superior numbers. It was a difficult problem, and it was only solved by a trick too elaborate and lengthy to be detailed here. The plan was very clever, so clever as almost to transcend the bounds of what is fair in war, but it was completely successful; and on the 4th of February Peterborough marched into Valencia without firing a shot.

1706. January.

January 24 / February 4.

He now cultivated the friendship of the priests and something more than the friendship of the ladies of Valencia, thereby combining pleasure with business and obtaining the best of information. Las Torres, who had once more superseded Los Arcos, presently appeared on the scene again, bringing four thousand men by land and a powerful siege-train by sea for the reduction of the city. Peterborough pounced upon the train directly after it had been landed and captured the whole of it; then sending twelve hundred men against the four thousand he surprised them, routed them, and took six hundred prisoners. But the pleasant and exciting life at Valencia was interrupted by an urgent summons to assist in the defence of Barcelona. King Lewis, at the entreaty of his grandson Philip, had resolved to make a great effort to recover it; and thus it was that at the beginning of April Marshal Tessé appeared before the city with twenty-five thousand men, and three days later began the siege in form. The garrison consisted of less than four thousand regular troops, the backbone of which were eleven hundred British of the Guards and the Thirty-fourth Foot.

March 23 / April 3.

1706. April. Weak as it was this little force made a gallant resistance, but the odds were too great against it, and but for the arrival of Peterborough it could not have held out for more than a fortnight. Even after his coming it was well-nigh overpowered; for of the three thousand troops that he brought with him the most part were employed chiefly in harassing Tessé's communications from the rear. The siege was finally raised on the April 30/May 11. advent of a relieving squadron under Admiral Leake, which so much discouraged Tessé that he abandoned the whole of his siege-train and retired once more over the French frontier.

Nothing now remained but to take advantage of this piece of good fortune. Peterborough had always favoured a dash on Madrid, and had twice urged this course upon King Charles in vain. He now pressed it for a third time with success, and presently sailed for Valencia with eleven thousand men. With immense trouble he procured horses and accoutrements to convert some of his infantry into dragoons, and then pushing forward a detached force of English he succeeded by the beginning of July in capturing Requena and Cuença and opening the road for King Charles to Madrid.

Meanwhile, after enormous delay, the English and Portuguese had actually begun operations from the side March 20/31. of Portugal against Marshal Berwick. On the 31st of March Lord Galway and General das Minas left Elvas with nineteen thousand men[1] and advanced slowly northward, forcing back Berwick, whose army was much inferior in number, continually before them. Alcantara, Plasencia, and Ciudad Rodrigo yielded to May 27/June 7. them after slight resistance; and by the 7th of June the Allied army had reached Salamanca, a country which two regiments, the Second and the Ninth, were to know better a century later. Then turning east it marched straight upon Madrid and entered the city on June 16/27. the 27th of June. So far all was well. The advance

[1] 2200 of them British, 2nd Dragoon Guards, 2nd, 9th, 17th, 33rd, and Brudenell's Foot.

from Portugal had been singularly slow, but the capital 1706. had been reached. King Philip had retired to Burgos, and King Charles had been proclaimed in Madrid. The object of the War of the Succession seemed to have been fulfilled in Spain.

At this juncture, however, the operations for no particular reason came to an end. Galway, without a thought apparently of following up Berwick, halted for a fortnight in Madrid, where the Portuguese troops behaved disgracefully, and then moving a short distance north-eastward took up a strong position at Guadalaxara. July 4/15. King Charles after immense delay suddenly altered the route which Peterborough had marked out for him and insisted on marching to Madrid through Arragon, even so not reaching Saragossa till the 18th of July. Mean- July 7/18. while the whole of the country through which Galway had marched rose in revolt against the House of Austria. Berwick, reinforced from France to twice the strength of Galway, cut him off from Madrid, and reproclaimed King Philip; and when Charles and Peterborough with three thousand men at last joined Galway on the 6th of August, the Archduke found that July 26/August 6. he must prepare not for triumphant entry into Madrid, but for what promised to be a difficult and perilous retreat.

Peterborough was for a sudden spring at Alcala and so on Madrid, but being over-ruled retired to Italy to raise a loan for the army. Galway, whose army had been so much reduced by sickness as to number, with Peterborough's reinforcement, but fourteen thousand men, still lingered close to Madrid for nearly a month in the vain hope of seeing the tide turn in his favour. Finally, being cut off from his base in Portugal, he marched for Valencia and the British fleet, Berwick troubling him no further than by occasional harassing of his rearguard. On crossing the Valencian frontier Sept. 17/28. he distributed his force into winter quarters; an example which, after the reduction of Carthagena and of sundry small strongholds, was imitated by Berwick at the end of November.

1706. So closed the year 1706, memorable for two of the most brilliant, even if in some respects disappointing, campaigns ever fought simultaneously by two British generals.

1707. Unexpected reinforcements from Britain came opportunely to revive the hopes of the Archduke Charles at the opening of the new year. It will be remembered that in the summer of 1706 a project for a descent on the Charente had been matured in England, for which Marlborough had detached certain of his battalions after Ramillies. The plan being considered doubtful of success, the destination of the expedition was altered to Cadiz. A storm in the Bay of Biscay, however, dispersed the fleet, which was only reassembled at Lisbon after very great delay, and after waiting in that port for two months was directed to place its force at the disposal of Galway.[1] In December 1706 Peterborough returned from Italy to Valencia to attend the councils of war respecting the next campaign. The general outlook in the Peninsula was not promising. Marlborough indeed opined that nothing could save Spain but an offensive movement against France from the side of Italy, and Peterborough, adopting the same view, strongly advocated a defensive campaign. He was overruled, and since his endless squabbles with his colleagues and his military conduct in general had been called in question in England, he was shortly after relieved of his command and returned to England.

March. After his departure the Archduke Charles and the English commanders fell at variance over their alternative plans, with the result that Charles withdrew with the whole of the Spanish troops to Catalonia. Galway and Das Minas then decided first to destroy Berwick's magazines in Murcia, and this done to march up the

[1] The total force comprehended 6900 men. Two squadrons each of the 3rd and 4th Dragoons (now Hussars) and seven squadrons of foreigners; the 28th, 29th, Hill's, Watkins's, Mark Kerr's, Macartney's Foot, two battalions of Marines, one of Germans and six of Huguenots.

Guadalaviar, turn the head-waters of the Tagus, and so 1707. move on to Madrid. Though the reinforcements had reached the Valencian coast in January it was not until the 10th of April that Galway crossed the Murcian March 30/April 10. frontier and after destroying one or two magazines laid siege to Villena. While thus engaged he heard April. that Berwick having collected his army was advancing towards Almanza, some five and twenty miles to the north-east, and that the Duke of Orleans was on his way to join him with reinforcements. Thereupon Galway and Das Minas resolved to advance and fight him at once, apparently without taking pains to ascertain what the numbers of his army might actually be. Berwick had with him twenty-five thousand men, half French, half Spanish, besides a good train of artillery. Galway, owing to the frightful mortality on board the newly-arrived transports, had but fifteen thousand, of which a bare third were British, half were Portuguese, and the remainder Dutch, German, and Huguenot. Considering how poorly the Portuguese had behaved on every occasion so far, the result of an open attack against such odds could hardly be doubtful.

Berwick on his side drew up his army in the usual April 14/25. two lines on a plain to the south of Almanza, his right resting on rising ground towards Montalegre, his left on a height overlooking the road to Valencia, while his right centre was covered by a ravine which gradually lost itself on level ground towards his extreme right flank. The force was formed according to rule with infantry in the centre and cavalry on each flank, the Spaniards taking the right and the French the left. At midday, after a march of eight miles, Galway approached to within a mile of the position, and formed his line of battle according to the prescribed methods. The Portuguese, with poor justice, claimed the post of honour on the right wing, so that the British and Dutch took the left, though with several Portuguese squadrons among them in the second line. But finding himself weak in cavalry Galway made good

1707. the deficiency, after the manner of Gustavus Adolphus, by interpolating battalions of foot among his horse.[1]

April $\frac{14}{25}$. At three o'clock in the afternoon Galway opened the attack without preliminary fire of artillery by leading an advance of the horse on his left wing. He was driven back at first by sheer weight of numbers; but the Sixth and Thirty-third Foot, which were among the interpolated battalions, came up, and by opening fire on the left flank of the Spanish horse gave the English squadrons time to rally and by an effective charge to drive the Spaniards back in confusion. Meanwhile, the rest of the English foot on the left centre fell, heedless of numbers, straight upon the hostile infantry and drove them back in confusion upon their second line. The Guards and the Second Foot following up their success broke through the second line also and pursued the scattered fugitives to the very walls of Almanza. So far as the Allied left was concerned the battle was going well.

But meanwhile the Portuguese on the right remained motionless; and Berwick lost no time in launching his left wing of horse upon them. Then the first line of Portuguese horse turned and ran, the second line also turned and ran, and the first line of infantry was left to bear the brunt alone. For a time the battalions stood up gallantly enough, but the odds were too great, and they were presently overwhelmed and utterly dispersed. Then Berwick brought up his French, both horse and foot, against the victorious British on his right. The British cavalry had suffered heavily in the first attack, all four regiments having lost their commanding officers, and in spite of all their efforts they were borne back and swept away by the numbers of the French squadrons. The infantry, surrounded on all sides, fought desperately and repeatedly repulsed the enemy's onset, but being overpowered by

[1] Colonel Parnell calls this a novelty and approves it; Colonel Frank Russell condemns it. The practice was not proscribed, but it was recognised as extremely hazardous (see Kane's *Campaigns*, ed. 1757, pp. 69-70), and received its final condemnation at the hands of Napoleon. *Campagnes de Turenne*.

numbers, were nearly all of them, English, Dutch, and Germans, cut down or captured. By great exertions Galway, who was himself wounded, brought off some remnant of them in good order and retreated unpursued to Ontiniente, some twenty miles distant. The guns also were saved; but a party of two thousand infantry which had been brought off the field by General Shrimpton was surrounded on the following day and compelled to lay down its arms.

1707. April $\frac{14}{25}$.

In this action, which lasted about two hours, Galway lost about four thousand killed and wounded and three thousand prisoners. The British alone lost eighty-eight officers killed, and two hundred and eighty-six captured, of whom ninety-two were wounded. The Sixth regiment had but two officers unhurt out of twenty-three, the Ninth but one out of twenty-six, and other regiments[1] suffered hardly less severely. The simple fact was that, as the bulk of the Portuguese would not fight, the action resolved itself into an attack of eight thousand British, Dutch, and Germans upon thrice their number of French and Spaniards, in an open plain; and the defeat, though decisive, was in no sense disgraceful except to the Portuguese. The most singular circumstance in this fatal day was that the French were commanded by an Englishman, Berwick, and the English by a Frenchman, the gallant but luckless Ruvigny. The battle of course put an end to further operations on the side of the Allies. Galway, with such troops as he could collect, retired to the Catalonian frontier, and set himself to reorganise a force to defend the lines of the Segre and Ebro, while Berwick methodically pursued the reduction of Valencia and in December retired, according to rule, into winter quarters. So

[1] The British regiments present were the Queen's Bays, 3rd, 4th, and 8th Dragoons (now Hussars), Peterborough's and Pearce's Dragoons, Guards (mixed battalion); 2nd, 6th, 9th, 11th, 17th, 28th, 33rd, 35th, 36th, Mountjoy's, Macartney's, Breton's, Bowles's, Mark Kerr's Foot. List of casualties of officers will be found in the *Postboy*, 26th June 1707. See order of battle on next page.

1707. swiftly did disaster follow on the first brilliant successes in the Peninsula.

Since we shall not again see Peterborough in the field this chapter should not be closed without a few sentences as to his peculiar methods. These were outwardly simple enough. Good information to discover his enemy's weak points, deception to put him off his guard, the deepest secrecy lest that enemy should grow suspicious, most careful thinking out of details so that every unit of an insignificant force should know its duty precisely and do it, exact divination of the probable results of each successive step, and extreme suddenness and rapidity in execution; such were, so far as they can be set down, the secrets of his success. In a word, his was the principle of making war by moral rather than by physical force,

ORDER OF BATTLE. ALMANZA.

LEFT WING ONLY.

Left. — Right.

1st Line.

1st Royal Dragoons.
7th Dragoons (Hussars).
Essex's Dragoons (4th Hussars).
Guiscard's Dragoons.

Wade's Brigade.
6th ,,
33rd Foot.
8th Dragoons (Hussars).
Peterborough's Dragoons.
17th Foot.
Mountjoy's Foot.

Two regiments of Dutch horse.
Queen's Bays.
Four Dutch regiments of horse

Macartney's Brigade.
1 Batt. English Guards.
35th Foot.
Macartney's ,,
Mordaunt's Foot.

Two Dutch Brigades.

2nd Line.

Portuguese Dragoons.
Four Squadrons.

Hill's Brigade.
9th ,,
36th Foot.
Three Portuguese Squadrons.
Mark Kerr's Foot.
11th Foot.

Four Portuguese Squadrons.

2nd Foot.
Bretton's.
Nassau's.
Bowles's.

Postboy, 5th-7th June 1707.

by scaring men into the delusion that they were 1707. beaten rather than by actually beating them. It is a difficult art, of which the highest exponent was produced by the Navy a century later in the person of Lord Dundonald; and it is curious to note that both men were troubled by exactly the same defects. Peterborough was difficult, cantankerous, quarrelsome and eaten up by exaggerated appreciation of self. His letters were so interminably long and tedious, containing indeed little besides abuse of his colleagues, that they exhausted the patience even of Marlborough. In fact, it seems to be impossible for this type of man to work harmoniously with his equals, however he may be adored by his subordinates. The Duke of Wellington summed up Peterborough as a brilliant partisan, but his contemporaries thought more highly of him. Eugene declared that he thought like a general, and Marlborough himself acknowledged that he had predicted the ill consequences of the operations which, contrary to his advice, were undertaken in Spain. But whatever his merit as a general and a leader, he, like all of his kind, is a man of whom we take leave without regret, turning gladly from the fitful, if dazzling flashes of his eccentric genius, to the steady glowing light which illuminates every action of the great Duke of Marlborough.

AUTHORITIES.—It is well known that the exploits of Peterborough rest principally on Carleton's *Memoirs*, and that the authority of these *Memoirs* is disputed. Colonel Frank Russell in his *Life of Peterborough* of course makes him a hero, Colonel Arthur Parnell in his *War of the Succession in Spain* refuses to allow him any merit. Mr. Stebbing in his *Peterborough* (Men of Action Series) treats the controversy with strong good sense, and I have not hesitated to follow his view. I must none the less acknowledge my obligations to all three of these writers, and particularly to Colonel Parnell, who has gone deeply into the history of the war, taken immense pains to ascertain which British regiments were engaged at every action, and has furnished a most copious list of authorities. The *Mémoires de Berwick* are most trustworthy on the French side, and the *Richards Papers* (Stowe Coll. B.M.), as Colonel Parnell says, most important.

CHAPTER VII

1707. ALMANZA was a bad opening for the new year, but worse was to follow. Throughout the winter Marlborough had, as usual, been employed in diplomatic negotiations, which nothing but his skill and fascination could have carried to a successful issue. But on one most important point the Duke was foiled by the treachery of the Emperor, who, to further his own selfish designs on Naples, secretly concluded a treaty with France for the neutrality of Italy, and thus enabled the whole of the French garrisons in Italy to be withdrawn unmolested. The forces thus liberated were at once brought up to the scene of action on the Rhine and in Flanders, and the French were enabled to bring a superior force in the field against Marlborough. Again the Duke had hoped to save Spain by an invasion of France from the side of Savoy, but this project again had been deferred until too late, owing to the Emperor's cupidity for the possession of Naples. Finally, though Prince Lewis of Baden had died during the winter, he had been replaced on the Rhine by a still more incompetent prince, the Margrave of Bayreuth, who, far from making any diversion in the Duke's favour, never ceased pestering him to come to his assistance. So flagrant was this deplorable person's incapacity that he too was superseded before the close of the campaign, though too late for any effective purpose. His successor, however, deserves particular notice, being none other than the Elector of Hanover, afterwards our own King George the First, no genius in the field, but, as shall be seen in due time, an extremely sensible and clear-headed soldier. 1707.

The result of these complications was that Marlborough spent the greater part of the summer encamped, in the face of a superior French force, at Meldert, on a branch of the Great Geete, to cover his conquests in Flanders and Brabant. At last the Emperor, having accomplished his desires in Naples, made a diversion towards Provence which drew away a part of the French force to that quarter and enabled the Duke to move. But then bad weather intervened to prevent any successful operations. Twice Marlborough was within an ace of surprising Vendôme, who had superseded Villeroy in Flanders, and twice the marshal decamped in haste and confusion only just in time to save his army. Even so the Duke would have struck one heavy blow but for the intervention of the Dutch deputies. But fortune favoured the French; the rain came down in torrents, and the country was poached into such a quagmire by the cavalry that many of the infantry were fairly swallowed up and lost.[1] Thus tamely ended the campaign which should have continued the work of Ramillies.[2]

[1] Parker.

[2] ORDER OF BATTLE. CAMPAIGN OF 1707.

RIGHT WING ONLY.

Left. 1st Line. — Right.

Lord North and Grey's Brigade.	Temple's Brigade.	Meredith's Brigade.			Palmer's Brigade.	Stair's Brigade.
3rd Buffs.	2nd Batt. Royal Scots.	1 Batt. 1st Guards.	Foreign horse.	Orrery's Foot.	1st Dragoon Guards.	Scots Greys.
21st Royal Scots Fusiliers.	18th Royal Irish.	1 " Royal Scots.		Evans's "	5th " "	5th Royal Irish Dragoons.
37th Foot.	Temple's Foot.	16th Foot.			7th " "	
26th Cameronians.	24th Foot.	23rd Royal Welsh.			6th " "	
15th Foot.	10th "	8th Foot.			3rd " "	
Gore's "						

No British in the Second Line.

Postboy, 26th June 1707.

1707. Returning home in November Marlborough found difficulties almost as great as he had left behind him in Flanders. There were quarrels in the Cabinet, already foreboding the time when the Queen and the people should turn against him. The Court of France was reverting to its old methods and endeavouring to divide England by providing the Pretender with a force for invasion. Again the hardships of the campaign in Flanders and the defeat of Almanza had not only created discontent, but had enormously increased the demand for recruits. The evil work of the Dutch deputies and the incorrigible selfishness and jealousy of the Empire had already prolonged the war beyond the limit assigned by the short patience of the English people.

Happily Parliament was for the present still loyal to the war, and voted not only the usual supplies but money for an additional ten thousand men. Five new battalions[1] were raised, and three more of the old establishment were detailed for service in Flanders.[2] But far more satisfactory was the fact that in 1708 all regiments took the field with new colours, bearing the cross of St. Andrew blended with that of St. George, pursuant to the first article of the Treaty of Union, passed in the previous year between England and Scotland.

1708. The early spring of 1708 was wasted by the French in a futile endeavour to set the Pretender afoot in March 29/April 9. Scotland with a French force at his back; nor was it until the 9th of April that Marlborough sailed for the Hague, where Eugene was already awaiting him. There the two agreed that the Duke should as usual command in Flanders, while Eugene should take charge of an army on the Moselle, nominally for operations on that river, but in reality to unite with

[1] Slane's, Brazier's, Delaune's, Jones's, Carles's, all raised in September.

[2] Mixed battalion of Guards, 19th Foot, Prendergast's (late Orrery's).

Marlborough by a rapid march and give battle to the French before they could call in their remoter detachments. There was a considerable difficulty with the Elector of Hanover, who was to command on the Rhine, owing to his jealousy of Eugene, but this trouble was satisfactorily settled, as were all troubles of the time, by the intervention of Marlborough. Thereupon the Electoral Prince, true to the quarrelsome traditions of his family, at once insisted on taking service with Eugene, simply for the sake of annoying his father; thus adding one more to the many causes of friction which, but for Marlborough, would soon have brought the Grand Alliance to a standstill. This Electoral Prince will become better known to us as King George the Second. 1708.

The French on their part had made extraordinary exertions in the hope of a successful campaign. Since Ramillies they had drawn troops from all quarters to Flanders; and from thenceforth the tendency in every succeeding year grew stronger for all operations to centre in that familiar battle-ground. On the Rhine the Elector of Bavaria held command, with Berwick, much exalted since Almanza, to help him. The French main army in Flanders numbered little less than a hundred thousand men, and was under the orders of Vendôme, with the Duke of Burgundy in supreme command. The presence of the heir to the throne, of his brother the Duke of Berry, and of the Chevalier de St. George, as the Pretender called himself, all portended an unusual effort.

Marching up at the end of May from their rendezvous on the south of the Haine, the French army moved north to the forest of Soignies. Marlborough thereupon at once concentrated at Hal and summoned Eugene to him with all haste. His own army numbered but eighty thousand men, and though as usual he showed a bold front he knew that such disparity of numbers was serious. The French then manœuvred towards Waterloo as if to threaten Louvain, a move- May.

1708. ment which the Duke met by a forced march to Park May 24/June 4 on the Dyle. Here he remained perforce inactive for a whole month, waiting for Eugene, who was delayed by June 4 to June 24/July 5. some petty formalities which were judged by the Imperial Court to be far more important than military operations. Suddenly, on the night of the 4th of July, June 23/July 4. the French broke up their camp, marched westward to cross the Senne at Hal and detached small corps against Bruges and Ghent. Unable to meet the Allies with the sword, the French had substituted gold for steel and had for some time been tampering with the new authorities in these towns. The gold had done its work. Within twenty-four hours Ghent and Bruges had opened their gates, and the keys to the navigation of the Scheldt and Lys were lost.

June 24/July 5. Marlborough, who was quite ready for a march, was up and after the French army immediately. At two o'clock in the morning his army was in motion, streaming off to pass the Senne at Anderlecht. The march was long and severe, the roads being in so bad a state that the right wing did not reach its halting-ground until six o'clock in the evening, nor the left wing till two o'clock on the following morning; but this great effort brought the Allies almost within reach of the June 25/July 6. French army. In the night intelligence was brought to Marlborough that the enemy was turning back to fight him. He was in the saddle at once, to form his line of battle; but the news was false. The French in reality were making off as fast as they could; and before the truth could reach Marlborough they were across the Dender. Marlborough's cavalry was instantly on their track, but could do no more than capture a few hundred prisoners together with most of the French baggage. That same day came definite information of the loss of Ghent and Bruges, and of the investment of the citadel of Ghent. Brussels took the alarm at once. The French, as they feared, had for once got the better of the Duke. The French army was encamped at Alost, where, like a king between two pieces at draughts, it threatened both the citadel of Ghent and Brussels; and all was panic in the capital. The Duke was fain to move on to Assche, midway between Alost and Brussels, to restore the confidence of the fearful city. 1708.

Here Eugene joined him. Finding it hopeless to arrive in time with his army, he had pushed on alone; nor could he have arrived more opportunely, for the Duke was so much weakened by an attack of fever that he was hardly fit for duty. It was indeed a trying moment. The next design of the French was evidently aimed at Oudenarde for the recovery of the line of the Scheldt. They were already across the Dender and ahead of Marlborough on the road to it, and moreover had broken down the bridges behind them; yet Marlborough dared not move lest he should expose Brussels. He sent orders to the Governor of Ath to collect as many troops as he could and throw himself into Oudenarde, which that officer punctually did; and then there was nothing to be done but to wait. Two days sufficed to place the citadel of Ghent in the hands of the French, and to set their army free for further operations. Accordingly on the 9th of July Vendôme sent forward detachments to invest Oudenarde, and moved with the main army up the Dender to Lessines, from which point he intended to cover the siege. Great was his astonishment on approaching the town on the following day to find that Marlborough had arrived there before him, and was not only within reach of Oudenarde but interposed between him and his own frontier. June 28/July 9.

For at two o'clock on the morning of the 9th of July the Allied army had marched off in beautiful order in five columns, and by noon had covered fifteen miles to Herfelingen on the road to the Dender. Four hours later Cadogan was sent forward with eight battalions and as many squadrons to occupy Lessines and throw bridges over the Dender; and when tattoo beat that night the army silently entered on a march of thirteen further miles to the same point. Before dawn came the welcome intelligence that Cadogan had reached June 29/July 10.

1708. his destination at midnight, laid his bridges, and made his disposition to cover the passage of the troops. The army tramped on, always in perfect order, crossed the river and was taking up its camping-ground, when the heads of the enemy's columns appeared on the distant heights and were seen first to halt and then to retire. Marlborough on the curve of the arc had outmarched Vendôme on the chord.

The French, finding the whole of their plans disconcerted, now wheeled about north-westward towards Gavre on the Scheldt, to shelter themselves behind the river and bar the advance of the allies on Bruges. But the Duke had no intention to let them off so easily. Burgundy and Vendôme were not on good terms; their differences had already caused considerable confusion in the army; and Marlborough was fully aware of the fact. At dawn on the morning of the 11th the unwearied Cadogan started off with some eleven thousand men[1] and twenty-four guns to prepare the roads, construct bridges, and make dispositions to cover the passage of the Scheldt below Oudenarde. By half-past ten he had reached the river, just above the village of Eyne, and on ascending the low heights above the stream and looking westward he saw before him a kind of shallow basin or amphitheatre, seamed by little ditches and rivulets, and broken by hedges and enclosures. To the south the rising ground on which he stood swept round almost to the glacis of Oudenarde, thence curved westward from the village of Bevere into another broad hill called the Boser Couter to the village of Oycke and beyond, thence round northward across the valley of the river Norken to Huysse, whence trending still to northward it died away in the marshes of the Scheldt. Near Oycke two small streams rise which, after pursuing for some way a parallel course,

June 30 / July 11.

[1] 16 battalions and 30 squadrons. In these were included the brigades of Sabine, viz., 8th, 18th, 23rd, 37th; of Evans, viz., Orrery's, Evans's and two foreign battalions; and of Plattenberg, which included the Scotch regiments of the Dutch service.

unite to run down into the Scheldt at Eyne; beyond 1708.
them the Norken runs beneath the heights of Huysse June 30 / July 11
in a line parallel to the Scheldt.

Presently parties of French horse appeared on the ground to the north. Vendôme's advanced-guard, under the Marquis of Biron, had crossed the Scheldt leisurely at Gavre, six miles farther down the river, and was now moving across his front with foragers out, in happy unconsciousness of the presence of an enemy. A dash of Cadogan's squadrons upon the foragers quickly brought Biron to Eyne and beyond it, where he caught sight of Cadogan's detachment of scarlet and blue battalions guarding the bridge, and presently of a body of cavalry in the act of crossing; for Marlborough, uneasy while his advanced-guard was still in the air, had caught up a column of Prussian horse and galloped forward with it in all haste. Biron at once reported what he had seen to Vendôme, who, perceiving that the mass of the Allied army was still on the wrong side of the Scheldt, gave orders to take up a position parallel to the river; the line to rest its left on the village of Heurne and extend by Eyne and Beveren to Mooregem on the right. In pursuance of his design he directed seven battalions to occupy Heurne forthwith; but at this point the Duke of Burgundy interposed. The heights of Huysse in rear of the Norken from Asper to Wannegem formed in his judgment a preferable position; and there, two miles from the Scheldt, he should form his line of battle, facing south-east. So the army was guided to the left bank of the Norken, while the seven battalions, obeying what they conceived to be their orders, marched down to the village not of Heurne but of Eyne, and backed by a few squadrons, took up the position assigned to them by Vendôme.

Meanwhile, responding to urgent messages from Marlborough, the main body of the Allies was hurrying forward, and by two o'clock the head of the infantry had reached the Scheldt. Part of the cavalry passed through Oudenarde to take advantage of the town

1708. June 30 / July 11. bridge; the foot began to cross by the pontoons, and Cadogan, whose eye had marked the march of the French into Eyne, at once summoned the whole of his advanced-guard across to the left bank. Sabine's brigade supported by the other two crossed the rivulet against Eyne, while the Hanoverian cavalry moved up to the rear of the village and cut off all hope of retreat. Presently Sabine's British were hotly engaged; but the French made but a poor resistance. It is the weakness of the French soldier that he apprehends too quickly when his officers have not given him a fair chance. Three battalions out of the seven were captured entire, the remaining four were killed or taken piecemeal in their flight. The cavalry, flushed by their success, then advanced under Prince George against the few French squadrons in rear of the village, charged them, routed them, and drove them across the Norken. The Prince had his horse shot under him in this encounter, for his family has never wanted for courage, and he remembered the day of Oudenarde to the end of his life.

The Duke of Burgundy now made up his mind to a general action, and made every preparation for defence of the position behind the Norken. But when four o'clock came and the Allied army was not yet in order of battle, he changed his plan, pushed a body of cavalry from his right across the stream, and set the whole of his centre and right in motion to advance likewise. Marlborough, perceiving the movement, judged that the attack would be directed against his left, in the hope that Cadogan's battalions about Eyne would be left isolated and open to be crushed by an advance of the French left. Two of Cadogan's regiments, Prussians, which had been pushed forward half a mile beyond Eyne to Groenewald were at once reinforced by twelve more of the advanced guard; the British cavalry was formed up on the heights at Bevere, and the Prussian horse further to the Allied right near Heurne. No more could be done until the rest of the army should gradually cross the river which divided it from the battlefield.

At length about five o'clock thirty French battalions debouched upon Groenewald, which was as yet held only by Cadogan's two advanced regiments, and began the attack. The Prussians stuck to their post gallantly and held their own among the hedges, until presently Cadogan's reinforcement, and later on twenty more battalions under the Duke of Argyll,[1] came up to their assistance. Forming in succession on the left of the Prussians as they reached the fighting line, these regiments extended the field of action as far south as Schaerken; and the combat was carried on with great spirit. The ground was so strongly enclosed that the fight resolved itself into duels of battalions, the cream of the infantry on both sides being engaged. At one moment the French outflanked the left of the Allies and drove them back, but fresh battalions of Marlborough's army kept constantly streaming into action, which recovered the lost ground and prolonged the line of fire always further to the south.

1708. June 30 / July 11.

Marlborough and Eugene, who had hitherto remained together, now parted, and the Duke handing over eighteen battalions to the Prince entrusted him with the command of the right. This accession of strength enabled Eugene to relieve Cadogan's corps, which had been forced to give way before Groenewald, and even to pierce through the first line of the enemy's infantry. General Natzmar thereupon seized the moment to throw the Prussian cavalry against the second line. His squadrons were received with a biting fire from the hedges as they advanced; and the French Household Cavalry watching the favourable moment for a charge drove back the Prussians with very heavy loss.

Meanwhile Marlborough with the Hanoverian and Dutch infantry was pressing forward slowly on his left, the French fighting with great stubbornness and gallantry, and contesting every inch of ground from hedge to hedge. At last the enemy being forced back to Diepenbeck, a few hundred yards in rear of Schaerken, stood

[1] Among them the Royal Scots and Buffs.

1708. June 30/July 11. fast, and refused despite all the Duke's efforts to give way for another foot. But Marlborough had still twenty battalions of Dutch and Danes with almost the entire cavalry of the left at his disposal, and he had noticed that the French right flank rested on the air. He now directed Marshal Overkirk to lead these troops under cover of the Boser Couter round the French right and to fall with them upon their rear. The gallant old Dutchman, though infirm and sick unto death, joyfully obeyed. Two brigades were thrown at once on the flank of the troops that were so stoutly opposing Marlborough; while the cavalry advanced quickly on the reverse slope of the Boser Couter,[1] and then wheeling to the right fell on the rear of the unsuspecting French. A part of the Household Cavalry and some squadrons of dragoons tried bravely to stand their ground, but they were borne back and swept away. Overkirk's troops pressed rapidly on; and the French right was fairly surrounded on all sides.

Now at last an effort was made to bring forward the French left, which through Burgundy's perversity or for some inscrutable reason, had been left motionless on the other side of the Norken; but it was too late. The infantry, though led by Vendôme himself, failed to make the slightest impression, and the cavalry dared not advance. The ground before them was intricate and swampy, and the whole of the British cavalry, withdrawn from their first position by Eugene, stood waiting to plunge down upon them directly they should move. The daylight faded and the night came on, but the musketry flashed out incessantly in an ever narrowing girdle of fire, as the Allies wound themselves closer and closer round the enveloped French right. At length at nine o'clock Marlborough and Eugene, fearful lest their own troops should engage each other in the darkness, with some difficulty enforced the order to halt and cease firing. Vast numbers of the French seized the

[1] That is to say, on the western side of the road from Oudenarde to Deynze.

moment to escape, but presently all the drums of the Allies began with one accord to beat the French retreat, while the Huguenot officers shouted "A moi, Picardie! A moi, Roussillon!" to gather the relics of the scattered regiments of the enemy around them. In this way some thousands of prisoners were gleaned, but the harvest which would have been reaped in another hour of daylight was lost. In the French army all was confusion. Vendôme tried in vain to keep the troops together till the morning, but Burgundy gave the word for retreat; and the whole ran off in disorder towards Ghent. 1708. June 30 / July 11.

So ended the battle of Oudenarde, presenting on one side a feature rare in these days, namely, a general engagement without an order of battle.[1] It was undoubtedly the most hazardous action that Marlborough ever fought. His troops were much harassed by forced marches. They had started at two o'clock on Monday morning and had covered fifty miles, including the passage of two rivers, when they came into action at two o'clock on Wednesday afternoon. It would be reckoned no small feat in these days to move eighty thousand men over fifty miles in sixty hours, but in those days of bad roads and heavy packs the effort must have been enormous. Finally, the army had to pass the Scheldt in the face of the enemy, and ran no small risk of being destroyed in detail. Yet the hazard was probably less than it now seems to us, and generals in our own day have not hesitated to risk similar peril with success. The French commanders were at variance; the less competent of them, being heir-apparent, was likely to be toadied by officers and supported by them against their better judgment; and finally the whole French army was very much afraid of Marlborough. Notwithstanding their slight success in Ghent and Bruges, their elation had evaporated speedily when they found Marlborough before them at Lessines.

[1] The ground, though drained and built over about Bevere, seems to have lost little of its original character, and is worth a visit.

1708. All this Marlborough knew well, and knew also that if an impromptu action, if one may use the term, must be fought, there was not a man on the other side who had an eye for a battlefield comparable to Eugene's and his own. The event justified his calculations; for the victory was one of men who knew their own minds over men who did not. Another hour of daylight, so Marlborough declared, would have enabled him to finish the war. The total loss of the Allies in the battle was about three thousand killed and wounded, the British infantry though early engaged suffering but little, while the cavalry, being employed to watch the inactive French left, hardly suffered at all.[1] The French lost six thousand killed and wounded and nine thousand prisoners only, but they were thoroughly shaken and demoralised for the remainder of the campaign. The wearied army of the Allies lay on its arms in the battlefield, while Marlborough and Eugene waited impatiently for the dawn. As soon as it was light forty squadrons, for the most part British, were sent forward in pursuit, while Eugene returned to his own army to hasten its march and to collect material for a siege. The main army halted to rest for two days where it lay, during
July 1/12. which time the intelligence came that Berwick had been summoned with his army from the Moselle, and was marching with all haste to occupy certain lines constructed by the French to cover their frontier from
July 2-3/13-14. Ypres to the Lys. At midnight fifty squadrons and thirty battalions under Count Lottum, a distinguished Prussian officer, started for these lines; the whole army followed at daybreak, and while on the march the Duke received the satisfactory news that Lottum had captured the lines without difficulty. Next day the whole of Marlborough's army was encamped along the Lys between Menin and Commines, within the actual territory of France.

Detached columns were at once sent out to forage

[1] British losses: 4 officers and 49 men killed, 17 officers and 160 men wounded.

and levy contributions. The suburbs of Arras were burnt, and no effort was spared to bring home to the French that war was hammering at their own gates. But the Allies were still doubtful as to the operations that they should next undertake. So long as the French held Bruges and Ghent they held also the navigation of the Scheldt and Lys, so that it was of vital importance to tempt Vendôme, if possible, to evacuate them. The British Government was preparing a force[1] under General Erle for a descent upon Normandy by sea, and Marlborough was for co-operating with this expedition, masking the fortress of Lille, and penetrating straight into France — a plan which the reader should, if possible, bear in mind. But the proposal was too adventurous to meet with the approval of the Dutch, and was judged impracticable even by Eugene unless Lille were first captured as a place of arms. Ultimately it was decided, notwithstanding the closing of the Scheldt and Lys, to undertake the siege of Lille; and all the energies of the Allies were turned to the collection of sixteen thousand horses to haul the siege-train overland from Brussels. 1708. July.

During the enforced inaction of the army for the next few weeks, the monotony was broken only by the arrival of a distinguished visitor, Augustus the Strong, Elector of Saxony and King of Poland, together with one of his three hundred and sixty-four bastards, a little boy of twelve named Maurice, who had run away from school to join the army. We shall meet with this boy again as a man of fifty, under the name of Marshal Saxe, at a village some twenty miles distant called Fontenoy.

At length the preparations for the siege were complete, and the huge convoy set out from Brussels for its long march. Now, if ever, was the time for the French to strike a blow. Vendôme in the north at

[1] The force consisted of detachments of the 3rd and 4th Dragoons (now Hussars), 12th, 29th, Hamilton's, Dormer's, Johnson's, Moore's, Caulfield's, Townsend's, Wynne's Foot.

1708. Ghent and Berwick in the south at Douay had, between them, one hundred and ten thousand men: the distance to be traversed by the convoy was seventy-five miles, and the way was barred by the Dender and the Scheldt. Such, however, was the skill with which the march was conducted that the French never succeeded even in threatening the vast, unwieldy columns, which duly
Aug. $\frac{1}{12}$. reached their destination without the loss even of a single waggon. Of all the achievements of Marlborough and Eugene, this seems to have been judged by contemporary military men to be the greatest.[1]

Lille, the capital of French Flanders, was one of the early conquests of Lewis the Fourteenth, and, if the expression may be allowed, the darling town of the Court of Versailles. Situated in a swampy plain and watered by two rivers, the Deule and Marque, its natural position presented difficulties of no ordinary kind to a besieging force, and, in addition, it had been fortified by Vauban with his utmost skill. The garrison, which had been strengthened by Berwick, amounted to fifteen thousand men, under the command of brave old Marshal Boufflers, who had solicited the honour of defending the fortress. To the north, as we have seen, lay Vendôme, and to the south Berwick, with a joint force now amounting to about ninety-four thousand men.[2] It was for Marlborough and Eugene with an inferior strength of eighty-four thousand men[3] to hold them at bay and to take one of the strongest fortresses in the world before their eyes.

A detailed account even of so famous a siege would be wearisome, the more so since the proportion of British troops detailed for regular work in the trenches was but five battalions,[4] but there are a few salient features which cannot be omitted. The point selected

[1] See, for instance, the commendations of Feuquières.

[2] 135 battalions, 260 squadrons.

[3] 122 battalions, 230 squadrons.

[4] These were, according to a contemporary plan (Fricx), the 16th, 18th, 21st, 23rd, 24th Foot.

for attack was the north side, the first advance to which was opened by a single English soldier, Sergeant Littler of the First Guards,[1] who swam across the Marquette to a French post which commanded the passage of the stream and let down the drawbridge. Two days later the town was fully invested, and Marlborough took post with the covering army at Helchin on the Scheldt. 1708. Aug. 2/13.

The investment had not been accomplished for more than a fortnight when the Duke was informed that Berwick and Vendôme were advancing towards the Dender to unite their forces at Lessines. After manœuvring at first to hinder the junction Marlborough finally decided to let it come to pass, being satisfied that, if the French designed to relieve Lille, they could not break through in the face of his army on the east side, but must go round and approach it from the south. In this case, as both armies would move in concentric circles around Lille as a centre, Marlborough being nearer to that centre could be certain of reaching any given point on the way to it before the French. Moreover, the removal of the enemy from the east to the south would free the convoys from Brussels from all annoyance on their march to the siege.

As he had expected, the French moved south to Tournay, and then wheeling northward entered the plain of Lille, where they found Marlborough and Eugene drawn up ready to receive them.[2] Vendôme and Berwick had positive orders to risk a battle; and Aug. 22/Sept. 2.

[1] He is claimed as a Guardsman by General Hamilton (*Hist. Grenadier Guards*), though Millner assigns him to the 16th Foot. This is the only name of a man below the rank of a commissioned officer that I have encountered in any of the books on the wars of Marlborough, not excluding the works of Sergeants Deane and Millner. Littler was deservedly rewarded with a commission.

[2] The Allied order of battle was peculiar. The artillery was all drawn up in front, in rear of it came a first line of 100 squadrons, then a second line of 80 squadrons, then a third line of 104 battalions, with wings of 14 squadrons more thrown out to the right and left rear. *Daily Courant*, 6th September 1708.

1708. there had been much big talk of annihilating the Allies. Yet face to face with their redoubtable enemies they hesitated. Finally, after a week's delay, which enabled Marlborough greatly to strengthen his position by entrenchment, they advanced as if to attack in earnest, but withdrew ignominiously after a useless cannonade without accepting battle. Had not Marlborough and Eugene been restrained by the Dutch deputies, the marshals would have had a battle forced on them whether they liked it or not, but, as things were, they were permitted to retire. To such depth of humiliation had Marlborough reduced the proud and gallant French army.

Aug. 27-28 / Sept. 7-8. The retreat left Eugene free to press the siege with vigour; but a great assault, which cost him three thousand men,[1] failed to give him the advantage for which he had hoped, and a week later Marlborough was called in from the covering army to give assistance. For the next assault, on the counterscarp, Sept. 19-20 / 20-21. the Duke lent the Prince five thousand English, and it is said that English and French never fought more worthily of their reputation than on that day; but the assault was thrice repelled, and it was only through the exertions of Eugene himself that a portion of the works was at last captured after a desperate effort and at frightful cost of life. Altogether the siege was not going well. The engineers had made blunders; a vast number of men had been thrown away to no purpose; and ammunition and stores were beginning to run short. Lastly, Boufflers maintained always a very grand and extremely able defence.

Vendôme and Berwick could now think of no better expedient than to throw themselves into strong positions along the Scarpe and Scheldt, from Douay to Ghent, in order to cut off all convoys from Brussels. But Marlborough was prepared for this, and had not

[1] The five English regiments lost about 350 killed and wounded in this assault. This would mean probably from a fifth to a sixth of their numbers. *Daily Courant*, 6th September 1708.

captured Ostend after Ramillies for nothing. England held command of the sea; and Erle's expedition, which had effected little or nothing on the coast of Normandy, was at hand to help in the transport of supplies from the new base. Erle, who had considerable talent for organisation, soon set Ostend in order, seized two passages over the Newport Canal at Leffinghe and Oudenburg and prepared to send off his first convoy. As its arrival was of vital importance to the maintenance of the siege, the French were as anxious to intercept as the English to forward it. Vendôme accordingly sent off Count de la Mothe with twenty-two thousand men to attack it on its way, while Marlborough despatched twelve battalions and fifteen hundred horse to Ostend itself, twelve battalions more under General Webb to Thourout, and eighteen squadrons under Cadogan to Roulers, at two different points on the road, to help it to its destination.

1708.

Sept. $\frac{16}{27}$.

The convoy started at night, and in the morning Cadogan sent forward Count Lottum with a hundred and fifty horse to meet it. At noon Lottum returned to Thourout with the intelligence that he had struck against the advanced-guard of a French force at Ichtegem, two miles beyond Wynendale and some four miles from Thourout on the road to Ostend. Webb at once collected every battalion within his reach, twenty-two in all, and marched with all speed for Ichtegem, with Lottum's squadron in advance. The horse, however, on emerging from the defile of Wynendale, found the enemy advancing towards them into the plains that lay beyond it. Lottum retired slowly, skirmishing, while Webb pushed on and posted his men in two lines at the entrance to the defile. The strait was bounded on either hand by a wood, and in each of these woods Webb stationed a battalion of Germans to take the French in flank. The dispositions were hardly complete when the enemy came up and opened fire from nineteen pieces of artillery. Lottum and his handful of horse then retired, while just in the nick of

Sept. $\frac{17}{28}$.

1708. time three more battalions reach Webb from the rear and formed his third line.

Sept. 17/28. The French cannonade was prolonged for nearly two hours, but with little effect, for Webb had ordered his men to lie down. At length at five o'clock the French advanced in four lines of infantry backed by as many of horse and dragoons. They came on with great steadiness and entered the space between the two woods, their flank almost brushing the covert as they passed, serenely unconscious of the peril that awaited them. Then from right and left a staggering volley crashed into them from the battalions concealed in the woods. Both flanks shrank back from the fire, and huddled themselves in confusion upon their centre. De la Mothe sent forward some dragoons in support; and the foot, recovering themselves, pressed on against the lines before them. So vigorous was their attack that they broke through two battalions of the first line, but the gap being instantly filled from the second, they were forced back. Again they struggled forward, trusting by the sheer weight of eight lines against two to sweep their enemy away. But the eternal fire on front and flank became unendurable, and notwithstanding the blows and entreaties of their officers the whole eight lines broke up in confusion, while Webb's battalions, coolly advancing by platoons "as if they were at exercise," poured volley after volley into them as they retired. Cadogan, who had hastened up with a few squadrons to the sound of the firing, was anxious to charge the broken troops, but his force was considered too weak; and thus after two hours of hot conflict ended the combat of Wynendale. The French engaged therein numbered almost double of the Allies, and lost close on three thousand men, while the Allies lost rather less than a thousand of all ranks. The signal incapacity displayed by the French commander did not lessen the credit of Webb, and Wynendale was reckoned one of the most brilliant little affairs of the whole war.[1]

[1] I have failed, in spite of much search, to identify the British

The safe arrival of the convoy before Lille raised 1708. the hopes of the besiegers; and Vendôme, now fully alive to the importance of cutting off communication with Ostend, marched towards that side with a considerable force, and opening the dykes laid the whole country under water. Marlborough went quickly after him, but the marshal would not await his coming; and the Duke by means of high-wheeled vehicles and punts contrived to overcome the difficulties caused by the inundation. At last, after a siege of sixty days the Oct. 11/22. town capitulated; and the garrison retired into the citadel, where Eugene proceeded to beleaguer it anew.

While the new siege was going forward the Elector of Bavaria arrived on the scene from the Rhine, from whence the apathy of the Elector of Hanover had most unpardonably allowed him to withdraw, and laid siege Nov. 13/24. to Brussels with fifteen thousand men. This was an entirely new complication; and since the French held the line of the Scheldt in force, it was difficult to see how Marlborough could parry the blow. Fortunately the garrison defended itself with great spirit, the English regiments[1] setting a fine example, and the Duke, in no wise dismayed, laid his plans with his usual secrecy and decision. Spreading reports, which he strengthened by feint movements, that he was about to place his troops in cantonments, he marched suddenly and silently eastward on the night of the 26th of November, crossed Nov. 15/26. the Scheldt at two different points before the enemy

regiments present, excepting one battalion of the 1st Royals. Marlborough, as Thackeray has reminded us by a famous scene in *Esmond*, attributed the credit of the action in his first despatch to Cadogan. Another letter, however, which appeared in the *Gazette* three days later (23rd September), does full justice to Webb, as does also a letter from the Duke to Lord Sunderland of 18th-29th September (*Despatches*, vol. iv. p. 243). Webb's own version of the affair appeared in the *Gazette* of 9th October, but does not mention the regiments engaged. Webb became a celebrated bore with his stories of Wynendale, but the story of his grievance against Marlborough would have been forgotten but for Thackeray, who either ignored or was unaware of the second despatch.

[1] Notably Prendergast's. *Gazette*, 25th November.

1708. knew that he was near them, took a thousand prisoners, and then remitting the bulk of his force to the siege of Lille, pushed on with a detachment of cavalry and two Nov. 17/28. battalions of English Guards to Alost. On his arrival he learned that the Elector had raised the siege of Brussels and marched off with precipitation. The bare name of Marlborough had been sufficient to scare him away.

Meanwhile Eugene's preparations before the citadel of Lille were in rapid progress, and Marlborough was already maturing plans for a further design before the close of the campaign. It had been the earnest desire of both commanders to reduce Boufflers to unconditional Nov. 28/Dec. 9. surrender; but time was an object, so on the 9th of December the gallant old marshal and his heroic garrison marched out with the honours of war. So ended the memorable siege of Lille. It had cost the garrison eight thousand men, or more than half of its numbers, and the Allies no fewer than fourteen thousand men. The honours of the siege rested decidedly with Boufflers, and were paid to him by none more ungrudgingly than Marlborough and Eugene. Yet as an operation of war, conducted under extraordinary difficulties in respect of transport, under the eyes of a superior force and subject to diversions, such as that of the Elector of Bavaria, it remains one of the highest examples of consummate military skill.

The fall of Lille was a heavy blow for France, but it was not the last of the campaign. Within eight days Marlborough and Eugene had invested Ghent, which after a brief resistance surrendered with the honours of war. The capitulation of Bruges quickly followed, and the navigation of the Scheldt and Lys having been regained, the two commanders at last sent their troops into winter quarters.

But even this did not close the sum of English successes for 1708, for from the Mediterranean had come news of another conquest, due to the far-seeing eye and far-reaching hand of Marlborough. Early in the year Galway had withdrawn from Catalonia to Lisbon, and the command in Catalonia had been given at Marlborough's instance to Field-Marshal von Staremberg, an Imperial officer of much experience and deservedly high reputation. Staremberg, however, could do little with but ten thousand men against the Bourbon's army of twice his strength, so by Marlborough's advice the troops were used to second the operations of the Mediterranean squadron. Sardinia, the first point aimed at, was captured almost without resistance, and the fleet then sailed for Minorca. Here somewhat more opposition was encountered; but after less than a fortnight's work, creditably managed by Major-General Stanhope, the Island was taken at a trifling cost of life.[1] Thus the English gained their first port in the Mediterranean; and the news of the capture of Minorca reached London on the same day as that of the fall of Lille.

1708.

Sept. 13/24.

[1] The British troops employed were the 6th Foot, 600 marines, and a battalion of seamen.

NOTE.—I have been unable to discover any Order of Battle for the campaign of 1708. The regiments that bear the name of Oudenarde on their appointments are the 1st, 3rd, 5th, 6th, and 7th Dragoon Guards, the 2nd Dragoons, 5th Lancers, Grenadier Guards, Coldstream Guards, 1st, 3rd, 8th, 10th, 15th, 16th, 18th, 21st, 23rd, 24th, 26th, 37th Foot.

CHAPTER VIII

1708. The successes of the past campaign were sufficient to set the British Parliament in good humour, and to prompt it to vote a further increase of ten thousand German mercenaries for the following year. Nevertheless political troubles were increasing, and there were already signs that the rule of Godolphin and Marlborough was in danger. The death of the Prince Consort had been a heavy blow to the Duke. Prince George may have deserved Lord Macaulay's character for impenetrable stupidity, but there can be little doubt that his heavy phlegmatic character was of infinite service to steady the weak and unstable Queen Anne.

In the spring of 1709, however, it seemed reasonable to hope that peace, which would have set all matters right, was well-nigh assured. France, already at the last gasp through the exhaustion caused by the war, was weakened still further by a severe winter which had added famine to all her other troubles; and Lewis sought anxiously, even at the price of humiliation, for
1709. peace. He approached Marlborough, reputed the most avaricious and corruptible of men, with a gigantic bribe to obtain good terms, but was unhesitatingly rebuffed. The Duke stated the conditions which might be acceptable to England; and had the negotiations been trusted to him, there can be little doubt but that he would have obtained the honourable peace which he above all men most earnestly desired. He was, however, overruled by instructions from home imposing terms which Lewis could not be expected to grant; the war was continued;

and Marlborough, who had striven his hardest to bring 1709. it to an end, was of course accused of prolonging it deliberately for his own selfish ends.

The French, now menaced with an invasion and a march of the Allies to Paris, had strengthened their army enormously by withdrawing troops from all quarters to Flanders, and had set in command their only fortunate general, that very able soldier and incomparable liar, Marshal Villars. To cover Arras, the north-western gate of France, Villars had thrown up a strong line of entrenchments from the Scarpe at Douay to the Lys, which were generally known, after the name of his headquarters, as the lines of La Bassée. There he lay, entrenched to the teeth, while Marlborough and Eugene, after long delay owing to the lateness of the spring, encamped with one hundred and ten thousand men to the south of Lille, between two villages, with which the June. reader will in due time make closer acquaintance, called Lincelles and Fontenoy. Thence they moved south straight upon Villars' lines with every apparent preparation for a direct attack upon them and for forcing their way into France at that point. The heavy artillery was sent to Menin on the Lys; report was everywhere rife of the coming assault, and Villars lost no time in summoning the garrison of Tournay to his assistance. On the 26th of June, at seven in the evening, Marl- June $\frac{15}{26}$. borough issued his orders to strike tents and march; and the whole army made up its mind for a bloody action before the lines at dawn. To the general surprise, after advancing some time in the direction of the French, the columns received orders to change direction to the left. After some hours' march eastward they crossed a river, but the men did not know that the bridge lay over the Marque and that it led them over the battlefield of Bouvines; nor was it until dawn that they saw the gray walls and the four spires of Tournay before them and discovered that they had invested the city.

Tournay had been fortified by Vauban and was one

1709. of the strongest fortresses in France,[1] but its garrison had been weakened by the unsuspecting Villars, and there was little hope for it. The heavy artillery of the Allies, which had been sent to Menin, went down the Lys to Ghent and up the Scheldt to the besieged city, June 26/July 7. the trenches were opened on the 7th of July, and after three weeks, despite the demonstrations of Villars and of incessant heavy rain, Tournay was reduced to sur- July 19/23. render.[2] Then followed the siege of the citadel, the most desperate enterprise yet undertaken by the Allied troops, inasmuch as the subterraneous works were more numerous and formidable than those above ground. The operations were, therefore, conducted by mine and countermine, with destructive explosions and confused combats in the darkness, which tried the nerves of the soldiers almost beyond endurance. The men did not object to be shot, but they dreaded to be buried alive by the hundred together through the springing of a single mine.[3] Four English regiments[4] bore their share in this work and suffered heavily in the course of it, Aug. 23/Sept. 3. until on the 3rd of September the citadel capitulated.

[1] There are still some remains of the old walls of Tournay on the south side of the town, and the ruins of Vauban's citadel close by, from which the extent of the works may be judged.

[2] The British regiments employed in the siege were the 1st Royals (2 battalions), 3rd Buffs, 37th, Temple's, Evans's and Prendergast's Foot.

[3] The following description written from the trenches gives some idea of the work: "Now as to our fighting underground, blowing up like kites in the air, not being sure of a foot of ground we stand on while in the trenches. Our miners and the enemy very often meet each other, when they have sharp combats till one side gives way. We have got into three or four of the enemy's great galleries, which are thirty or forty feet underground and lead to several of their chambers; and in these we fight in armour by lanthorn and candle, they disputing every inch of the gallery with us to hinder our finding out their great mines. Yesternight we found one which was placed just under our bomb batteries, in which were eighteen hundredweight of powder besides many bombs: and if we had not been so lucky as to find it, in a very few hours our batteries and some hundreds of men had taken a flight into the air." *Daily Courant*, 20th August.

[4] 8th, 10th, 15th, 16th.

Before the close of the siege Marlborough and Eugene, leaving a sufficient force before Tournay, had moved back with the main army before the lines at Douay. They had long decided that the lines were far too formidable to be forced, but they saw no reason for communicating this opinion to Villars. On the 31st of August Lord Orkney, with twenty squadrons and the whole of the grenadiers of the army, marched away silently and swiftly eastward towards St. Ghislain on the Haine. Three days later, immediately after the capitulation of the citadel of Tournay, the Prince of Hessen-Cassel started at four o'clock in the afternoon in the same direction; at nine o'clock Cadogan followed him with forty squadrons more, and at midnight the whole army broke up its camp and marched after them. Twenty-six battalions alone were left before Tournay to superintend the evacuation and to level the siege works, with orders to watch Villars carefully and not to move until he did.

1709. Aug. $\frac{20}{31}$. Aug. 23 / Sept. 3.

The Prince of Hessen-Cassel soon overtook Orkney, from whom he learned that St. Ghislain was too strongly held to be carried by his small force. The Prince therefore at once pushed on. Rain was falling in torrents, and the roads were like rivers, but he continued his advance eastward behind the woods that line the Haine almost without a halt, till at length at two o'clock on the morning of the 6th of September he wheeled to the right and crossed the river at Obourg three miles to the north-east of Mons. Before him lay the river Trouille running down from the south through Mons, and in rear of it a line of entrenchments, thrown up from Mons to the Sambre during the last war to cover the province of Hainault. A short survey showed him that the lines were weakly guarded; and before noon he had passed them without opposition. His force, despite the weather and the state of the roads, had covered the fifty miles to Obourg in fifty-six hours.

Aug. 26 / Sept. 6.

Too late Villars discovered that for the second time he had been duped, and that Marlborough had no

1709. intention of forcing his way into France through the lines of La Bassée and the wet swampy country beyond them, when he could pass the lines of the Trouille without loss of a man. He was in a difficult position, for Mons was but slenderly garrisoned and difficult of access, while, if captured, it would be a valuable acquisition to the Allies. The approach to it from the westward was practically shut off by a kind of natural barrier of forest, running, roughly speaking, from St. Ghislain on the Haine on the north to Maubeuge on the Sambre to the south. In this barrier there were but two openings, the Trouée de Boussut between the village of that name and the Haine, and the Trouées d'Aulnois and de Louvière, which are practically the same, some miles further to the south. These will be more readily remembered, the northern entrance by the name of Jemappes, the southern by the name of Malplaquet. Villars no sooner knew what was going forward than he pushed forward a detachment with all speed upon the northern entrance, which was the nearer Aug. 27/Sept. 7. to him. The detachment came too late. The Prince of Hessen-Cassel was already astride of it, his right at Jemappes, his left at Ciply. The French thereupon fell back to await the approach of the main army of the Allies.

Meanwhile that army had toiled through the sea of Aug. 26/Sept. 6. mud on the northern bank of the Haine, and crossing the river had by evening invested Mons on the eastern side. Aug. 27/Sept. 7. On the following day Villars and his whole army also arrived on the scene and encamped a couple of miles to westward of the forest-barrier from Montreuil to Athis. Here he was joined by old Marshal Boufflers, who had volunteered his services at a time of such peril to France. The arrival of the gallant veteran caused such a tumult of rejoicing in the French camp that Marlborough and Eugene, not knowing what the clamour might portend, withdrew all but a fraction of the investing force from the town, and advancing westward into the plain of Mons caused the army to bivouac between Ciply and Quévy in order of battle.

Villars meanwhile had not moved, being adroit enough to threaten both passages and keep the Allies in doubt as to which he should select. While therefore the mass of the Allied army was moved towards the Trouée d'Aulnois, a strong detachment was sent up to watch the Trouée de Boussut. That night Villars sent detachments forward to occupy the southern passage, and by midday of the morrow his whole army was taking up its position across the opening. Marlborough at once moved his army forward, approaching so close that his left wing exchanged cannon shot with Villars's right. Everything pointed to an immediate attack on the French before they should have time to entrench themselves. Whether the Dutch deputies intervened to stay further movements is uncertain. All that is known is that a council of war was held, wherein, after much debate, it was resolved to await the arrival of the detachment from the Trouée de Boussut and of the troops that had been left behind at Tournay, and that in the meanwhile eighteen battalions should be sent north to the capture of St. Ghislain and the investment of Mons turned into a blockade. Evidently in some quarter there was reluctance to hazard a general action.

1709. Aug. 28/Sept. 8. Aug. 29/Sept. 9.

Villars now set himself with immense energy to strengthen his position; and, when Marlborough and Eugene surveyed the defences at daybreak of the following morning, they were astonished at the formidable appearance of the entrenchments. Marlborough was once more for attacking without further delay, but he was opposed by the Dutch deputies and even by Eugene. The attack was therefore fixed for the morrow; and another day was lost which Villars did not fail to turn to excellent account.

Aug. 30/Sept. 10.

The entrance from the westward to the Trouée d'Aulnois or southern entrance to the plain of Mons is marked by the two villages of Campe du Hamlet on the north and Malplaquet on the south. About a mile in advance of these villages the ground rises to its highest elevation, the opening being about three thousand

1709. paces wide, and the ground broken and hollowed to right and left by small rivulets. This was the point selected by Villars for his position. It was bounded on his right by the forest of Laignières, the greatest length of which ran parallel to the Trouée, and on the left by a forest, known at different points by the names of Taisnières, Sart and Blangies, the greatest length of which ran at right angles to the Trouée. Villars occupied the forest of Laignières with his extreme right, his battalions strengthening the natural obstacles of a thick and tangled covert by means of abattis. From the edge of the wood he constructed a triple line of entrenchments, which ran across the opening for full a third of its width, when they gave way to a line of nine redans. These redans in turn yielded place to a swamp backed by more entrenchments, which carried the defences across to the wood of Taisnières. Several cannon were mounted on the entrenchments and a battery of twenty guns before the redans. On Villars's left the forests of Taisnières and Sart projected before the general front, forming a salient and re-entering angle. Entrenchments and abattis were constructed in accordance with this configuration, and two more batteries were erected on this side, in addition to several guns at various points along the line, to enfilade an advancing enemy. Feeling even thus insecure Villars threw up more entrenchments at the villages of Malplaquet and Chaussée du Bois in rear of the wood of Sart, and was still hard at work on them to the last possible moment before the action. Finally in rear of all stood his cavalry, drawn up in several lines. The whole of his force amounted to ninety-five thousand men.

The position was most formidable, but it had its defects. In the first place the open space before the entrenchments was broken at about half a mile's distance by a small coppice, called the wood of Tiry, which could serve to mask the movements of the Allied centre. In the second place the forest of Sart ran out beyond the fortified angle in a long tongue, which would effectually conceal any troops that might be directed against the extreme left flank. Finally the French cavalry, being massed in rear of the entrenchments, could take no part in the action until the defences were forced, and was therefore incapable of delivering any counterstroke. Marlborough and Eugene accordingly decided to make a feint attack on the French right and a true attack on their left front and flank. Villars would then be obliged to reinforce his left from his centre, which would enable the defences across the open to be carried, and the whole of the allied cavalry to charge forward and cut the French line in twain. 1709.

The dawn of the 11th of September broke in dense heavy mist which completely veiled the combatants from each other. At three o'clock prayers were said in the Allied camp, and then the artillery was moved in position. Forty pieces were massed in a single battery in the open ground against the French left, and were covered with an epaulment for defence against enfilading fire; twenty-eight more were stationed against the French right, and the lighter pieces were distributed, as usual, among the different brigades. Then the columns of attack were formed. Twenty-eight battalions under Count Lottum were directed against the eastern face of the salient angle of the forest of Taisnières, and forty battalions of Eugene's army under General Schulemberg against the northern face, while a little to the right of Schulemberg two thousand men under General Gauvain were to press on the French left flank in rear of their entrenchments. In rear of Schulemberg fifteen British battalions under Lord Orkney were drawn up in a single line on the open ground, ready to advance against the centre as soon as Schulemberg and Lottum should have done their work. Far away beyond Gauvain to the French left General Withers with five British and fourteen foreign battalions and six squadrons was to turn the extreme French left at the village of La Folie. Aug. 31 / Sept. 11.

For the feint against the French right thirty-one

1709. Aug. 31 / Sept. 11. battalions, chiefly Dutch, were massed together under the Prince of Orange. The cavalry was detailed in different divisions to support the infantry. The Prince of Orange was backed by twenty-one Dutch squadrons under the Prince of Hesse, Orkney by thirty more under Auvergne, Lottum by the British and Hanoverian cavalry, and Schulemberg by Eugene's horse. The orders given to the cavalry were to sustain the foot as closely as possible without advancing into range of grape-shot, and as soon as the central entrenchments were forced to press forward, form before the entrenchments and drive the French army from the field. The whole force of the Allies was as near as may be equal to that of the French.

At half-past seven the fog lifted and the guns of both armies opened fire. Eugene and Marlborough thereupon parted, the former taking charge of the right, the latter of the left of the army. Then the divisions of Orange and of Lottum advanced in two dense columns up the glade. Presently the Dutch halted, just beyond range of grape-shot, while Lottum's column pushed on under a terrific fire to the rear of the forty-gun battery and deployed to the right in three lines. Then the fire of the cannon slackened for a time, till about nine o'clock a salvo of the forty guns gave the signal for attack. Lottum's and Schulemberg's divisions thereupon advanced perpendicularly to each other, each in three lines, Gauvain's men crept into the wood unperceived, and Orkney extended his scarlet battalions across the glade.

Entering the wood Schulemberg's Austrians made the best of their way through marshes and streams and fallen trees, nearer and nearer to the French entrenchments. The enemy suffered them to approach within pistol-shot and delivered a volley which sent them staggering back; and though the Austrians extended their line till it joined Gauvain's detachment, yet they could make little way against the French fire. Lottum's attack was little more successful. Heedless of the tempest of shot in their front and flank the Germans pressed steadily on, passed a swamp and a stream under a galling fire, and fell fiercely upon the breastwork beyond; but being disordered by the ground and thinned by heavy losses they were forced to fall back. Schulemberg then resumed the attack with his second line, but with all his exertions could not carry the face of the angle opposed to him. Picardie, the senior regiment of the French Line, held this post and would not yield it to the fiercest assault. The utmost that Schulemberg could accomplish was to sweep away the regiments in the wood, and so uncover its flank. 1709. Aug. 31 / Sept. 11.

Lottum, too, extended his front and attacked once more, Orkney detaching three British battalions, the Buffs, Sixteenth, and Temple's, to his assistance, while Marlborough took personal command of Auvergne's cavalry in support. The Buffs on Lottum's extreme left found a swamp between them and the entrenchments, so deep as to be almost impassable. In they plunged, notwithstanding, and were struggling through it when a French officer drew out twelve battalions and moved them down straight upon their left flank. The British brigade would have been in a sorry plight had not Villars caught sight of Marlborough at the head of Auvergne's horse and instantly recalled his troops. So the red-coats scrambled on, and turning the flank of the entrenchment while Lottum's men attacked the front, at length with desperate fighting and heavy loss forced the French back into the wood. Thus exposed to the double attack of Lottum and Schulemberg Picardie at last fell back, but joined itself to Champagne, the next regiment in seniority; and the two gallant corps finding a rallying-point behind an abattis turned and stood once more. Their comrades gave way in disorder, but the wood was so dense that the troops on both sides became disjointed, and the opposing lines broke up into a succession of small parties fighting desperately from tree to tree with no further guidance than their own fury.

1709. Aug. 31/Sept. 11. The entrenchments on the French left had been forced; and Villars sent urgent messages to his right for reinforcements. But Boufflers could spare him none. After Schulemberg and Lottum had been engaged for half an hour, the Prince of Orange lost patience and, without waiting for orders, opened not a false but a real attack against the French right. On the extreme left of Orange's division were two Highland regiments of the Dutch service, Tullibardine's and Hepburn's, and next to them King William's favourite Blue Guards. These were to attack the defences in the forest of Laignières, while the rest fell upon the entrenchments in the open; and it was at the head of the Highlanders and of the Blue Guards that Orange took his place. A tremendous fire of grape and musketry saluted them as they advanced, and within the first few yards most of the Prince's staff were struck dead by his side. His own horse fell dead beneath him, but he disentangled himself and continued to lead the advance on foot. A few minutes more brought his battalions under the fire of a French battery on their left flank. Whole ranks were swept away, but still the Prince was to be seen waving his hat in front of his troops; and Highlanders and Dutchmen pressing steadily on carried the first entrenchment with a rush. They then halted to deploy, but before they could advance further Boufflers had rallied his men, and charging down upon his assailants drove them back headlong. On Orange's right, success as short-lived was bought at as dear a price. The Prince still exerted himself with the utmost gallantry, but his attack was beaten back at all points. The loss of the Dutch amounted to six thousand killed and wounded; the Blue Guards had been annihilated, and the Hanoverian battalions, which had supported them, had suffered little less severely. In fact, the Prince's precipitation had brought about little less than a disaster.

The confusion in this part of the field called both Marlborough and Eugene to the Allied left to restore order. Further useless sacrifice of life was checked, for enough and more than enough had been done to prevent Boufflers from detaching troops to Villars. But soon came an urgent message requiring their presence once more on the right. Schulemberg and Lottum had continued to push their attack as best they could; and red-coated English, blue-coated Prussians, and white-coated Austrians were struggling forward from tree to tree, tripping over felled trunks, bursting through tangled foliage, panting through quagmires, loading and firing and cursing, guided only by the flashes before them in the cloud of foul blinding smoke. But now on the extreme right Withers was steadily advancing; and his turning movement, though the Duke and Eugene knew it not, was gradually forcing the French out of the wood. Villars seeing the danger called the Irish Brigade and other regiments from the centre, and launched them full upon the British and Prussians. Such was the impetuosity of the Irish that they forced them back some way, until their own formation was broken by the density of the forest. Eugene hastened to the spot to rally the retreating battalions and though struck by a musket ball in the head refused to leave the field. Then up came Withers, just when he was wanted. The Eighteenth Royal Irish met the French Royal Regiment of Ireland, crushed it with two volleys by sheer superiority of fire, drove it back in disorder, and pressed on.[1] Eugene also advanced and was met by Villars, who at this critical moment was bringing forward his reinforcements in person. A musket shot struck the Marshal above the knee. Totally unmoved the gallant man called for a chair from which to continue to direct his troops, but presently fainting from pain was carried insensible from the field. The French, notwithstanding his fall, still barred the advance of the Allies, but they had been driven from their entrenchments and from the wood on the left, and only held their own by the help of the troops

1709.
Aug. 31
Sept. 11.

[1] Parker.

1709. Aug. 31 / Sept. 11. that had been withdrawn from the centre. The moment for which Marlborough had waited was now come.

The forty-gun battery was moved forward, and Orkney leading his British battalions against the redans captured them, though not without considerable loss, at the first rush. Two Hanoverian battalions on their left turned the flank of the adjoining entrenchments, and Orange renewing his attack cleared the whole of the defences in the glade. The Allied cavalry followed close at their heels. Auvergne's Dutch were the first to pass the entrenchments, and though charged by the French while in the act of deploying succeeded in repelling the first attack. But now Boufflers came up at the head of the French Gendarmerie, and drove them back irresistibly to the edge of the entrenchments. Here, however, the French were checked, for Orkney had lined the parapet with his British, and though the Gendarmerie thrice strove gallantly to make an end of the Dutch, they were every time driven back by the fire of the infantry. Meanwhile the central battery, which had been parted right and left into two divisions, advanced and supported the infantry by a cross-fire, and Marlborough coming up with the British and Prussian horse fell upon the Gendarmerie in their turn. Boufflers, however, was again ready with fresh troops, and coming down upon Marlborough with the French Household Cavalry crashed through his two leading lines and threw even the third into disorder. Then Eugene coming up with the Imperial horse threw the last reserves into the melée and drove the French back. Simultaneously the Prince of Hesse hurled his squadrons against the infantry of the French right, and with the help of the Dutch foot isolated it still further from the centre. Then Boufflers saw that the day was lost and ordered a general retreat to Bavay, while he could yet keep his troops together. The movement was conducted in admirable order, for the French though beaten were not routed, while the Allies were too much exhausted to pursue. So Boufflers retired unmolested, though it was not yet three o'clock, honoured alike by friend and foe for his bravery and his skill. 1709.

Thus ended the battle of Malplaquet, one of the bloodiest ever fought by mortal men. Little is known of the details of the fighting, these being swallowed up in the shade of the forest of Taisnières, where no man could see what was going forward. All that is certain is that neither side gave quarter, and that the combat was not only fierce but savage. The loss of the French was about twelve thousand men, and the trophies taken from them, against which they could show trophies of their own, were five hundred prisoners, fifty standards and colours and sixteen guns. The loss of the Allies was not less than twenty thousand men killed and wounded, due chiefly to the mad onset of the Prince of Orange. The Dutch infantry out of thirty battalions lost eight thousand men, or more than half of their number; the British out of twenty battalions lost nineteen hundred men,[1] the heaviest sufferers being the Coldstream Guards, Buffs, Orrery's and Temple's.[2]

[1] A nominal list in the *Postboy* of 1st October gives 36 officers killed and 46 wounded. An earlier list of 17th September gives 40 officers and 511 men killed, 66 officers and 1020 men wounded; but this is admittedly imperfect.

[2] ORDER OF BATTLE. CAMPAIGN OF 1709.

Left. RIGHT WING ONLY. Right.

1st Line.

				Two Foreign Brigadiers.	Orrery's Brigade.	Kelburn's Brigade.	Sybourg's Brigade.
8th Foot.	3rd Buffs.	2nd Batt. Royal Scots.	1 Batt. 1st Guards.	Twenty-seven squadrons of foreign dragoons.	26th Cameronians.	1st Dragoon Guards, 2 squadrons.	Scots Greys, 3 Squadrons.
24th Foot.	Temple's Foot.	23rd Royal Welsh.	1 " Coldstream Guards.		Two foreign battalions.	5th " " 2 "	5th Royal Irish Dragoons, 2 squadrons.
21st Royal Scots Fusiliers.	Evans's Foot.	Orrery's Foot.	1 " Royal Scots.		Prendergast's Foot.	7th " " 2 "	
18th Royal Irish.	16th Foot.		37th Foot.			6th " " 1 "	
			10th Foot.			3rd " " 2 "	

No British troops in the second line; but the 15th and 19th Foot were also present at the action of Malplaquet.

1709. The more closely the battle is studied the more the conviction grows that no action of Marlborough's was fought less in accordance with his own plans. We have seen that he would have preferred to fight it on either of the two preceding days, and that he deferred to Eugene against his own judgment in suffering it to be postponed. Then again there was the almost criminal folly of the Prince of Orange, which upset all preconcerted arrangements, threw away thousands of lives to no purpose, and not only permitted the French to retreat unharmed at the close of the day but seriously imperilled the success of the action at its beginning. Nevertheless there are still not wanting men to believe the slanders of the contemptible faction then rising to power in England, that Marlborough fought the battle from pure lust of slaughter.

Still, in spite of all blunders, which were none of Marlborough's, Malplaquet was a very grand action. The French were equal in number to the Allies and occupied a position which was described at the time as a fortified citadel. They were commanded by an able general, whom they liked and trusted, they were in good heart, and they looked forward confidently to victory. Yet they were driven back and obliged to leave Mons to its fate; and though Villars with his usual bluster described the victory as more disastrous than defeat, yet French officers could not help asking themselves whether resistance to Marlborough and Eugene were not hopeless. Luxemburg with seventy-five thousand men against fifty thousand had only with difficulty succeeded in forcing the faulty position of Landen; yet the French had failed to hold the far more formidable lines of Malplaquet against an army no stronger than their own. Say Villars what he might, and beyond all doubt he fought a fine fight, the inference could not be encouraging to France.

It was not until the third day after the fight that the Allies returned to the investment of Mons. Eugene was wounded, and Marlborough not only worn out by

fatigue but deeply distressed over the enormous sacrifice 1709.
of life. The siege was retarded by the marshy nature of the ground and by heavy rain; but on the 9th of Sept. 28/Oct. 9.
October the garrison capitulated, and therewith the campaign came to an end. Tournay had given the Allies firm foothold on the Upper Scheldt, and Mons was of great value as covering the captured towns in Flanders and Brabant. The season's operations had not been without good fruit, despite the heavy losses at Malplaquet.

CHAPTER IX

1708. Once more I return to Spain, where the armies of the Bourbons had recommenced operations in the winter of 1708. At the end of October General d'Asfeld having first captured Denia after a short siege had advanced against Alicante, which was garrisoned by eight hundred British[1] and Huguenots, under Major-General John Richards. The siege of Alicante is memorable chiefly for the manner of Richards's death. The castle was built on the solid rock, and the only possible method of destroying its defences was by means of mining. After three months of incessant work d'Asfeld hewed a gallery through the rock beneath the castle, charged it with seventy-five tons of powder, and then summoned Richards to surrender, inviting him at the same time to send two officers to inspect the mine. Two officers accordingly were sent, who returned with the report that the explosion of the mine would doubtless be destructive, but not, in their judgment, fatal to further defence. Richards therefore rejected the summons, nor, though d'Asfeld thrice repeated it, would he return any other answer.

Immediately over the gallery were two guards, each
1709. of thirty men, which could not be withdrawn without peril to the safety of the castle. Early in the morning
Feb. 20/March 3. fixed for the springing of the mine, the sentries were posted as usual, pacing up and down in the keen morning air, when General Richards and all the senior officers of the garrison who were off duty came

[1] Hotham's regiment and artillery.

and joined them. They were come to stand by their men in the hour of trial. A little before six a thin column of blue smoke came curling up the rock, and a corporal of the guard reported that the match had been fired. Richards and his officers remained immovable, the guard stood under arms, and the sentries stuck to their posts. Presently the whole rock trembled again; the ground beneath their feet was rent into vast clefts which yawned for a moment with a hideous hollow roar and instantly closed. When the rumbling had ceased there were still eighteen men left on the rock, but Richards with eleven other officers and forty-two of their comrades had been swallowed up like the company of Korah. Yet Richards was right, for when Admiral Byng and General Stanhope arrived six weeks later the garrison still remained unconquered in the castle. But it was thought best to evacuate it, so the little force was carried away to Mahon, leaving Richards and his brave companions asleep in the womb of the rock. Among the forgotten graves of British soldiers that are sown so thickly over the world, one at least is safe from the ravages of time, the living tomb over which John Richards and his comrades stood, waiting undismayed till it should open to engulf them at Alicante. 1709. Feb. 20 / March 3.

Shortly after the removal of the garrison from the castle Lord Galway and the Portuguese opened the campaign on the side of Portugal near Campo Mayor. Their total force consisted of about fifteen thousand men, including barely three thousand British infantry[1] and artillery; but its weakest point was that it was commanded by a Portuguese officer, the Marquis de Fronteria. Opposed to it were five thousand Spanish horse and ten thousand Spanish foot under the Marquis de Bay, who advanced with his cavalry only to the plain of Gudina on the left bank of the Caya, in order to entice Fronteria across the river. Galway entreated Fronteria not to think of attacking Bay, but the

[1] 5th, 13th, 20th, 39th, Paston's, Stanwix's.

1709. April 26/May 7. Portuguese commander, disregarding his advice, sent the whole of his horse together with the Fifth, Twentieth, Thirty-ninth and Paston's regiments of British Foot across the Caya, and drew them up, rather less than five thousand men in all, on the plain beyond.

Bay at once sent for his infantry, but without waiting for them boldly attacked the Portuguese horse on Fronteria's right wing. Before the Spanish cavalry could reach them the Portuguese turned and fled, leaving the flank of the British infantry uncovered. The four regiments, however, stood firm, and having repulsed three charges formed a hollow square and made a steady and orderly retreat. Meanwhile Galway had sent forward Brigadier Sankey with the Thirteenth, Stanwix's and a Catalan regiment in support, but before they could reach their comrades Bay charged the other wing of Portuguese horse, which fled as precipitately as the former, and turning the whole of his force against Sankey's brigade isolated it completely and compelled it to surrender. The whole of the loss, as usual, fell on the British; and Galway, none too soon, vowed that they would never fight in company with the Portuguese again.

The action on the Caya practically ended the campaign in Portugal for 1709. The operations in Catalonia during the same year call for little notice;
1710. nor was it until July of the following year that Staremberg, reinforced by British[1] and Germans to a strength of twenty thousand foot and five thousand horse, was able to take the field with activity. He lay at the time at Agramont on the Segre, the Spanish army under Villadarias, the unsuccessful besieger of Gibraltar, being a couple of marches to south of him at Lerida. Staremberg resolved to take the offensive forthwith and to carry the war into Aragon.

[1] 2nd Dragoon Guards, Royal Dragoons, 8th Hussars, Nassau's and Rochford's Dragoons. Scots Guards, 6th, 33rd, Bowles's, Dormer's, Munden's, Dalzell's, Gore's. Together 4200 men, under General Stanhope.

Crossing the Segre he sent forward General Stanhope with a small force of dragoons and grenadiers to seize the pass of Alfaraz, before the Spaniards could reach it. Stanhope executed his task with his usual diligence; and the arrival of the Spanish army a few hours after him led to a brilliant little combat of cavalry at Almenara. The odds against the Allies were heavy, for they had but twenty-six squadrons against forty-two of the enemy. Both sides, each drawn up in two lines, observed each other inactive for some time, Staremberg hesitating to permit Stanhope to charge. At length, however, he let him go. The first line, wherein all the British were posted, sprang forward with Generals Stanhope and Carpenter at their head against the Spanish horse, and after a sharp engagement drove them back. The second line followed and forced them back still further upon their infantry. Panic set in among the Spaniards, and presently the whole of the Spanish army was in full retreat to Lerida. The loss of the enemy was thirteen hundred killed and wounded; that of the Allies did not exceed four hundred, half of whom were British.[1]

1710. July $\frac{16}{27}$.

After more than a fortnight's stay at Lerida King Philip summoned Bay to supersede Villadarias, but finding it impossible to advance in face of Staremberg retreated in the direction of Saragossa. Staremberg at once started in pursuit, overtook Bay under the walls of Saragossa and totally defeated him.[2] Contrary to his own better judgment he then marched for Madrid, and led the Archduke Charles for the second time into his capital. The bulk of the army was quartered in the suburbs, but a strong detachment was sent away under Stanhope to occupy Toledo, and, this done, to follow the Tagus to the bridge of Almaraz, where it

Aug. $\frac{7}{18}$. Sept. $\frac{17}{28}$.

[1] 2 brigadiers, 5 other officers and 73 men killed. 2 lieutenant-generals, 12 other officers and 113 men wounded.

[2] Having failed to ascertain the share of the British in this action, I omit it altogether. All that is sure is that they did their duty and that the cavalry suffered severely.

1710. should join hands with a force that was to advance from Portugal.

Sept. The plan was hardly formed before it was broken to pieces. On receiving the news of the defeat at Saragossa Lewis the Fourteenth at once formed an army of his garrisons on the frontier and sent it southward under the command of Vendôme. By the end of September he had united his force with Bay's at Aranda on the Douro and was drawing in fresh troops from all sides. The whole population being in his favour kept him well supplied with intelligence. Before either Stanhope or the Portuguese could reach Almaraz, Vendôme had pounced upon it and destroyed the bridge. Stanhope perforce retired to Toledo, and Vendôme, having by this time collected a force superior to that of the Allies, moved up the Tagus and encamped on the historic field of Talavera.

Staremberg now found it necessary to evacuate Madrid. The Archduke Charles had been coldly received, supplies were failing, and the army was much
Nov. 22/Dec. 3. weakened by sickness. Recalling Stanhope, therefore, from Toledo, he retired up the left bank of the Tajuña; the army, for convenience of forage and supplies, marching in five columns of different nations—Germans, Dutch, Spanish, Portuguese, and British. The third day's march brought the first four columns to Cifuentes, the British who formed the rearguard diverging across the river to Brihuega some fourteen miles from the rest.
Nov. 25/Dec. 6. Stanhope had observed a large body of horse following close at his heels during the march, and had reported the fact to Staremberg, but none the less received orders to halt for another day and to collect provisions.
Nov. 27/Dec. 8. Next morning the enemy's horse appeared on the hill in force, and was joined after a few hours, to the great astonishment of Stanhope, by its infantry. His efforts to obtain intelligence had been foiled by the hostility of the peasants, and neither he nor Staremberg had the faintest idea that there was any infantry within fifty miles of them. In truth this body of foot had,

under Vendôme's direction, covered one hundred and seventy miles in seven days, a march of incredible speed, which, in Stanhope's own words, was his undoing. By five o'clock in the evening Brihuega was fully invested by nine thousand men, and the escape of the British was impossible. 1710. Nov. 27/Dec. 8.

Stanhope's position was desperate. He had but eight battalions and eight squadrons, all so much weakened as to number together but two thousand five hundred men. The town, which was of considerable extent, had no defences but an old Moorish wall, too narrow in most places to afford a banquette for musketeers. Further, the streets were narrow and commanded on all sides by hills within range of artillery and even of musketry. Nevertheless he might hold out till Staremberg came to his relief; so rejecting the summons to surrender, he barricaded the gates, threw up entrenchments as well as he could, and at nightfall sent away his aide-de-camp, who at great risk passed through the enemy's lines, to Staremberg's camp.

At midnight King Philip and Vendôme arrived with the rest of the army, horse, foot, and artillery, increasing the investing force to over twenty thousand men. Before morning two batteries had already been erected, which opened fire at nine o'clock. Two breaches were speedily made in the wall, which the British could not repair except under fire, and a mine was dug to make a third. At three o'clock in the afternoon an assault was delivered at both breaches, and was met by a vigorous resistance. While the combat was raging around them, the mine was fired and a third breach was formed, through which large bodies of the enemy effected an entrance before they were perceived. The British however turned upon them and beat them out again. Finally, the first attack was totally repulsed; and the French entrenched themselves in the breaches to await reinforcements. Again the assault was renewed and again it was driven back with heavy loss by the deadly English fire. Ammunition now began to fail, but Nov. 28/Dec. 9.

1710. Nov. 28 / Dec. 9. the little garrison held its own with the bayonet, contesting every inch of ground, horse and dragoons fighting dismounted by the side of the foot, and every man doing his utmost. Forced back at length from their entrenchments the British set fire to the houses which had been gained by the enemy, and after four hours of hard fighting still held the best part of the town. But their ammunition by this time was almost exhausted, and there was no sign of Staremberg's appearance; so at seven o'clock Stanhope, unwilling uselessly to sacrifice the lives of his men, capitulated, and he and his gallant little force became prisoners of war. Never did British troops fight better than at Brihuega; but even where all were so much distinguished Stanhope could not refrain from giving special praise to the Scots Guards. The total loss of the British was six hundred killed and wounded. That of the enemy was nearly three times as great.

Nov. 29 / Dec. 10. It was not until the morning of the next day that Staremberg approached Brihuega, and meeting the advanced squadrons of Vendôme's, drew up his army for battle in the plains of Villa Viciosa. He had but thirteen thousand men against twenty thousand, but he made skilful dispositions, posting his left behind a deep ravine and strengthening his right, which lay on the open plain, by interlacing the battalions with his few feeble squadrons of horse. The British troops present, Lepell's dragoons, Dubourgay's and Richard's foot, were stationed on the left. The action opened with a long cannonade, after which Vendôme's horse of the right crossed the ravine, and coming down with great spirit and in overwhelming numbers on Staremberg's left swept it after a short resistance completely away. The English dragoons were very heavily punished and the two battalions were cut to pieces. The centre also was broken; and the victorious Spaniards at once fell on the baggage beyond it and began to plunder. But the right of the Allies had held its own, and Staremberg, taking advantage of the disorder among the Spaniards,

contrived with great coolness and skill to convert the 1710.
action into a drawn battle. The whole engagement, indeed, reproduces curiously the features of the early battles of our own Civil War. On the next day, however, Staremberg was compelled to retreat, leaving Nov. 30 / Dec. 9.
his artillery to the enemy; and though Barcelona, Tarragona, and Balaguer were still kept for the Austrian side, the campaign closed with the loss to the Allies of the whole of Spain.

I shall not trouble the reader with the petty operations of the following year, for the war in the Peninsula was practically closed by the battles of Brihuega and Villa Viciosa. The spasmodic nature of the operations has made them difficult and, I fear, wearisome to the reader to follow, quite apart from the dissatisfaction that necessarily attends a long tale of failure. Disunion of purpose and the extreme inefficiency of the Portuguese were the principal infirmities of the Allies throughout the war; the long distance from their true bases at Portsmouth and at Brill their principal disadvantage. Again and again the French were able to retrieve a defeat by sending their garrisons from the frontier-towns across the Pyrenees. Too late, on the appointment of Staremberg, the Allies decided that it would be better to fight the war in the Peninsula with Germans, who could march over Italy and cross the Mediterranean to Catalonia, instead of with English and Dutch, who must make the long and dangerous passage across the Bay of Biscay and through the Straits. But the true secret of the success of the Bourbons, as Lord Macaulay long ago pointed out, lay in the fact that the general sentiment of Spain was on their side, a force which, after another century, shall be seen working to make the fame of a great English commander in another and greater Peninsular war.

Unfortunately the disasters of the year 1710 were not confined to Spain. Up to the autumn of 1709 it seemed that England was still bent on prosecuting the war till the ends of the Grand Alliance should have been

1710. attained. Seven new regiments[1] at any rate had been formed during the year, which might be taken as an earnest of serious intentions. But ever since 1707 Robert Harley, who will be remembered as the proposer of the imbecile motion for disbandment which nearly drove King William from England, had been working with all the resources of a weak, crafty, and dishonest nature to undermine the Government that had so far carried the country triumphantly through the struggle. It was the misfortune of Great Britain at this time to lie at the mercy principally of three women, Queen Anne, the Duchess of Marlborough, and Mrs. Masham. Of these the Duchess alone had any ability, which ability, however, was greatly discounted by her meddlesome and imperious disposition. So long as she retained her ascendency over Anne, things went unpleasantly for the Queen but on the whole well for the country; when her ungovernable temper drove Anne into the arms of Mrs. Masham, the Queen led a quieter life, but the country suffered. Marlborough, who was aware of his wife's waning influence and foresaw the consequences, tried hard on his return from the campaign of 1709 to assure himself a permanent station of power by asking to be made commanner-in-chief for life. The request was tactless as well as unprecedented. Anne, greatly offended, replied by a positive refusal, which Marlborough, for once forgetting his usual serenity, received with culpably ill grace.

So far the Queen was undoubtedly right and Marlborough undoubtedly wrong; but at the beginning of the new year the situation was reversed. The colonelcy of a regiment fell vacant and was filled up by the Queen on the nomination not of the commander-in-chief but of Mrs. Masham by the appointment of her brother, Colonel Hill. Marlborough naturally resolved to resign at once, while the wise and sagacious Somers

[1] Desbordes's, Gually's, Sarlandes's, Magny's, 'Assa's dragoons, all composed of Huguenots but borne on the English establishment; Dalzell's and Wittewrong's foot.

remonstrated most strongly with the Queen against this 1710.
foolish step, as subversive of all discipline and injurious to the army. Unfortunately the Duke, instead of insisting that either he or Mrs. Masham must go, was persuaded to consent to a compromise, which the Queen regarded as a victory for herself and rejoiced over with all the fervour of a weak nature. In the intense personal bitterness of the struggle no one but Somers, outside the military profession, paused for a moment to reflect on its consequences to the Army.

The next object of the opposing faction was to get Marlborough out of England to the Low Countries as soon as possible, which was duly effected, at Harley's instance, by ordering him to take a part in the negotiations for a peace. These negotiations coming to naught, he opened the campaign in April by a rapid movement, which brought him safely over the lines of La Bassée, April $\frac{11}{22}$
and laid siege to Douay. The town made a firm defence for two months, but fell on the 26th of June; and June $\frac{15}{26}$.
Marlborough now proposed to himself either to invest Arras or to advance further into France and cross the Somme. Villars, however, though he had failed to relieve Douay, had made excellent dispositions for the defence of the frontier, and was lying unassailable behind a new series of lines, which he had drawn, as he said later, to be the *ne plus ultra* of Marlborough. The Duke therefore turned to the siege of Bethune, which surrendered on the 28th of August, and thereafter to Aug. $\frac{17}{28}$.
the sieges of Aire and St. Venant on the Upper Lys, which closed the campaign. Each one of these fortresses was strong and made a spirited resistance, costing the Allies altogether some fifteen thousand men killed and wounded. The operations, though less brilliant than those of other campaigns, completed the communication with Lille, opened the whole line of the Lys, and increased the facilities for joint action with an expedition by sea, landing at Calais or Abbeville. Another such blow as Ramillies would have gone near to bring the Allies before the walls of Paris. Throughout the

1710. campaign, however, Marlborough acted always with extreme caution, abandoning the plans which he had once favoured for concerted operations with the fleet. He knew that the slightest failure would lay him open to overwhelming attack from his enemies at home, whose triumph would mean not only his own fall but, what he dreaded much more, the ascendency of unscrupulous politicians who would sacrifice the whole fruits of the war to factious ends, and bring disgrace, perhaps ruin, upon England.

Meanwhile the Queen, with all the pettiness of a weak nature, kept parading her power by foolish interference with matters which she did not undestand. Marlborough had submitted a list of colonels for promotion to general's rank, but as the name of Colonel Hill was not among them she insisted on promoting every colonel of this year, regardless of expense, propriety, justice, or discipline, merely for the sake of including him. In August came a heavier blow in the dismissal of Godolphin and the appointment of Harley as Lord Keeper in his place, which accomplished the long-threatened downfall of the Government. By a refinement of insult the Duke's Secretary-at-War, Adam Cardonnel, was also removed and replaced, without the slightest reference to Marlborough, by Mr. Granville. Finally, shortly after his return from the campaign the Queen, despite his entreaties, definitely dismissed the Duchess from all her posts, and even went the length of ordering the Duke to forbid the moving of any vote of thanks for his services by Parliament.

The example thus set in high places was quickly followed. A few even of the Duke's own officers, such as the Duke of Argyll, to the huge disgust and contempt of the Army, turned against him. The mouth of every libeller and slanderer was opened. Swift and St. John, the only two Englishmen whose intellect entitled them to be named in the same breath with Marlborough, vied with each other in blackening his character. Nothing was too vile nor too extravagant to be insinuated against the greatest soldier, statesman, and diplomatist in Europe. He was prolonging the war for his own ends; he could make peace if he would, but he would not; he delighted in the wanton sacrifice of life; finally, he had neither personal courage nor military talent. "I suppose," wrote Marlborough bitterly, "that I must every summer venture my life in battle, and be found fault with in the winter for not bringing home peace, though I wish for it with all my heart and soul." 1710.

He would fain have resigned but for the remonstrances of Godolphin and Eugene, who entreated him to hold the Grand Alliance together for yet a little while, and gain for Europe a permanent peace. They might have spared their prayers had they known the secrets of the Cabinet, for Harley and his gang were already opening the secret negotiations with Lewis which were to dissolve the Alliance and grant to France all that Europe had fought for ten years to withhold from her. For these men, who accused Marlborough of wilful squandering of life, thought nothing of sending brave soldiers forth to lose their lives for a cause which they had made up their minds to betray. But it is idle to waste comment on such creatures, long dead albeit unhanged; though the fact must not be forgotten in the history of the relations of the House of Commons towards the Army. It will be more profitable to accompany the great Duke to his last campaign.

CHAPTER X

1711. THE French, fully aware of the political changes in England, had during the winter made extraordinary exertions to prolong the war for yet one more campaign, and to that end had covered the northern frontier with a fortified barrier on a gigantic scale. Starting from the coast of Picardy the lines followed the course of the river Canche almost to its source. From thence across to the Gy or southern fork of the Upper Scarpe ran a line of earthworks, extending from Oppy to Montenancourt. From the latter point the Gy and the Scarpe were dammed so as to form inundations as far as Biache, at which place a canal led the line of defence from the Scarpe to the Sensée. Here more inundations between the two rivers carried the barrier to Bouchain, whence it followed the Scheldt to Valenciennes. From thence more earthworks prolonged the lines to the Sambre, which carried them at last to their end at Namur.

This was a formidable obstacle to the advance of the Allies, but no lines had sufficed to stop Marlborough yet, and with Eugene by his side the Duke did not despair. Before he could start for the campaign, however, the news came that the Emperor Joseph was dead of smallpox, an event which signified the almost certain accession of the Archduke Charles to the Imperial crown and the consequent withdrawal of his candidature for the throne of Spain. Eugene was consequently detained at home; and worse than this, a fine opportunity was afforded for making a breach in the Grand Alliance. To render the Duke's difficulties still greater, though his force was already weakened by the necessity 1711. of finding garrisons for the towns captured in the previous year, the English Government had withdrawn from him five battalions[1] for an useless expedition to Newfoundland under the command of Mrs. Masham's brother, General Hill; an expedition which may be dismissed for the present without further mention than that it was dogged by misfortune from first to last, suffered heavy loss through shipwreck, and accomplished literally nothing.

Nevertheless the Imperial army was present, though without Eugene. The whole of the forces were assembled a little to the south of Lille at Orchies, and on the 1st of May Marlborough moved forward to a April 20/May 1. position parallel to that of Villars, who lay in rear of the river Sensée with his left at Oisy and his right at Bouchain. There both armies remained stationary and inactive for six weeks. Eugene came, but presently received orders to return and to bring his army with him. On the 14th of June Marlborough moved away June 3/14. one march westward to the plain of Lens in order to conceal this enforced diminution of his army. The position invited a battle, but Villars only moved down within his lines parallel to the Duke; and once more both armies remained inactive for five weeks. After the departure of Eugene the French commander detached a portion of his force to the Rhine, but even so he had one hundred and thirty-one battalions against ninety-four, and one hundred and eighty-seven squadrons against one hundred and forty-five of the Allies.

We now approach what is perhaps the most remarkable and certainly the most entertaining feat of the Duke during the whole war. Villars, bound by his instructions, would not come out and fight; his lines could not be forced by an army of inferior strength,

[1] 11th, 37th, Kane's, Clayton's, and one foreign battalion of foot. The losses of the expedition were 29 officers and 676 men drowned.

1711. and they could therefore be passed only by stratagem. The inundation on the Sensée between Arras and Bouchain could be traversed only by two causeways, the larger of which was defended by a strong fort at Arleux, the other being covered by a redoubt at Aubigny half a mile below it. Marlborough knew that he could take the fort at Arleux at any time and demolish it, but he knew also that Villars would certainly retake it and rebuild as soon as his back was turned. He therefore set himself to induce Villars to demolish it himself. With this view he detached a June 25/July 6. strong force under General Rantzau to capture the fort, which was done without difficulty. The Duke then gave orders that the captured works should be greatly strengthened, and for their further protection posted a large force under the Prussian General Hompesch on the glacis of Douay, some three miles distant from the fort.

As fate ordained it Hompesch, thinking himself secure under the guns of Douay, neglected his outposts and even his sentries, and was surprised two days later June 28/July 9. by a sudden attack from Villars, which was only repulsed with considerable difficulty and not a little shame. Villars was in ecstasies over his success, and Marlborough displayed considerable annoyance. However, the Duke reinforced Hompesch, as if to show the value which he attached to Arleux, and pushed forward the new works with the greatest vigour. Finally, when all was completed, he threw a weak garrison into the fort and led the rest of the army away two marches westward, encamping opposite the lines between the July 10/21. Canche and the Scarpe. Villars likewise moved westward parallel to him; but before he started he detached a force to attack Arleux. The commander of the fort sent a message to Marlborough that he could not possibly hold it, and the Duke at once despatched Cadogan with a strong force to relieve it. It was noticed, however, that Cadogan made no such haste as the urgency of the occasion would have seemed to require;

and indeed before he had gone half way he returned 1711.
with the intelligence that Arleux had surrendered.

Villars was elated beyond measure; and Marlborough for the first time in his life seemed to be greatly distressed and cast down. Throwing off his usual serenity he declared in public with much passion that he would be even with Villars yet, and would attack him, come what might of it, where he lay. Then came the news that Villars had razed the whole works of Arleux, over which he had spent such pains, entirely to the ground. This increased the Duke's ill-temper. He vowed that he would avenge this insult to his army, and renewed his menace of a direct attack on the entrenchments. Villars now detached a force to July $\frac{15}{26}$.
make a diversion in Brabant; and this step seemed to drive Marlborough distracted. Vowing that he would check its march he sent off ten thousand men under Lord Albemarle to Bethune, and the whole of his baggage and heavy artillery to Douay. Having thus July $\frac{17}{28}$.
weakened an army already inferior to that of the French, he repaired the roads that led towards the enemy's entrenchments, and with much display of vindictiveness, sulkiness, and general vexation advanced one march nearer to the lines. His army watched his proceedings with amazement, for it had never expected such proceedings from Corporal John.

Villars meanwhile was in a transport of delight. He drew every man not only from all parts of the lines but also from the neighbouring garrisons towards the threatened point, and asked nothing better than that Marlborough should attack. In the height of exultation he actually wrote to Versailles that he had brought the Duke to his *ne plus ultra*. Marlborough's strange manner still remained the same. On the 2nd of $\frac{\text{July 22}}{\text{August 2}}$.
August he advanced to within a league of the lines, and during that day and the next set the whole of his cavalry to work to collect fascines. At nightfall of the $\frac{\text{July 23}}{\text{August 3}}$.
3rd he sent away all his light artillery, together with every wheeled vehicle, under escort of a strong detach-

1711. July 24 / August 4. ment, and next morning rode forward with most of his generals to reconnoitre the lines. Captain Parker of the Eighteenth Royal Irish, who had obtained permission to ride with the Staff, was amazed at the Duke's behaviour. He had now thrown off all his ill-temper and was calm and cool as usual, indicating this point and that to his officers. "Your brigade, General, will attack here, such and such brigades will be on your right and left, such another in support, and you will be careful of this, that, and other." The generals listened and stared; they understood the instructions clearly enough, but they could not help regarding them as madness. So the reconnaissance proceeded, drearily enough, and was just concluding when General Cadogan turned his horse, unnoticed, out of the crowd, struck in his spurs and galloped back to camp at the top of his speed. Presently the Duke also turned, and riding back very slowly issued orders to prepare for a general attack on the morrow.

At this all ranks of the army, from the general to the drummer, fell into the deepest depression. Not a man could fail to see that direct attack on the lines was a hopeless enterprise at the best of times, and doubly hopeless now that half of the army and the whole of the artillery had been detached for other service. Again the violent and unprecedented outburst of surliness and ill-temper was difficult to explain; and the only possible explanation was that the Duke, rendered desperate by failure and misfortunes, had thrown prudence to the winds and did not care what he did. A few only clung faintly to the hope that the chief who had led them so often to victory might still have some surprise in store for them; but the most part gave themselves up for lost, and lamented loudly that they should ever have lived to see such a change come over the Old Corporal.

So passed the afternoon among the tents of the Allies; but meanwhile Cadogan with forty hussars at his heels had long started from the camp and was galloping hard across the plain of Lens to Douay, five leagues away. There he found Hompesch ready with his garrison, now strengthened by detachments from Bethune and elsewhere to twelve thousand foot and two thousand horse, and told him that the time was come. Hompesch thereupon issued his orders for the troops to be ready to march that night. Still the main army under Marlborough knew nothing of this, and passed the day in dismal apprehension till the sun went down, and the drummers came forward to beat tattoo. Then a column of cavalry trotted out towards the Allied right, attracting every French eye and stirring every French brain with curiosity as to the purport of the movement. Then the drums began to roll; and the order ran quietly down the line to strike tents and make ready to march immediately.

1711. July 24 / August 4.

Never was command more welcome. Within an hour all was ready and the army was formed into four columns. The cavalry having done their work of distracting French vigilance to the wrong quarter returned unseen by the enemy; and at nine o'clock the whole army faced to its left and marched off eastward in utter silence, with Marlborough himself at the head of the vanguard.

The night was fine, and under the radiant moonlight the men swung forward bravely hour after hour over the plain of Lens. The moon paled; the dawn crept up into the east throwing its ghastly light on the host of weary, sleepless faces; and presently the columns reached the Scarpe. So far the march had lasted eight hours, and fifteen miles had been passed. Pontoon-bridges were already laid across the river, and on the further bank, punctual to appointment, stood Brigadier Sutton with the field-artillery. The river was passed, and presently a messenger came spurring from the east with a despatch for the Duke of Marlborough. He read it; and words were passed down the columns of march which filled them with new life. "Generals Cadogan and Hompesch" (such was their purport) "crossed the causeway at Arleux

July 24-25 / August 4-5.

1711. July 25 / August 5. unopposed at three o'clock this morning, and are in possession of the enemy's lines. The Duke desires that the infantry will step out." The right wing of horse halted to form the rearguard and bring up stragglers, while a cloud of dust in the van told that the Duke and fifty squadrons with him were pushing forward at the trot. Then the infantry shook themselves up and stepped out with a will.

Villars had received intelligence of Marlborough's march only two hours after he had started, but he was so thoroughly bewildered by the Duke's intricate manœuvres that he did not awake to the true position until three hours later. Then, quite distracted, he put himself at the head of the Household Cavalry and galloped off at full speed. So furiously rode he that he wore down all but a hundred of his troopers and pushed on with these alone. But even so Marlborough was before him. At eight o'clock he crossed the lower causeway at Aubanchoeuil-au-bac and passing his cavalry over the Scarpe barred the road from the west by the village of Oisy. Presently Villars, advancing reckless of all precautions, blundered into the middle of the outposts. Before he could retire his whole escort was captured, and he himself only by miracle escaped the same fate.

The Marshal now looked anxiously for the arrival of his main body of horse; but the Allied infantry had caught sight of them on the other side of the Sensée, and weary though they were had braced themselves to race them for the goal. But now the severity of the march and the burden of their packs began to tell heavily on the foot. Hundreds dropped down unconscious and many died there and then, but they were left where they lay to await the arrival of the rearguard; for no halt was called, and each regiment pushed on as cheerfully as possible with such men as still survived. Thus they were still ahead of the French when they turned off to the causeway at Arleux, and, Marlborough having thrown additional bridges

over the Scarpe, they came quickly into their positions. 1711.
The right wing of infantry crossed the river about four July 25/August 5.
o'clock in the afternoon, having covered close on forty miles in eighteen hours; and by five o'clock the whole force was drawn up between Oisy and the Scheldt within striking distance of Arras, Cambrai, and Bouchain. So vanished the *ne plus ultra* of Villars, a warning to all generals who put their sole trust in fortified lines.

Marlborough halted for the next day to give his troops rest and to allow the stragglers to come in. Fully half the men of the infantry had fallen out, and there were many who did not rejoin the army until the third day. Villars on his side moved forward and offered Marlborough battle under the walls of Cambrai; but the Duke would not accept it, though the Dutch deputies, perverse and treacherous to the last, tried hard to persuade him. Had the deputies marched in the ranks of the infantry with muskets on their shoulders and a kit of fifty pounds' weight on their backs, they would have been less eager for the fray. Marlborough's own design, long matured in his own mind, was the capture of Bouchain, and his only fear was lest Villars should cross the Scheldt before him and prevent it. Then the deputies, who had been so anxious to hurry the army into an engagement under every possible disadvantage, shrank from the peril of a siege carried on by an inferior under the eyes of a superior force. But Marlborough, even if he had not been able to adduce Lille as a precedent, was determined to have his own way, and carried his point. At noon on the July 27/August 7.
7th of August he marched down almost within cannon-shot of Cambrai, ready to fall on Villars should he attempt to cross the Scheldt, halted until his pontoon-bridges had been laid a few miles further down the stream, and then gradually withdrawing his troops passed the whole of them across the river unmolested.

It is hardly credible that a vast number of foolish civilians, Dutch, Austrian, and even English, blamed Marlborough for declining battle before Cambrai, and

1711. that he was actually obliged to explain why he refused to sacrifice the fruit of his manœuvres by attacking a superior force in a strong position with an army not only smaller in numbers at its best, but much thinned by a forced march and exhausted by fatigue. "I despair of being ever able to please all men," he wrote. "Those who are capable of judging will be satisfied with my endeavours: others I leave to their own reflections, and go on with the discharge of my duty."

It is possible that Villars only refrained from hindering Marlborough's passage of the Scheldt in deference to orders from Versailles, of which the Duke was as well aware as himself; but it is more than doubtful whether he ever intended him to capture Bouchain. Though inferior in numbers, however, Marlborough covered himself so skilfully with entrenchments that Villars could not hinder him, and met all attempts at diversion so readily that not one of them
Sept. 2/13. succeeded. Finally, the garrison surrendered as prisoners of war under the very eyes of Villars. The Duke would have followed up his success by the siege of Quesnoi, the town before which English troops first came under the fire of cannon in the year of Creçy; but by this time Lewis, with the help of the contemptible Harley, had succeeded in detaching England from the Grand Alliance. Though, therefore, the English ministers continued to encourage Marlborough in his operations in order to conceal their own infamous conduct from the Allies, yet they took good care that those operations should proceed no further. So with the capture of Bouchain the last and not the least remarkable of Marlborough's campaigns came, always victoriously, to an end.

The most brilliant manifestation of military skill was, however, powerless to help him against the virulence of faction in England. The passage of the lines was described as the crossing of the kennel, and the siege of Bouchain as a waste of lives. In May the House of Commons had addressed the Queen for inquiry into abuses in the public expenditure, and 1711. when the Duke arrived at the Hague in November he found himself charged with fraud, extortion, and embezzlement. The ground of the accusation was that he had received in regular payment from the bread-contractors during his command sums amounting to £63,000. Marlborough proved conclusively that this was a perquisite regularly allowed to the commander-in-chief in Flanders as a fund for secret service, and he added of his own accord that he had also received a deduction of two and a half per cent from the pay of the foreign troops, which had been applied to the same object. But this defence, though absolutely valid and sound, could avail him little. His reasons were disregarded, and on the 31st of December he was dismissed from all public employment.

Three weeks later the House of Commons voted that 1712. his acceptance of these two perquisites was unwarrantable and illegal, and directed that he should be prosecuted by the Attorney-General. This done, the Ministry appointed the Duke of Ormonde to be commander-in-chief in Marlborough's place, and confirmed to him the very perquisites which the House had just declared to be unwarrantable and illegal. Effrontery and folly such as this are nothing new in representative assemblies, but it is significant of the general attitude of English civilians towards English soldiers, that not one of Harley's gang seems to have realised that this vindictive persecution of Marlborough was an insult to a brave army as well as a shameful injustice to a great man, nor to have foreseen that the insult might be resented by the means that always lie ready to the hand of armed and disciplined men.

It is not necessary to dwell on the operations, if such they may be called, of the Duke of Ormonde. He did indeed take the field with Eugene, but under instructions to engage neither in a battle nor a siege, but virtually to open communications with Villars. By July the subservience of the British Ministry to

1712. Lewis the Fourteenth had been so far matured that Ormonde was directed to suspend hostilities for two months, and to withdraw his forces from Eugene. Then the troubles began. The auxiliary troops in the pay of England flatly refused to obey the order to leave Eugene, and Ormonde was compelled to march away with the British troops only. Even so the feelings of anger ran so high that a dangerous riot was only with difficulty averted. The British and the auxiliaries were not permitted to speak to each other, lest recrimination should lead either to a refusal of the British to leave their old comrades or to a free fight on both sides. The parting was one of the most remarkable scenes ever witnessed. The British fell in, silent, shamefaced, and miserable; the auxiliaries gathered in knots opposite to them, and both parties gazed at each other mournfully without saying a word. Then the drums beat the march and regiment after regiment tramped away with full hearts and downcast eyes, till at length the whole column was under way, and the mass of scarlet grew slowly less and less till it vanished out of sight.

At the end of the first day's march Ormonde announced the suspension of hostilities with France at the head of each regiment. He had expected the news to be received with cheers: to his infinite disgust it was greeted with one continuous storm of hisses and groans. Finally, when the men were dismissed they lost all self-control. They tore their hair and rent their clothes with impotent rage, cursing Ormonde with an energy only possible in an army that had learned to swear in the heat of fifty actions. The officers retired to their tents, ashamed to show themselves to their men. Many transferred themselves to foreign regiments, many more resigned their commissions; and it is said, doubtless with truth, that they fairly cried when they thought of Corporal John.

More serious consequences followed. The march was troublesome, for the Dutch would not permit the retiring British to pass through their towns, and 1712. the troops were consequently obliged to cross every river that barred their way on their own pontoons. Again, all the old contracts for bread had been upset by Harley and his followers through their prosecution of Marlborough: it was nothing to them that an army should be ill-fed, so long as they gained power and place. St. John, it must be noted, was a principal accomplice in this rascality—St. John, who alone of living Englishmen had intellect sufficient to measure the gigantic genius of Marlborough; who, moreover, as Secretary-at-War during the greatest of the Duke's campaigns, had gained some insight into those prosaic details of supplies and transport which are all in all to the organisation of victory. Ormonde, a thoroughly mediocre officer, was not a man to grapple with such difficulties. Bad bread heightened the ill-feeling of the soldiers towards him. Agitators insinuated to the worst characters in the army that they would lose all the arrears of pay that were due to them; and the story found ready and reasonable credence from recollection of the scandals that had followed the Peace of Ryswick. The good soldiers, then as always a great majority, refused to have anything to do with a movement so discreditable, and reported what was going forward to their officers; but either their tale was disbelieved or, as is more likely, apathy and general disorganisation prevented the nipping of the evil in the bud. Finally, three thousand malcontents slipped away from the camp, barricaded themselves in a defensive position, and sent a threatening message to the commander-in-chief demanding good bread and payment of arrears. Then discipline speedily reasserted itself. The mutineers were surrounded and compelled to surrender. A court-martial was held; ten of the ringleaders were executed on the spot and the mutiny was quelled once for all. Fortunate it was that the outbreak took place while the troops were still abroad, or the House of Commons might have learned by a second bitter experience that the patience of the British soldier, though very great, is not inexhaustible.[1]

1713. March 31/April 11. The negotiations so infamously begun with King Lewis shortly after found as infamous an end in the Peace of Utrecht, which not only sacrificed every object for which the war had been fought, but branded England with indelible disgrace. Five months earlier Marlborough had left England, to all intent a banished man. Before his departure he had endured incredible insults in the House of Lords, the worst and falsest of them from one of his own officers, the Duke of Argyll. The defection and ingratitude of Argyll, however, only brought out the more strongly the general loyalty of the Army towards its great chief. Marlborough's most prominent officers were of course subjected to the same degradation as himself. Cadogan, for instance, was removed from the Lieutenancy of the Tower to make room for Brigadier Hill; and even the Duke's humble secretary, Adam Cardonnel, was not too small an object for the malignant spite of the House of Commons. But honourable men, such as Lord Stair, the colonel of the Scots Greys, threw up their commissions in disgust; and plain, honest officers, such as Kane and Parker, have left on record the immense contempt wherein Argyll, brave soldier though he was, was held in the Army. The Dutch also rose, though too late, to the occasion. When Marlborough sailed into Ostend at the end of November, 1712, the whole garrison was under arms to receive him, and when he left it, it was under a salute of artillery. At Antwerp, in spite of his protests, his reception was the same; the cannon thundered in his honour, and all ranks of the people

[1] Strangely enough it was in these very weeks (13th July) that Richard Cromwell, the ex-protector, died, at the age of eighty-seven; one of the very few men who had seen the rise of the New Model, the culmination of Oliver Cromwell's military work in the hands of Marlborough, and the fall of Marlborough himself.

turned out to meet him with joyful acclamations. He took the most secluded road to Maestricht, but go whither he would, fresh parties of horse always appeared to escort him. Above all, he was comforted by the unchanging confidence and sympathy of Eugene. 1713.

There for the present we must leave him till the time, not far distant, shall come to tell of his restoration. That the welcome given to him by the Dutch may have been a consolation to him we can hardly doubt, and yet he cannot but have felt that these same Dutch had been his undoing. For, despite the shameful perfidy of the English politicians who drove Marlborough from England and concluded the Treaty of Utrecht, the main responsibility for the catastrophe rests not with them but with those unspeakable Dutch deputies who, by wrecking the Duke's earlier campaigns, prolonged beyond the limits of the patience of the House of Commons the War of the Spanish Succession.

Authorities.—The literature of the War of the Spanish Succession is, as may be guessed, not slender. On the English side there are the lives of Marlborough by Lediard and Coxe, as well as the French life, in three volumes, which was written by Napoleon's order. There are also the journals of Archdeacon Hare for the campaign of Blenheim, and a valuable letter from him respecting Oudenarde; the narratives of General Stearne, of Kane, Parker, and Sergeant Millner, all unfortunately of one regiment, the 18th Royal Irish; and, for the campaign of 1708 only, the journal of Private John Deane of the 1st Guards (privately printed 1846). Dumont's *Histoire Militaire* gives admirable maps and plans. Many curious items are also to be found in Lamberti. I have not failed to study the archives of the War Office preserved at the Record Office, with results that will be seen in the next chapter, and I have been carefully through the contemporary newspapers. Minor authorities, such as Tindal's *History* and the like, are hardly worth mention. Marlborough's *Despatches*, though decried by Lord Mahon (Preface to *History of England*), I have found most valuable. On the French side Quincy remains the chief authority, together with the *Archives Militaires* in the printed collection. The *Mémoires* of St. Simon, Villars, Millot, and others have also been consulted, and good and pertinent comment is always to be found in Feuquières.

For the war in Spain see at the close of Chapter VI.

CHAPTER XI

1702-1713. ALTHOUGH the narrative of the War of the Spanish Succession has not infrequently been interrupted in order to give the reader an occasional glimpse of the progress and difficulties of the military administration at home, yet much has been of necessity omitted, lest the strand, enwoven of too many and too distinct threads, should snap with the burden of its own weight and unravel itself into an inextricable tangle. I propose therefore at this point to summarise the orders, regulations, and enactments of the War Office and of the House of Commons during the reign of Queen Anne to the Peace of Utrecht, so as, if possible, to convey some notion of the legacies, other than those of glory and prestige, that were bequeathed to the Army by this long and exhausting war.

The reader will, I think, have gathered at least that the extension of operations and the consequent increase of the British forces during the war was almost portentously rapid. A few figures will make this more apparent. In 1702 and 1703 Flanders was practically the only scene of active operations, the raid on Cadiz being of too short duration and too little account to be worthy of serious mention. In both of these years the British troops with Marlborough were set down at eighteen thousand men. In 1704 to 1706 they rose to twenty-two thousand, and in 1708 to 1709 to twenty-five thousand men, reverting once again to twenty-two thousand from 1711 to 1712. Concurrently with the first increase of 1704 came the first despatch of eight thousand troops to the Peninsula, rising to nine thousand in 1705, ten thousand in 1706, and twenty-six thousand[1] from 1707 to 1709, relapsing between 1710 and 1712 to rather over twenty thousand. The total number of forces borne on the list of the British Army at its greatest was six troops of Household Cavalry, eleven regiments of horse, sixteen of dragoons, and seventy-five of foot, comprehending in all seventy-nine battalions.[2] The nominal war strength of a battalion in Flanders was, as a rule, in round numbers nine hundred and forty of all ranks, in the Peninsula from seven hundred and fifty to eight hundred and eighty, a diversity of establishments which gave rise to much trouble and confusion. It would not be safe to reckon the British infantry at any period during the war as exceeding fifty thousand men. The regiments of dragoons again varied from a normal strength of four hundred to four hundred and fifty, rising in occasional instances to six hundred; but they cannot reasonably be calculated at a higher figure than six thousand men. The regiments of horse were subject to similar variations, but their total strength, even including the six strong troops of Household Cavalry, cannot be counted as more than seven thousand men. There then remains the artillery, of which, from want of data as well as from vagueness of organisation, it is impossible to make any accurate calculation. Speaking generally, the highest strength actually attained by British troops at home and abroad during the war may be set down at seventy thousand men.[3] 1702-1713.

[1] Nominally 30,000, but 4000 are deducted for Huguenot regiments.

[2] Including Huguenot regiments the numbers would be 22 regiments of dragoons and 81 of foot. The three regiments of Guards, though varying greatly in strength, may be reckoned practically at two battalions apiece; the Royal Scots had also two battalions, both on active service.

[3] These figures are based principally on the estimates submitted to the House of Commons, which are printed in the journals, but can only be approximately accurate. The confusion in the statement is worthy of the War Office. First, there is the establishment

1702-1713. The defect that will seem most flagrant, according to modern ideas, in the scheme above sketched is the multiplicity of distinct units that go to make up so small a force. The French had long abandoned the system of single battalions, and indeed given to their regiments the name of brigades. In the British Army the Guards and the Royal Scots alone had two battalions; and though we know by actual information that, in the case of the former, the battalions at home were used to feed those abroad, yet it is indubitable that both battalions of the Royal Scots took the field and kept it from beginning to end of the war. For this, however, the principles that then governed the conduct of a war and the maintenance of an army sufficiently account. The year was divided for military purposes into two parts—the campaigning season, which lasted roughly from the 1st of April to the 1st of October, and the recruiting season, which covered the months that remained over. Directly the campaign was ended and the troops distributed into winter quarters, a sufficient number of officers returned home to raise for each regiment the recruits that were needed. In strictness no officer enlisted a man except for his own corps; and it was only occasionally that a regiment, having enlisted more recruits than were required for its own wants, transferred its superabundance to another.

But apart from this, we find throughout the reign of Queen Anne a resolute and healthy opposition to the principle of completing one regiment by drafts from

for England (after 1707 for Great Britain), including colonial garrisons. Next, establishment for Flanders and augmentation for Flanders; establishment for Portugal and augmentation for Portugal; establishment for Catalonia and augmentation for Catalonia, making, with Ireland, eight different establishments, involving transfers and changes and explanations without end. The House of Commons (see Journals, January 1708) was puzzled and dissatisfied, but obtained small satisfaction. Probably the Treasury was partly to blame as well as the War Office.

The estimates for 1709 provide for 69,000 men, exclusive of the Irish establishment and of Artillery. *Commons Journals.*

another. At the beginning of the war the ranks of the Army were, thanks to the wanton imbecility of the House of Commons, so empty that it was impossible to send any appreciable number of regiments abroad without depletion of those that were left at home. As an exceptional favour therefore the first troops sent to Spain and to the West Indies were completed by drafts; but at that point the practice was checked.[1] Marlborough had early set his face against so vicious a system, and although once, under pressure of orders from the Queen herself, he directed it to be enforced, yet it is sufficiently clear from his language and from his ready deference to the protest of the officer concerned, that he fully recognised the magnitude of its evil.[2] After the disaster of Almanza the War Office appears to have been urged in many quarters to resort to drafting, but St. John told the House of Commons outright that the practice had been found ruinous to the service, prejudicial alike to the corps that furnished and that received the draft. As Marlborough's influence declined, the mischievous system seems to have been revived, and although in more than one case colonels flatly declined to part with their men,[3] yet at the close of the war we find garrisons denuded by drafts to an extent that was positively dangerous.[4] The same objectionable practice, as is well known, is still rampant among us; would that the authority of Marlborough could help to break it down. 1702-1713.

There remains the question why, instead of raising new regiments, the authorities did not raise additional battalions to existing regiments? The reply is that they doubtless knew their own business, and adopted the best plan that lay open to them. Englishmen have

[1] *Commons Journals*, 3rd and 18th February 1708.

[2] *Despatches*, vol. ii. p. 460.

[3] *Secretary's Common Letter Book*, 26th May 1709. *S. P., Dom.*, vol. xvii. p. 85.

[4] Thus in August 1710 the garrison of Portsmouth was reduced by drafts to 360 men. *S. P., Dom.*, vol. xvii. p. 19.

1702-1713. a passion for independent command. To this day, as the history of the volunteers shows, there are many men who, though unwilling to serve in any existing corps, would cheerfully expend ten times the care, trouble, and expense on a regiment, or even on a troop or company, of their own. It must be remembered, too, that a regiment in those days was not only a command but a property, that it afforded to officers opportunities for good and for evil such as are now undreamed of, that, lastly, it was in the vast majority of cases called by its colonel's name.

Let us now, before examining the measures taken for the supply of recruits, glance briefly at the principal centres and causes of consumption and of demand. The inquiry must not be considered superfluous, for the primary force in the maintenance of a voluntary army is attraction, and it is only after full knowledge of the elements of repulsion which work counter to it that the failure of the attractive force, and the necessity for substituting coercion in its place, can be rightly understood. The theatres of war claim first attention, and of these Flanders claims the precedence. It is well known that sickness or fatigue are more destructive in war than bullet and sword, and Marlborough's campaigns can have been no exception to the rule. Yet it is remarkable that the British were never so much thinned as after the campaign of Blenheim, wherein they bore the brunt of two severe actions. The march to the Danube was of course severe, but the men stood it well; nor do we hear of extraordinary sickness on the return march. All that we know is that when the British regiments reached the Rhine they were too weak to be fit for further work. We never hear the like in subsequent campaigns, in spite of severe marching and sieges. Yet the capture of one of Vauban's fortresses was always a long and murderous piece of work, while, if the trenches were flooded by heavy rain or the natural oozing of marshy ground, an epidemic of dysentery was sure to follow. We have no returns of the losses from sickness in Flanders, but it is certain that the operations in that field were by no means the most deadly to the troops, nor the most exhausting to England. This must be ascribed almost entirely to the care and forethought of the great Duke. Marlborough knew the peculiar weaknesses as well as the peculiar value of his own countrymen, and was careful to keep them always well fed. In the second place, and this was most important, the theatre of war was but a few hours distant from England, so that a force once fairly set on foot could be maintained with comparative ease. Recruits, too, did not feel that they were going to another part of the world, and would never return home. Moreover, a bounty had been granted for Blenheim, there was some prospect of plunder,[1] and there was the glory of marching to certain victory with Corporal John. 1702-1713.

It was far otherwise in the Peninsula. There a campaign was broken not only by winter-quarters, but also by summer-quarters in the hot months of July and August. Again, the voyage to Lisbon, and still more to Catalonia, to say nothing of the risk of storm and shipwreck, occupied days and weeks, whereas the passage to Flanders was reckoned by hours. The transport-service, too, had a bad name. Although after 1702 the official complaints of bad and insufficient food ceased, yet the mortality on board the troop-ships sent to the Peninsula shows that the sickness and misery must have been appalling. The reinforcements despatched to Lisbon in the summer of 1706 with a total strength of eight thousand men were reduced to little more than half of their numbers when they landed in Valencia in February 1707. They had suffered from bad weather and long confinement, it is true, but theirs was no exceptional case.[2] In 1710, of a detachment of three hundred men that were landed, only a hundred ever

[1] The men, as is plain from the pages of Parker, Kane, and Millner, looked forward to a wealth of spoil as soon as they should penetrate into the heart of France.

[2] *Commons Journals*, 18th February 1708.

1702-1713. reached their regiments.[1] In 1711 five weak regiments lost sixty men dead, and two hundred disabled from sickness in a voyage of ten days.[2] A private of the First Guards summed up his experience of a month in a transport as "continual destruction in the foretop, the pox above board, the plague between decks, hell in the forecastle, and the devil at the helm."[3]

This was one great discouragement to recruits; and others became quickly known to them. The Peninsula was ill-supplied, transport was difficult, the quarters of the troops were very unhealthy, and the Portuguese unfriendly even to brutality.[4] Altogether, though steel and lead played their part in the destruction of the British in the Peninsula, the havoc that they wrought was trifling compared with that of privation and disease. Prisoners of course were never lost for long, as Marlborough had always abundance of French to give in exchange for them; but in spite of this, the waste in Portugal and Spain was terrible, and the service proportionately unpopular.

So much for the two theatres of war; but the sphere of foreign service was not bounded by these. New York, Bermuda, and Newfoundland each possessed a small garrison; and the West Indies, as we have seen, claimed from four to six battalions. This colonial service was undoubtedly the most unpopular of all. When the single company that defended Newfoundland left England in 1701, their destination was carefully concealed from the men lest they should desert. The most hardened criminal could hope for pardon if he enlisted for Jamaica. Once shipped off to the West Indies, the men seem to have been totally forgotten. No proper provision was made for paying them;

1 *Cal. Treas. Papers*, 18th November 1710.

2 *S. P., Dom.*, vol. xviii. p. 116.

3 Deane.

4 There is nothing more remarkable than the mortality among the British troops, in what town soever quartered, in the Peninsula. The complaints against the Portuguese will be found very bitter in the letters of Colonel Albert Borgard of the Artillery. *S. P. Spain.*

colonels who cared for their men were compelled to borrow money to save them from starvation; colonels who did not, came home, together with many of their officers, and left the men to shift for themselves.[1] Clothing, again, was entirely overlooked. The troops in Jamaica were reduced almost to nakedness; and when finally their clothing, already two years overdue, was ready for them, it was delayed by a piece of bungling such as could only have been perpetrated by the War Office.[2] Another great difficulty was that, there being no regular system of reliefs, colonels never knew whether to clothe their men for a hot or a temperate climate. Recruits were consequently most difficult to obtain, although owing to the unhealthiness of the climate they were in great request. The result was that old men and boys were sent across the Atlantic only to be at once discharged, at great pecuniary loss, by the officers, who were ashamed to admit creatures of such miserable appearance into their companies.[3] 1702-1713.

Again, during the course of the war, two new acquisitions demanded garrisons of three or four battalions apiece. Minorca appears to have given no very serious trouble; but Gibraltar having been reduced virtually to ruins by the siege was, owing to the lack of proper habitations, a hot-bed of sickness. The authorities seem in particular to have neglected the garrison of Gibraltar, though they took considerable pains for the fortification of the Rock. In 1706 more than half of the garrison was disabled through disease brought on by exposure,[4] yet it was not until four years later that

[1] *Cal. Treas. Papers*, 18th June and 18th November 1706.

[2] The regiment being in the Irish establishment the clothing was ordered in Ireland. When, after long delay, the clothing arrived at Bristol, it was discovered that, being of Irish manufacture, it could not be discharged without the Treasurer's warrant; which, of course, entailed the delay, appreciable enough in those days, of a journey to London and back.

[3] *Cal. Treas. Papers*, 18th November 1707.

[4] *S. P., Dom.*, vol. viii. 81.

1702-1713. orders were given for the construction of barracks,[1] while even in 1711 the men were obliged to burn their own miserable quarters from want of fuel.[2]

These lapses in countries beyond sea might possibly find some excuse in the plea of inexperience, though this should not be admitted in a country which for nearly four centuries had continually sent expeditions across the Channel, and for more than two centuries across the Atlantic also. Yet there were similar faults at home which show almost incredible thoughtlessness and neglect. Thus in 1709 many soldiers at Portsmouth perished from want of fire and candle,[3] while the garrison of Upnor Castle was required to supply a detachment of guards in the marshes three miles from any house or shelter, where the men on duty stood up to their knees in water.[4] No one had thought that they might want a guard-room or at least tents. Again, it was not until a ship's load of men invalided from Portugal had been turned adrift in the streets of Penrhyn, penniless and reduced to beg for charity, that any provision was made for the sick and wounded. Then at last, in the fourth campaign of the war, commissioners were appointed to make them their special care. So far no one had been responsible for them, the duty having been thrust provisionally upon the commissioners of transport.[5] In a word, no forethought nor care was to be found beyond the reach of Marlborough's own hand; all administration on the side of the War Office, even under the secretaryship of so able a man as Henry St. John, was marked by blindness and incompetence.

The ground being now cleared, and the principal obstacles in the way of recruiting being indicated, it is

1 *S. P., Dom.*, vol. xvi. 92.
2 *Cal. Treas. Papers*, 15th August 1711.
3 *Ibid.*, 12th October 1709.
4 *Secretary's Common Letter Book*, 20th September and December 1705.
5 *S. P., Dom.* (12th March 1711), vol. xix. 21.

time to examine the means employed by Parliament to overcome them. We may properly confine ourselves to England, since she with her population of five and a quarter millions was necessarily the main source for the supply of men. Ireland was not yet the recruiting-ground that she became at a later day, for the simple reason that none but Protestants could be enlisted. She had, however, her five distinctly national regiments,[1] a small proportion which enabled her to provide a dozen or fifteen more in the course of the war. Protestant Ireland, in fact, still under the spell of William of Orange, played her part very fully and generously during these years. Scotland, as became a country of great military traditions, maintained a larger number of national regiments than her sister,[2] but being thinly populated, inaccessible in many districts and already engaged to furnish troops to the Dutch service, was unable to provide more than three additional battalions. The greatest stress therefore fell, and fell rightly, upon England. 1702-1713.

Transporting ourselves therefore for a moment to the opening of the war, when the Army was still smarting under its shameful treatment by Parliament after the Peace of Ryswick, we find without surprise that the strain of providing recruits made itself felt very early. The Mutiny Act of 1703 shows this by a clause empowering the Queen to order the delivery from gaol of capital offenders who had been pardoned on condition of enlistment. This enactment was of course something like a reversion to the methods of Elizabeth; but although this class of recruit does not sound desirable, yet the competition for it was so keen that a regular roster was kept to ensure that every regiment should profit by the windfall in its turn.[3] It must be remembered that many a man was then con-

[1] 5th, 6th, 8th Dragoons; 18th, 27th Foot.

[2] Two troops Household Cavalry, Scots Greys and 7th Dragoons, Scots Guards, and 1st Royals (each two battalions), 21st, 25th, 26th Foot.

[3] *Secretary's Common Letter Book*, 22nd May 1704.

1702-1713. demned to death who would now be released under the First Offenders' Act; but apart from this, criminals were welcome to the recruiting officer, first, because they cost nothing, and secondly, because they were often men of fine physique.[1] In the later years of the war the sweepings of the gaols were in particular request, and the multiplication of petitions from the condemned shows that the fact was appreciated within the walls of Newgate.

In the session of 1703-4 an Act, for which there was a precedent in the days of King William, was passed to provide for the discharge of all insolvent debtors from prison, who should serve or procure another to serve in the fleet or Army. This probably brought some useful young recruits who enlisted to procure the release of their fathers; and there is evidence that the bankrupt was as much sought after by recruiting officers as the sheep-stealer. Another most important Act of the same session was the first of a long series of annual Recruiting Acts. Under this, a bounty of one pound[2] was offered for volunteers; and justices of the peace were empowered to levy as recruits all able-bodied men who had no visible employment or means of subsistence, and to employ the officers of borough and parish for the purpose. For each such recruit a bounty of ten shillings[3] was allowed for himself as well as a fee of ten shillings to the parish officer. To remove any temptation to malpractice, no officer of the regular Army was permitted to sit as a justice under the Act; and all voters were specially exempted from its operation, the possession of the franchise being apparently considered, as it probably was, a sufficiently visible means of subsistence.

This latter measure brought with it a considerable crop of abuses. In the very next session it was found

[1] Not always, however, for among the capital offenders pardoned I find a boy of ten.

[2] Levy money of £2, of which one moiety for the recruit.

[3] Levy money of £1.

necessary to give special protection to harvest-labourers, many having been already impressed, while many more had hidden themselves from fear of impressment. But this was by no means all. Voters occasionally shared the fate of their unenfranchised brethren, and required hasty deliverance with many apologies to the member for their borough.[1] The high bounty again gave a stimulus to wrongful impressment, fraudulent enlistment, and desertion. It was found necessary after a few months to restrain the zeal of parish officers, who enlisted men that were already soldiers. Again, there were recruiting-officers who would discharge the recruits brought to them for a pecuniary consideration, an occurrence which though not common was not unknown. Finally, recruits would occasionally try to break away in a body, which led to desperate fighting and to awkward complications. In one instance a large number of recruits made so determined an attempt to overpower the guard and escape that they were not quelled until two of them had been actually slain. The guard, who thought with justice that they had done no more than their duty, then found themselves threatened with an indictment for murder; and the War Office was obliged to call in the Attorney-General to advise how they should be protected.[2] Turbulent scenes with the rural population over the arrest of deserters and the impressment of idle fellows were by no means infrequent. We have, for instance, accounts of the whole town of Exminster turning out with flails and pitchforks against an officer who claimed a deserter, and of the mob of Abergavenny, mad for the rescue of an impressed recruit, driving the officers from house to house, and compelling them to fire in self-defence.[3] 1702-1713.

After the campaign of Blenheim, the heavy losses in the field, and the resolution to send a large force

1 Abundant instances in the *Secretary's Common Letter Book.*

2 *Ibid.*, 13th March 1704.

3 *S. P., Dom.*, vol. v. 135; vol. ix. 75.

1702-1713. to the Peninsula drove the military authorities to desperate straits. Suggestions of course came in from various quarters; among them a proposal from a gentleman of Amsterdam that every one who had two or more lacqueys should send one into the Army, the writer having observed that members of Parliament "abounded in that sort of person."[1] But the stress of the situation is shown by the fact that a Bill was actually introduced to compel every parish and corporation to furnish a certain number of recruits, though it was presently dropped as being an imitation from the French and unfit for a free country.[2] The authorities therefore contented themselves by ordering stricter enforcement of the Recruiting Act, and apparently with success.[3] During the next two years there was no change in the Act, excepting the addition, in 1706, of a penalty of five pounds against parochial officers who should neglect to execute it. But in 1707 the measure showed signs of failing, and was hastily patched up by increasing the bounty to two pounds[4] for volunteers enlisting during the recruiting season, and to one pound for such as enlisted after the campaign had been opened. Some effort was also made to systematize the power granted by the Act by convening regular meetings of justices at stated times and places.

The close of the year, however, found the Commons face to face with the disaster of Almanza, and with urgent need for close upon twenty thousand recruits. The Recruiting Act now assumed a new and drastic form. The authority to impress men of no employment was transferred from the justices to the commissioners of the land-tax, with full powers

1 *S. P., Dom.*, vol. v. 128.

2 Tindal.

3 A curious and, I imagine, illegal stretch of the Royal prerogative appears in the shape of a Royal warrant for the impressment of fifes, drums, and hautbois. *H. O. M. E. B.*, 1st Jan. 1705.

4 The levy money was £4 per man, of which it seems that half was bounty, and half for expenses of the recruiting officer.

to employ the parochial officers. The penalty on 1702-1713. these officers for neglect of duty was increased to ten pounds, while for diligent execution of the same a reward of one pound was promised them for every recruit, as well as sixpence a day for the expense of keeping him until he should be made over to his regiment. The parish likewise received three pounds for every man thus recruited, in order to quicken its zeal against the idle. Finally, as an entire novelty, borrowed be it noted from the French,[1] volunteers were enlisted at the same high rate of bounty for a term of three years, at the close of which they were entitled to claim their discharge. Great results were evidently expected from these provisions, for the standard of height for recruits was still maintained at five feet five inches,[2] men below that stature being accepted only for marines. So from this year until the close of the war it is possible to study the first trial of short service in England.

Unfortunately abuses seemed only to multiply under the new Act. The campaign of Oudenarde, prolonged as it was into December, drained Marlborough's army heavily, and the spring of 1709 found the forces in want of yet another fifteen thousand recruits. Moreover, from the moment when Marlborough's power began to decline the tone of the Army at home began to sink. The justices again were jealous of the commissioners of land-tax, and in some instances openly abused and reproached them.[3] In at least one case they were found conniving with officers to accept money for the discharge of impressed men.[4] Officers on their

[1] The system was introduced by Lewis XIV. in the autumn of 1703. The still earlier suggestion of a short-service system in the sixteenth century has already been related.

[2] The number of volunteers enlisted in March 1708 for the regiments in the Peninsula was something over 800, of which London and Middlesex supplied just twenty-three.

[3] Newspapers, 13th March 1709.

[4] *S. P., Dom.* (15th September 1708), vol. xiv.

1702-1713. side also began to misbehave, withholding the bounty from recruits and subjecting them to the gantlope if they complained, and in some instances not only withholding the bounty but demanding large bribes for their discharge.[1] As the war continued, matters grew worse and worse. Sham press-gangs established themselves with the object of levying blackmail;[2] and as a climax Army and Navy began to fight for the possession of impressed men.

At the opening of 1711 the first batch of men enlisted for three years completed their term, but found to their surprise that their discharge did not come to them automatically, as they had expected. The officers had no instructions. They were unwilling too to part with the sixty best soldiers in each regiment, for such these men of short service had proved to be, and could only promise to let them go as soon as orders should arrive from home. Harley's Secretary-at-War, with the characteristic ill faith of the politician towards the soldier, boldly proposed to pass an Act compelling them to serve for two years longer; but the Attorney-General, to whom the matter was referred, decided that the men were beyond all question entitled to their discharge.[3] Thereupon, rather late in the day, the Secretary-at-War hurriedly ordered the instant discharge of a man whose term had expired, in order to encourage others to enlist.[4] Finally, in 1711 abuses increased so rapidly under the new administration that the whole system of recruiting broke down.[5] The evils of Harley's short tenure of office were by no means bounded by the Peace of Utrecht.

[1] *E.g.*, *Secretary's Common Letter Book*, 21st September and 23rd December 1708.

[2] *S. P.*, *Dom.* (undated), vol. x.

[3] *Ibid.* (20th February 1711), vol. xviii.; (14th April 1712), vol. xxii.

[4] Lord Lansdowne. *Secretary's Common Letter Book*, 12th March 1712. The question had originally been brought up a year before.

[5] *Ibid.*, 23rd April 1711.

There remains a further question still to be dealt with, that namely of desertion, which directly and indirectly sapped the strength of the Army as much as any campaign. Let it not be thought that this evil was confined to England, for it was rampant in every army in Europe, and nowhere a greater scourge than in France. Nor let the deserter from the army in the field be too severely judged, for his anxiety was not to serve against his own countrymen but simply to get back to his own home. Some of the English deserters in Flanders were even cunning enough to pass homeward as exchanged prisoners belonging to the fleet.[1] But it was before starting for the seat of war that deserters gave most trouble, particularly if, as was often unavoidably the case, the regiments were kept waiting long for their transports.[2] No punishment seemed to deter others from abetting them.[3] If we may judge by the records of the next reign a thousand to fifteen hundred lashes was no uncommon sentence on a deserter, while not a few were actually shot in Hyde Park.[4] The only resource, therefore, was to check the evil as far as possible by prevention. Thus we constantly find large bodies of troops under orders for foreign service quartered in the Isle of Wight, from which they could not easily escape. This remedy was at least in one case found worse than the disease, for the numbers of the men being too great to be accommodated in the public houses, very many of them perished from exposure to the weather. Thereupon the Secretary-at-War made inquiry as to barns and empty houses for them, according to the traditions of his office, fatally too late.[5] 1702-1713.

1 *Secretary's Common Letter Book*, 6th July 1707.

2 Four regiments destined for the Peninsula in 1711 were kept waiting three months for their ships at Cork. In that time they lost 500 men by desertion, probably not much less than a fourth of their numbers.

3 A clause against concealment of deserters was inserted in the Mutiny Act of 1708-9.

4 Abundant instances in *Secretary's Common Letter Book*.

5 *Ibid.*, 18th October 1707.

1702-1713. Another practice, which from ignorance of its origin has been blindly followed till within the last few years, also took its rise from the prevalence of desertion at this period, namely that of shifting troops from quarter to quarter of England by sea. On the same principle men were frequently cooped up in the transport-vessels for weeks and even months before they sailed on foreign service, occasionally with frightful consequences. Thus in 1705 certain troops bound for Jamaica were embarked on transports on the 18th of May. They remained there for two months with fever and small-pox on board, until at last, the medical supplies being exhausted, the case was represented to the Secretary-at-War. The reply was that they were to receive such relief as was possible; but they remained in the same transports until October, when at last they were drafted off in parties of sixty on the West Indian packets to their destination. Forty-eight of them were lost through a storm in port long before October, but the number that perished from sickness is unknown, and was probably most sedulously concealed.[1]

Let us now turn to the pleasanter theme of the changes that were wrought for the benefit of the soldier. The first of these appears in the Mutiny Act of 1703, and was doubtless due in part to the scandals revealed in the office of the Paymaster-General. The rates of pay to all ranks below the status of commissioned officers are actually given in the Act, with express directions, under sufficient penalties, that the subsistence money shall be paid regularly every week, and the balance over and above it every two months. Further, all stoppages by the Paymaster-General, Secretary-at-War, commissaries, and muster-masters are definitely forbidden, and the legitimate deductions strictly limited to the clothing-money, one day's pay to Chelsea Hospital, and one shilling in the pound to the Queen. The continuance of this last tax was of course a crying

[1] *Secretary's Common Letter Book*, 25th, 27th July; 17th August; October 1705.

injustice, but the abolition of the other irregular claims 1702-1713. was distinctly a gain to the British soldier, due, as it is satisfactory to know, to the newly appointed Controllers of Accounts. Altogether the condition of the soldier as regards his pay seems decidedly to have improved, Marlborough's attention to this most important matter having evidently borne good fruit. It is true that in Spain and the colonies, to which he had not leisure nor opportunity to give personal attention, the neglect of the Secretary-at-War caused great grievances and much suffering; it is true also that even in England, when his influence was gone, there was a recurrence of the old scandals under the miserable administration of Harley;[1] yet on the whole the improvements in this province were at once distinct and permanent.

Another valuable reform in respect of clothing was due to the direct interposition of Marlborough himself. In 1706 the abuses in this department were, at his instance, made the subject of inquiry by Secretary St. John and General Charles Churchill, with the ultimate result that the pattern and allowance of clothing and the deduction of off-reckonings were laid down by strict rule, while the whole business of clothing, though still left to the colonels, was subjected to the control of a board of six General officers, whose sanction was essential to the validity of all contracts and to the acceptance of all garments. Thus was established the Board of General Officers,[2] whose minutes are still the great authority for the uniforms of the eighteenth century.

Unfortunately these benefits could weigh but little against the disadvantages already described. It is certain that despite the standard laid down by Act of

[1] See, for instance, the complaint of a regiment which had been paid in unsaleable tallies. Several officers had been arrested for debts contracted by their men for want of their pay. *Secretary's Common Letter Book*, 18th April 1711.

[2] Such a Board, or rather intermittent meeting of Generals, had been established in January 1706. For the report of St. John and Churchill and the new regulations, see *Miscellaneous Orders*, 4th February 1706; 14th January 1708.

1702-1713. Parliament, vast numbers of boys were enlisted as well as men of fifty and sixty years of age, who no sooner entered the field than they were sent back into hospital. Good regiments, however, then as now obtained good recruits, sometimes through the offer of extra bounty from the officers,[1] more often through the character of the officers themselves. The presence of thieves, pirates, and other criminals in the ranks must necessarily have introduced a certain leaven of ruffianism, yet neither in Flanders nor in the Peninsula do we find anything approaching to the outrageous bursts of indiscipline which were witnessed a century later at Ciudad Rodrigo and Badajoz. There was, it is true, the mutiny under the Duke of Ormonde, but it was of short duration and easily suppressed; and altogether, for reasons that shall presently be given, Marlborough's army seems to have been better conducted than Wellington's. Unfortunately, although two men who served in the ranks left us journals of a whole or part of the war, we remain still without a picture of the typical soldier of Marlborough. The one figure that emerges with any distinctness from the ranks is that of Christian Ross, a woman who served as a dragoon in several actions, was twice wounded before her sex was discovered, and ended her career as virago, sutleress, and out-pensioner of Chelsea Hospital.[2] The rest, with the exception of Sergeant Littler, Sergeant John Hall,[3] and Private Deane remain buried in dark oblivion, leaving a lamentable gap that can never be filled in our military history.

From the men I pass to the officers. Our information in regard to them is curiously mixed. Certain of the abuses that dishonoured them have already been revealed, nor can these be said to exhaust the list. There

[1] I can adduce only one instance in proof, that of the Duke of Schomberg, who offered £2 a man to old soldiers to join his regiment of dragoons (Newspaper Advertisement, 27th July 1703), but the fact is indubitable.

[2] There are two or three memoirs of her, attributed to Defoe and others.

[3] See Steele's *Tatler* (No. 87), 29th Oct. 1709.

were grave scandals in the Guards, which had the misfortune to possess one colonel, of a distinguished Scottish family, who revived the worst traditions of Elizabeth and Charles the Second. Not only did he systematically enlist thieves and other bad characters as "faggots,"[1] but he did not scruple to accept recruits who offered themselves for the sake of defrauding their creditors, to receive money from them for doing so, and to extort more money by threatening to withhold his protection or to ship them off to fight in Spain. These men did no duty,[2] wore no uniform and drew no pay, to the great profit of the colonel and the great disgrace of the regiment; and the evil grew to such a height that when the House of Commons finally took the matter in hand, the "faggots" were found to number one-fourth of the nominal strength of the regiment.[3] Such cases, however, as this of the infamous Colonel Chartres were rare; and the decrease of this particular vice of officers in Queen Anne's time presents a pleasing contrast to its prevalence in the time of King William. 1702-1713.

Another habit, which sounds particularly objectionable in modern ears, was the occasional unwillingness of officers to accompany their regiments, and their readiness to leave them, when employed on distasteful service. This was especially true of regiments on colonial stations, particularly in the West Indies,[4] and was by no means unknown of those actually on active service in Flanders and the Peninsula. Sometimes the offenders had received leave of absence, which the Secretary-at-War would willingly grant as a matter of jobbery in the case

1 *S. P., Dom.* (11th September 1705), vol. vi.

2 They went on guard once and were put in the guard-room once, that their names might appear on the list of prisoners.

3 *Commons Journals*, 5th, 13th, 22nd February; 8th, 26th May 1711.

4 See the case of Lillingston's regiment in Antigua, *Cal. Treas. Papers*, 18th November 1707: for the Mediterranean garrisons and Peninsula, *S. P., Dom.* (December 1705), vol. vii.; (19th June 1709), vol. xiv.

1702-1713. of a friend,[1] but more often they took leave without asking for it, occasionally for as much as five years together,[2] without objection from the colonel or rebuke from the War Office. One colonel took it as a great grievance when Marlborough insisted that he should sell his commission since he was unwilling to do duty;[3] and altogether the general connivance at shirking of this kind rendered the offence so little discreditable that it must not be judged by the standard of to-day. Speaking generally, however, the officers had far more grievances that command our pity than faults which provoke our indignation.

One hardship that bore on officers with peculiar severity was the expense of obtaining recruits. They received, of course, levy-money for the purpose, but this was frequently insufficient, while no allowance was made for recruits lost through desertion, sickness, or other misfortunes over which they had no control. Marlborough was most strict in discouraging, except in extreme cases, any attempts of officers to transfer their burdens from themselves to the State, though he freely admitted, not without compassion, that officers had been ruined by sheer bad luck with their recruits. We find bitter complaints from officers in the Peninsula that owing to the heavy mortality in the transports, their recruits, by the time that they reached them, cost them eight or nine pounds a head.[4] Indeed, if one may judge from contemporary newspapers, which are quite borne out by scattered evidence, the sufferings

[1] *E.g. Secretary's Common Letter Book*, 22nd December 1710.

[2] *Ibid.*, 22nd December 1708.

[3] *Despatches*, vol. v. pp. 21, 241. This colonel, Bennett by name, was an admirable officer at his work, and had done excellent service at Gibraltar.

[4] *Cal. Treas. Papers*, 18th November 1710, 6th January 1711. Recruits were practically bought and sold at from £2 to £3 a head at ordinary times, colonels receiving so much a man when they furnished drafts. In strictness one officer took a recruit from another, and paid to him the expenses of raising a substitute. See *Commons Journals*, 8th May 1711.

of officers on account of recruiting were almost unendurable.[1] 1702-1713.

Remounts again were a heavy tax upon the officer. An allowance of levy-money at the rate of twelve pounds a horse[2] was granted to officers for the purpose, but was complained of as quite inadequate to the charge,[3] in consequence of heavy losses through the epidemic of horse-sickness in Flanders. Carelessness in the hiring and fitting of transports also caused much waste of life among the horses,[4] until Marlborough, as his letters repeatedly show, took the matter into his own hands. It is interesting to learn that Irish horses, being obtainable for five pounds apiece,[5] were much used in Spain, though less in Flanders, Marlborough having a prejudice in favour of English horses as of English men, as superior to all others. This cheapness, however, was of little

1 See *Humours of a Coffee House* (a dialogue), 26th December 1707. *Guzzle.*—How go on your recruits this winter? *Levy* (an officer).—Very poorly. I am almost broke; they cost us so much to raise them, and run away so fast afterwards that, without the Government will consider us, we shall be undone, and the service will suffer into the bargain. . . . Some of us were forced to live on five shillings weekly; the rest was stopped by the Colonel for the charge we had been at in raising recruits; and after all they deserted from us and the service wanted what the nation paid for. . . . What recruits stayed with us, we were no better, for being most of them boys, they fell sick as soon as we got into the field. . . If our regiments were only complete as they ought to be, you would hear something to surprise you in a campaign.

See also *Secretary's Common Letter Book*, 23rd April 1711, wherein the Generals report that under the present system of mustering, recruiting is impossible, and recommend that if any men die, desert, or are discharged, their names may be kept on the rolls for the next two musters; and see Coxe's *Marlborough*, vol. vi. pp. 232, 233.

2 *Miscellaneous Orders* (*Guards and Garrisons*), 17th May 1707.

3 *Ibid.* (*Forces Abroad*), 5th March 1706.

4 Conyngham's regiment (8th Hussars) lost on passage to Portugal 27 chargers out of 70, and 141 troop horses out of 216, owing to the use of two such transports. The animals were beaten to pieces and stifled for want of room.

5 "Good squat dragoon horses," *S. P., Dom.*, 27th February, 10th August 1705.

ɔ2-1713. service to the officers. They were expected to pay for the transport of their horses at a fixed rate, and though at length in reply to their complaints free transport was granted for twenty-six horses to a battalion, yet this privilege was again withdrawn as soon as it was discovered that Irish animals were to be purchased at a low price.[1]

Again, the officers were always subject to extortion from civilians. Parish constables, to whom the law allowed sixpence a day for the subsistence of recruits, declined to deliver them unless they were paid eightpence a day.[2] But as usual the chief delinquents were the regimental agents. The Controllers of Accounts early made an attack on these gentry, but with little success, the fellows pleading that they were not public officials but private servants of the colonel, and therefore not bound to produce their accounts. The complaints of the officers against them were endless, and with good reason. Perhaps the most heartless instance of an agent's rascality was that of one who stole the small allowance made by a lieutenant on active service to his wife, and refused to pay it until ordered by the Queen.[3] Officers clamoured that the agents should be tried by court-martial, but this was not permitted, and perhaps wisely, for a court-martial would probably have sentenced a scoundrel to the gantlope, in which case the men would not have let him escape alive.

Yet another tax fell upon officers in the shape of contribution to pensions and regimental debts. In every regiment except those serving in Flanders a fictitious man was allowed in the roll of each troop or company, whose pay was taken to form a fund for the support of officers' widows;[4] but in Marlborough's army

1 *Secretary's Common Letter Book*, 27th February, 10th August 1705.

2 *Ibid.*, 19th February 1709.

3 *Ibid.*, 15th February 1712.

4 Hence the expression, once very common, of a widow's man. Readers of Marrayat will remember that when Peter Simple was searching the ship for Cheeks the marine, he was informed that Cheeks was a widow's man.

these widows were supported by a voluntary subscription 1702-1713. from the officers, without expense to the State. By some contrivance, which seems utterly outrageous and was presumably the work of the War Office or of the Treasury, this voluntary fund was saddled with the maintenance of widows who had lost their husbands in the previous war, so that in 1709 Marlborough was obliged to protest and to ask for the extension of "widows' men" to some at least of his own troops.[1] Again, some regiments appear to have been charged with pensions to particular individuals, though by what right or for what service it is impossible to say.[2] Yet again, by misfortune, carelessness, or roguery of a colonel, or more commonly of an agent, regiments found themselves burdened with debts amounting to several thousand pounds, as, for instance, through the loss of regimental funds by shipwreck or through mismangement of the clothing. In such cases the only possible relief was the sale, by royal permission, of the next company or ensigncy for the liquidation of the debt.[3]

Another form of pension which, though sometimes used for worthy objects, was at least as often perverted to purposes of jobbery, was the appointment of infant officers. In many instances children received commissions in a regiment wherein their fathers had commanded and done good service, either for the relief of the widows, if those fathers had fallen in action, or for a reward if they were still living. Sometimes these children actually took the field, for there is record of one who went to active service in Flanders at the age of twelve, "behaving with more courage and conduct than could have been expected from one of his years,"

[1] *Despatches*, vol. v. pp. 356, 412. A scale of widows' pensions from £50 a year for a colonel's to £16 for a cornet's or ensign's was fixed by regulation, 23rd August 1708. *Miscellaneous Orders (Guards and Garrisons)*, under date.

[2] *E.g*, Cadogan's regiment (5th Dragoon Guards). Marlborough tried to obtain relief for it. *Secretary's Common Letter Book*, 5th April 1705.

[3] W. O. *Miscellaneous Orders*. 17th April 1712.

1702-1713. and ruined his career at sixteen by killing his man in a duel.[1] But beyond all doubt in many instances the favour was granted without sufficient cause, while even at its best it was an abuse of public money and a wrong done to the regiment. This abuse was of course no new thing, and did not amount to an actual grievance; but it had fostered a feeling, that was already too strong, of the privileges conferred on colonels by their proprietary rights in their regiments.

The grant of commissions to children was forbidden by the Royal Regulations of 1st May 1711, a collection of orders which had at any rate for their ostensible object a considerable measure of reform, and therefore demands some notice here. Hereby the grant of brevets, which had given considerable trouble to Marlborough, and had already been forbidden in 1708, was again prohibited; and finally an attempt was made to limit the sale and purchase of commissions. To this end no sale of commissions whatever was permitted except by royal approbation under the sign manual, and then only to officers who had served for twenty years or had been disabled by active service. The announcement appears to have been treated as a joke;[2] and within six months the rule, in consequence of representations from Marlborough, was considerably modified.[3] If (so the Duke pointed out) subalterns who have been unlucky with their recruits may not sell their commissions, the debt will fall on the regiment: if, again, the successors to officers who die on service do not contribute something towards the dead man's wife and family, many widows and children must starve; lastly, colonels often wish to promote officers from other regiments to their own when they have no officer of their own fit for advancement, which is for the good of the service but must

[1] See account of Captain Richard Hill. *S. P., Dom.*, Anne, vol. x. (undated).

[2] *Miscellaneous Orders* (*Guards and Garrisons*), 19th October 1711.

[3] *Ibid.* (*Forces Abroad*), 1st May 1711.

become impracticable unless the superseded officer receive something in compensation.[1] His arguments were seen to be irresistible unless the State were prepared to incur large additional military expenditure, and the rules were shortly afterwards amended in the spirit of his recommendations and for the reasons that he had adduced.[2] 1702-1713.

Thus almost the final administrative act of Marlborough as Captain-General was to uphold the system of purchase then existing against the hasty reforms of civilian counsellors. Enough has been said to show that contemporary military policy in England, with which he was chiefly identified, tended always to make the regiment more and more self-contained and less dependent on the support of the State: it will be seen before long how regiments met the charge imposed on them by the institution of regimental funds in the nature of insurance. The drawback of such a system is obvious. Excess of independence in the members can hardly but entail weakening of central control, with incoherence and consequent waste of energy in the action of the entire body. Regimental traditions, regimental pride, are priceless possessions well worthy the sacrifice of ideal unity of design and perfect assimilation to a single pattern. But regimental isolation, fostered and encouraged on principle to the utmost, must inevitably bring with it a certain division of command, a want of subordination to the supreme authority, in a word that measure of indiscipline in high places which distinguishes an aggregation of regiments from an army.

1 *Despatches*, vol. v. p. 412. Amended regulations, *Miscellaneous Orders* (*Forces Abroad*), 7th September 1712. In the same letter Marlborough pleaded for the abolition of the 5 per cent purchase money paid to Chelsea Hospital, which was done by Order of 1st April 1712. *H. O. M. E. B.*, under date.

2 Even as things were, officers were occasionally obliged to accept a Chelsea pension; a captain of horse being admitted on the footing of a corporal of horse. *Secretary's Common Letter Book*, 10th January 1712.

1702-1713. Yet who can doubt but that Marlborough acted with his usual strong good sense as a soldier and his usual sagacity as a statesman? He had risked his popularity in the Army by his avowed severity towards officers in the matter of recruits,[1] because he knew that the slightest attempt to shift this burden upon the State would mean the refusal of Parliament to carry on the war, and a wholesale disbandment of the Army. He favoured the sale of commissions on precisely the same principle; for, as his letter clearly shows, he foresaw the growth of what is now called a non-effective vote, and doubted the willingness of Parliament to endure it. That which he dreaded has now come to pass, for better or worse; the country is saddled with a vast load of pensions, and the Commons grow annually more impatient over increase of military expenditure without corresponding increase of efficiency. Marlborough's choice lay between an aggregation of regiments and no army, and of two evils he chose the less. It still remains to be proved that he was wrong.

From the regimental I pass to the general administration. Herein the first noticeable feature is the amalgamation by the Act of Union of the English and Scotch establishments into a single establishment for Great Britain. Ireland of course still remained with a separate establishment of her own, and all the paraphernalia of Commander-in-Chief, Secretary-at-War, and Master-General of Ordnance. There continued always in Ireland as heretofore a different rate of pay for all ranks, which, owing to constant transfer of regiments from Ireland to England or abroad gave rise to great confusion in the accounts. The chief matter of interest in Ireland is the very reasonable jealousy of the Irish Commons for the retention within the kingdom of all regiments on the Irish establishment, or at least for the substitution of other regiments in their place if they should be withdrawn. Their intention was that Irish revenue should be spent in Ireland, and it is satisfactory

[1] Coxe's *Marlborough*, vol vi. p. 232, 233.

to note that it was rigidly and conscientiously respected by the authorities in England.[1] 1702-1713.

Another important matter was a first attempt to settle the position of the marines, who up to the middle of the reign were subject to a curious and embarrassing division of control. St. John early disclaimed all authority over them,[2] but they were evidently subject to the regulations of the army and suffered not a little in consequence. The rigid rule that regiments must be mustered before they were paid inflicted great hardship on marines, for it could not be carried out when a regiment was split up on half a dozen different ships, and the result was that the men were not paid at all. Even when ashore they were exposed to the same inconvenience owing to the inefficiency of the commissaries,[3] so that some regiments actually received no wages for eight years.[4] The inevitable consequence was hatred of the service and mutiny, which at one moment threatened to be serious.[5] Finally, on the 17th of December 1708, the marines were definitely placed under the jurisdiction of the Lord High Admiral.[6]

I come now to the most fateful of all changes in the administration, namely the rise to supreme importance of the Secretary-at-War. Attention has already been drawn to the duties and powers which silently accumulated in the hands of this civilian official after the death of Monk, owing to the lack of

1 Journals of Irish House of Commons. Speeches from the throne, 1703, 1707, 1710, 1713.

2 *Secretary's Common Letter Book*, 21st August 1704. "The marines are entirely under the Prince's (George of Denmark's) direction. You must apply to his secretary."

3 The Commissary of the Musters at Portsmouth was "a superannuated old man who was rolled about in a wheel-barrow." *Cal. Treas. Papers*, 15th November 1703.

4 *E.g.*, Caermarthen's and Shovell's, *ibid.*, 7th November 1706.

5 *S. P., Dom.* (29th March 1709), vol. xiv. Thirty-eight mutineers marched on London from Portsmouth in order to lay down their arms publicly at Whitehall. They were stopped at Putney. See also *Cal. Treas. Papers* of same date.

6 *H. O. M. E. B.*, under date.

1702-1713. efficient control by the Sovereign. The reigns of King William and Queen Anne, in consequence of the constant absence of the Captain-General on active service, did nothing to restore this lost control, and the almost unperceived change which released the Secretary-at-War from personal attendance on the Commander-in-Chief in the field virtually abolished it altogether. The terms of the Secretary-at-War's commission remained the same, "to obey such orders as he should from time to time receive from the Sovereign or from the General of the forces for the time being, according to the discipline of war;"[1] but the situation was in reality reversed. Even in King William's time the Secretary-at-War had countersigned the military estimates submitted to Parliament; from the advent of St. John he assumes charge of all military matters in the Commons, often taking the chair of the committee while they are under discussion. Thus he becomes the mouthpiece of the military administration in the House, and, since the Commander-in-Chief is generally absent on service he ceases to take his orders from him, but becomes, except in the vital matter of responsibility, a Secretary-of-State, writing in the name of the Queen or of her consort, or finally in his own name and by his own authority without reference to a higher power. Lastly, his office, thus exalted to importance, becomes the spoil of political party; Secretaries-at-War follow each other in rapid succession,— St. John, Walpole, Granville, Lord Lansdowne, Windham, Gwynne; and the Army is definitely stamped as a counter in the eternal game of faction.

The power of the Secretary-at-War in Queen Anne's time is sufficiently shown by his letter-books. In the Queen's name he gives orders for recruiting, for

[1] *H. O. M. E. B.*, St. John's Commission, 20th April 1704, 8th June 1707; Walpole's, 23rd February 1708; Granville's, 17th October 1710; Windham's, 28th June 1712; Francis Gwynne's, 31st August 1713.

drafting, for armament, for musters, for change of quarters, relief of garrisons, hire of transports, embarkation of troops, patrolling of the coast, escort of treasure, and in a word for all matters of routine. In the Duke of Marlborough's name also he directs men to be embarked, money to be advanced, and recruits to be furnished, and even criticises the execution of the orders issued by him on behalf of the Queen.[1] On his own authority he bids colonels to send him muster-rolls and lists of recruiting staff and to provide their regiments with quarters, regrets that he cannot strengthen weak garrisons, and lays down the route for all marches within the kingdom.[2] He corresponds direct with every rank of officer without the slightest regard for discipline or dignity. We find Walpole threatening a lieutenant with forfeiture of his commission for absence without leave, bidding a captain be thankful that owing to his own clemency he is not cashiered for fraud,[3] regretting that he cannot in conscience excuse one subaltern from attending his regiment on foreign service,[4] ordering another to pay for his quarters immediately,[5] summoning a third person to the War Office to account to him for wrongful detention of a recruit. Granville promises an officer leave of absence from foreign service, but must first, in common decency, apply to the General in command.[6] Lord Lansdowne begs the Governor of Portsmouth not to be too hard on a young regiment in the matter of guard-duties, orders the discharge of a soldier when three years of his service have expired, and writes to the Irish Secretary-at-War for leave of 1702-1713.

[1] Compare the Duke of Wellington's evidence in 1837 : "The Commander-in-Chief cannot at this moment move a corporal's guard (four men) from hence to Windsor without going to a civil department for authority."

[2] *Secretary's Common Letter Book*, 22nd December 1708.

[3] *Ibid.*, 29th January 1709.

[4] *Ibid.*, 7th March 1709.

[5] *Ibid.*, 14th May 1709.

[6] *Ibid.*, 22nd December 1710.

1702-1713. absence for a friend.[1] Finally, all ask favours of colonels on behalf of officers and men. One thing only they left for a time untouched, namely the sentences of court-martial, which St. John expressly abjured in favour of the Judge Advocate-General; but for the rest they issued orders, approbations, and reprimands with all the freedom of a Commander-in-Chief.

The Office of Ordnance remained as before independent of the War Office, though of course liable to fulfil its requisitions for arms and stores. It is remarkable that Marlborough, like Wellington a century later, no sooner became Master-General[2] than he restored the organisation of King James the Second. But the strain imposed upon the Department by the multitude of forces in the field was too severe for it. Two months before Blenheim was fought the supply of firelocks and socket-bayonets was exhausted; and in succeeding years, as disasters grew and multiplied in Spain, the Office was obliged frequently, and to the great indignation of English manufacturers, to purchase arms abroad.[3]

The subject of weapons leads us directly to the progress of the Army in the matter of armament, equipment, and training. The first point worthy of notice is the disappearance of the time-honoured pike. Pikes were issued to a battalion in the proportion of one to every five muskets as late as 1703, but were delivered back into store in the following year;[4] and in 1706 a letter from St. John announces that pikes are considered useless and that musket and bayonet must be furnished to every man.[5] The bayonet was, of

[1] *Secretary's Common Letter Book*, 1st and 3rd March, 24th May 1712.

[2] *H. O. M. E. B.*, 30th June 1702. Marlborough was appointed Master-General on 26th March.

[3] *Commons Journals*, 29th March 1707. The cost of Dutch muskets was £8000, and of English £11,000 per 10,000; but great superiority was claimed for the English.

[4] *H. O. M. E. B.*, 16th April 1703. April 1704 (arms of Evans's regiment).

[5] *Secretary's Common Letter Book*, 12th June 1706.

course, the socket-bayonet; and the musket, being of a new and improved model, was a weapon much superior to that issued in the days of King William.[1] Partly, no doubt, owing to the efficiency of this musket, which carried bullets of sixteen to the pound, as against the French weapon, which was designed for bullets of twenty-four to the pound, and still more owing to superiority of discipline and tactics, the fire of the British was incomparably more deadly than that of the French.[2] The secret, so far as concerned tactics, lay in the fact that the British fired by platoons according to the system of Gustavus Adolphus, whereas the French fired by ranks; and the perfection of drill and discipline was superbly manifested at Wynendale. For this, as well as for the better weapon, the Army had their great chief to thank, for the Duke knew better than any the value of fire-discipline, as it is called, and would put the whole army through its platoon-exercise by signal of flag and drum before his own eye.[3] Nevertheless, the cool head and accurate aim for which the British have always been famous played their part, and a leading part, in the victories of Marlborough. 1702-1713.

Of the drill proper there is little to be said, though some few changes are significant of coming reforms. The number of ranks was left unfixed, being increased or reduced according to the frontage required, but probably seldom exceeded three and was occasionally reduced to two. The old method of doubling ranks was still preserved; but the men no longer fell in by files, and the file may be said definitely to have lost its old position as a tactical unit. A company now fell in in single rank, was sorted off into three or more divisions and formed into ranks by the wheel of the divisions from line into column, which was a complete

[1] *H. O. M. E. B.*, 14th October 1704. *Commons Journals*, 19th March 1707.

[2] Parker. See the account of the meeting between the Royal Irish of England and of France at Malplaquet.

[3] Millner. 30th May, 1707.

1702-1713. novelty. The manual and firing exercise remained as minute and elaborate as ever; and a single word of command shows that the old exercise of the pike was soon to be adopted for the bayonet.[1] With these exceptions there was little deviation from the old drill of Gustavus Adolphus; but the real improvement, which made that drill doubly efficient, was in the matter of discipline. That the lash and the gantlope were unsparingly used in Marlborough's army there can be no doubt, and that they were employed even more savagely at home can be shown by direct evidence;[2] but the Duke, as shall presently be shown, understood how to make the best of his countrymen by other means besides cutting their backs to pieces.

For the cavalry, of which he was evidently very fond, Marlborough did very signal service by committing it definitely to action by shock. Again and again in the course of the war the French squadrons are found firing from the saddle with little or no effect, and the British crashing boldly into them and sweeping them away. There are few actions, too, in which the Duke himself is not found in personal command of the horse at one period or another of the battle—at Blenheim in the great charge which won the day, at Ramillies at the most critical moment, at Malplaquet in support of the British infantry, and most brilliantly of all at the passage of the lines at Landen. Yet he was too sensible not to imitate an enemy where he could do so with advantage. The French gendarmerie had received pistol-proof armour in 1703;[3] the British horse in Flanders, at Marlborough's suggestion, received a cuirass

[1] The Duke of Marlborough's new exercise of firelocks and bayonets, by an officer in the Foot Guards. London, N.D.

[2] The most appalling sentence was that given to a guardsman at home who had slaughtered his colonel's horse for lucre of the hide—seven distinct floggings of eighteen hundred lashes apiece, or twelve thousand six hundred lashes in all. His life was despaired of after the first flogging, and the Queen remitted the remaining six. *Secretary's Common Letter Book*, 12th Jan. 1712.

[3] Newspapers, 3rd March 1703.

in 1707, a reform which was copied by the Dutch and urged upon all the rest of the Allies.[1] It is characteristic of the Duke's never-failing good sense that the cuirasses consisted of breast-pieces only, so that men should find no protection unless their faces were turned towards the enemy. 1702-1713.

As to the artillery there is little to be said except that the organisation by companies appears to have been thoroughly accepted, and the efficiency of the arm thereby greatly increased. The Duke was never greater than as an artillerist. Every gun at Blenheim was laid under his own eye; and the concentration of the great central battery at Malplaquet and its subsequent advance shows his mastership in the handling of cannon. For the rest, the artillery came out of the war with not less, perhaps with even more, brilliancy than the other corps of the army; and, though no mention is made of the fact by the historian of the regiment, it is likely that no artillery officers ever worked more strenuously and skilfully in the face of enormous difficulties than the devoted men who brought their guns first down to the south side of the Danube and then back across the river to the battlefield of Blenheim.[2]

It is impossible to quit this subject without a few words on the great man who revived for England the ancient glory of Creçy, Poictiers, and Agincourt, the greatest, in the Duke of Wellington's words, who ever appeared at the head of a British Army. There are certain passages in his life which make it difficult sometimes to withhold from him hard names; but allowance should be made for one who was born in revolution, nurtured in a court of corruption, and matured in fresh revolution. Wellington himself admitted that he never understood the characters of that period, nor exercised due charity towards them, till he had observed the effects of the French Revolution on the minds and consciences

1 *Despatches*, vol. iii. pp. 309, 335, 461; *S. P., Dom.*, vol. xix. 23.

2 The testimony to these exertions is to be found only in Hare's Journal, but it is emphatic.

1702-1713. of French statesmen and marshals. Marlborough's fall was brought about by a faction, and his fame has remained ever since a prey to the tender mercies of a faction. But the prejudices of a partisan are but a sorry standard for the measure of one whose transcendent ability as a general, a statesman, a diplomatist, and an administrator, guided not only England but Europe through the War of the Spanish Succession, and delivered them safe for a whole generation from the craft and the ambition of France.

Regarding him as a general, his fame is assured as one of the great captains of all time; and it would not become a civilian to add a word to the eulogy of great soldiers who alone can comprehend the full measure of his greatness. Yet one or two small points are worthy of attention over and above the reforms, already enumerated, which were introduced by him in all three arms of the service. First, and perhaps most important, is the blow struck by Marlborough against the whole system, so much favoured by the French, of passive campaigns. It was not, thanks to Dutch deputies and German princelets, as effective as it should have been, but it still marked a step forward in the art of war. It must never be forgotten that we possess only the wreck of many of Marlborough's finest combinations, shattered, just as they were entering port, against the rocks of Dutch stupidity and German conceit. Next, there is a great deal said and written in these days about night marches and the future that lies before them. It will be well to glance also at the past that they have behind them, and to mark with what frequency, with what consummate skill, and what unvarying success they were employed under far greater than modern difficulties by Marlborough.

Next let it be observed how thoroughly he understood the British soldier. He took care to feed him well, to pay him regularly, to give him plenty of work, and to keep him under the strictest discipline; and with all this he cherished a genial feeling for the men, which showed itself not only in strict injunctions to watch over their comfort but in acts of personal kindness kindly bestowed. The magic of his personality made itself felt among his men far beyond the scope of mere military duty. His soldiers, as the Recruiting Acts can testify, were for the most part the scum of the nation. Yet they not only marched and fought with a steadiness beyond all praise, but actually became reformed characters and left the army sober, self-respecting men.[1] Marlborough, despite his lapses into treachery as a politician, was a man of peculiar sensitiveness and delicacy. He had a profound distaste for licentiousness either in language or in action, and he contrived to instil a like distaste into his army. His force did not swear terribly in Flanders, as King William's had before it, and although the annual supply of recruits brought with it necessarily an annual infusion of crime, yet the moral tone of the army was singularly high. Marlborough's nature was not of the hard, unbending temper of Wellington's. The Iron Duke had a heart so steeled by strong sense, duty, and discipline that it but rarely sought relief in a burst of passionate emotion. Marlborough was cast in a very different mould. He too, like Wellington, was endowed with a strong common sense that in itself amounted to genius, and possessed in the most trying moments a serenity and calm that was almost miraculous. But there was no coldness in his serenity, nothing impassive in his calm. He was sensitive to a fault; and though his temper might remain unchangeably sweet and his speech unalterably placid and courteous, his face would betray the anxiety and worry which his tongue had power to conceal.[2] With such a temperament there was a bond of humanity between him and his men that was lacking in Wellington. Great as Wellington was, the Iron Duke's army could never have nicknamed him the Old Corporal.

1702-1713.

[1] Lediard.

[2] "The Duke does not say much, but no one's countenance speaks more." Hare's Journal.

1702-1713. The epithet Corporal suggests comparison with the Little Corporal, who performed such marvels with the French Army. Undoubtedly the name was in both cases a mark of the boundless confidence and devotion which the two men could evoke from their troops, and which they could turn to such splendid account in their operations. Marlborough could make believe that he was meant to throw away his entire army and yet be sure of its loyalty; Napoleon could throw away whole hosts, desert them, and command the unaltered trust of a new army. In both the personal fascination was an extraordinary power; but here the resemblance ends. Napoleon, for all his theatrical tricks, had no heart nor tenderness in him, and could not bear the intoxication of success. Marlborough never suffered triumph to turn his head, to diminish his generosity towards enemies, to tempt him from the path of sound military practice, or to obscure his unerring insight into the heart of things. Twice his plans were opposed as too adventurous by Eugene, first when he wished to hasten the battle of Malplaquet, and secondly when he would have masked Lille and advanced straight into France; but even assuming, as is by no means certain, that in both instances Eugene was right, there is no parallel here to the gambling spirit which pervaded the latter enterprises of Napoleon. "Marlborough," said Wellington, "was remarkable for his clear, cool, steady understanding," and this quality was one which never deserted him. Nevertheless, if there be one attribute which should be chosen as supremely characteristic of the man, it is that which William Pitt selected as the first requisite of a statesman—patience; "patience," as the Duke himself once wrote to Godolphin, "which can overcome all things";[1] patience which, as may be seen in a hundred passages during the war, was possessed by him in such measure that it appears almost godlike. These are the qualities which mark the sanity of perfect genius, that distinguish a

[1] Mahon, *Hist. of England*, vol. iii. p. 368.

Milton from a Shelley, a Nelson from a Dundonald, 1702-1713. and a Marlborough from a Peterborough; and it is in virtue of these, indicating as they do the perfect balance of transcendent ability, that Marlborough takes rank with the mightiest of England's sons, with Shakespeare, with Bacon, and with Newton, as "the greatest statesman and the greatest general that this country or any other country has produced."[1]

[1] St. John.

END OF VOL. I

www.ingramcontent.com/pod-product-compliance
Ingram Content Group UK Ltd.
Pitfield, Milton Keynes, MK11 3LW, UK
UKHW020716041026
15590UKWH00002B/701